ASTEROIDS AND METEORITES

ASTEROIDS AND METEORITES

Editor

Gregor Pahlke

scitus academics

Asteroids and Meteorites

Edited by **Gregor Pahlke**

This Book is Printed on Acid-Free Paper. ♾

ISBN: 9781681178790
Library of Congress Control Number: 2017948906

SCITUS Academics LLC,
616, Corporate Way, Suite 2, 4766,
Valley Cottage, NY 10989

www.scitusacademics.com

Printed in United States of America.

Preface

Meteorites, and their fall to Earth, have the potential to inform studies of the asteroid impact hazard and of impact mitigation. An asteroid is a rocky object in space that's smaller than a planet — they're sometimes called minor planets or planetoids, according to NASA. Other sources refer to them loosely as "space debris," or leftover fragments from the formation of the solar system. There are millions of asteroids orbiting the sun, some 750,000 of which are found in the asteroid belt, a vast ring of asteroids located between the orbits of Mars and Jupiter. Asteroids can be as large as hundreds of kilometers wide: The asteroid Ceres, sometimes referred to as a dwarf planet, is 940 km (584 miles) wide. Meteorites are usually categorized as iron or stony. As the name implies, iron meteorites are composed of about 90 percent iron; stony meteorites are made up of oxygen, iron, silicon, magnesium and other elements. The collisions of these objects with Earth are basically random events, but still we have some idea how often they happen. Localized destruction happens every couple of hundreds of years and is somewhat equivalent to a hydrogen bomb. Last such event happened in 1908 near Tunguska river in Siberia. The number of casualties depends on the place of impact (the objects of this size usually explode in the air before reaching the ground, just like an atom bomb). If a city is struck, casualties could be close to a million, while Tunguska event had zero to one reported casualty (reports vary). An impact in the ocean would create a tsunami and definitely produce significant destruction on the nearby seaside. These events usually do not leave a crater and typically involve a 100-meter asteroid or comet.

This book presents the celestial phenomena of asteroids, meteoroids and meteorites, and comets. It illustrates the role that collisions with meteors, comets, and asteroids have played in the history of Earth and other planets in the solar system and examines what is being done to protect Earth from future collisions.

Table of Contents

CHAPTER 1

Asteroids and The Origin of Life—Two Steps of Chemical Evolution on The Surface of These Objects

M. B. Simakov

Group of Exobiology, Institute of Cytology RAS, Tikhoretsky Av., 4 St. Petersburg, 194064, Russia

ABSTRACT

It is now well-known that carbonaceous chondrites contain large quantities of prebiotic molecules, including amino acids, carbohydrates, and heterocyclic bases of nucleic acids. It has become evident that the chemical evolution proceeds under conditions of open space, which serves as a huge chemical reactor. The extraterrestrial origin of these prebiotic molecules has been established on the basis of diversity, molecular structure, and isotopic composition. The molecular composition of a meteorite's organics resembles abiotic processes governed by physico-chemical processes that occurred on the parent bodies, including asteroids. The synthesis of soluble compounds in carbonaceous chondrites consists of a multi-step process in which organic precursors were formed in the interstellar cloud, incorporated with many volatiles into icy planetesimals and, upon aqueous processing produced the variety of compounds which can form complexes under space energy sources. We show experimentally that the solid mixtures of biologically significant compounds (nucleosides and amino acids) produce more complex compounds when they are exposed to either vacuum UV photons or ionizing radiation.

Keywords : Chemical evolution abiogenic synthesis amino acids nucleotides

INTRODUCTION

It is well-known that carbonaceous chondrites contain large quantities of prebiotic molecules (Cronin et al., 1988), including amino acids (Botta and Bada, 2002), carbohydrates (Cooper et al., 2001), and heterocyclic bases of nucleic acids (Stocks and Schwartz, 1981). Chondrites contain a much wider variety of amino acids than is normally found in terrestrial organisms, and their isotopic composition suggests space origin. The amino acids share a deuterium enrichment found in other meteoritic organic extracts, such as kerogen, strongly hinting at synthesis by processes operating in the interstellar medium or the Solar nebula

UV-radiation probably causes significant photochemistry within the mixed-molecular ices found in space. This process has been simulated in the laboratory using high-tech gadgetry. By simulating the photochemistry of interstellar and cometary ices in the laboratory, we were able to produce a host of organic compounds (i.e. compounds composed primarily of carbon, the kinds of molecules from which we and all living things are made). We suggest that this process may be responsible for the richness of the organics seen in the diffuse and dense interstellar medium, comets, and meteorites.

Simple organic compounds were formed originally as products of low temperature reactions that occurred in the interstellar molecular clouds, under radiation processing of astrophysical ices, and by gas-phase reactions in the primordial solar nebula (Fisher-Tropsch-type reactions), and they entered the parent bodies at the time of their accretion. Such reactions may have been involved in the production of interstellar molecules, organic compounds in meteorites, and prebiotic organic matter on planets. The production of organic compounds on asteroids as parent bodies of the meteorites may have been catalyzed by the minerals found on the surface of these bodies.

There are many environments in space in which it is possible to find solid bodies subjected to UV-irradiation and particle bombardment. McDonald et al. (1996) reported the production of a large diversity of organic compounds by experimental simulation. Alkanes, ketones, esters, and other complex molecules have been detected by gas chromatography/mass spectrometry following the irradiation of the mixture H_2O:CH_3OH:CO_2:C_6H_6 (80:16:3.2:0.8) with helium plasma. A wide variety of the initial amino acids were also found following the irradiation of an ice mixture of propane (or methane, or carbon monoxide), ammonia, and water with 3 MeV protons. Dr. Kobayashi with co-workers identified such compounds as glycine (the most abundant), alanine and aminobutyric acid (Takahashi et al., 1999). Amino acids were formed from carbon monoxide, nitrogen, and water, which are possible constituents of extraterrestrial atmospheres, following irradiation with highenergy particles (protons, helium nuclei, and electrons, all components of cosmic rays). Glycine, aspartic acid, serine, alanine, threonine, glutamic acid, aminobutyric acids, aminoisobutyric acid, valine, norvaline, sarcosine and proline were detected (Kobayashi et al., 1998). These compounds were identified after irradiation of different gas mixtures (CH_4-N_2-H_2O; CO-NH_3-H_2O; CO-N_2-H_2O) by X-ray (Takahashi et al., 1999) and visible-ultraviolet (VUV) irradiation (Takahashi et al., 2005).

Among several energy sources available for abiogenic synthesis of biomolecules in space, UV-light with different wavelengths and cosmic rays are two of the most abundant. Some extraterrestrial material exhibits strong evidence that irradiation by H and He ions is an important process on solids throughout the interstellar and interplanetary medium (Bradley, 1994). Strazzulla and Johnson (1991) demonstrated the formation of non-volatile organic polymers from proton-irradiated simulated cometary ices. The synthesis of simple oligopeptides (Gly-Trp and Gly-Tyr) from a solid mixture of single amino acids under the action of vacuum UV-radiation ($\lambda <$ 200 nm) was reported by Simakov et al. (1996). Kobayashi et al. (1995) showed that some amino acids, such as glycine and alanine, were detectable in hydrolysates when a mixture of simple inorganic compounds (CO, NH_3 and H_2O) was irradiated in a cryostat at 10 K (simulated cometary ices) with high-energy protons. Our experiment indicates that oligopeptides are also produced from mixtures of amino acids irradiated with high-energy charged particles

In order to simulate the process of chemical evolution that takes place in the interstellar and interplanetary medium, some experimental mixtures have been placed into the space environment onboard the BION-type sputnik. The full range of the solar light spectrum and the primary components of cosmic rays are available in space at high density. In contrast, the flux and energy range of light radiation are limited in laboratory exposure experiments on the ground. UV-light and cosmic rays can work together on organic molecules in space, but it is difficult to apply these two energy sources together in laboratory. The scientific aim of this experiment was to expose simulated organic films to interstellar dust and meteoritic (asteroid) surfaces. To this end, (1) the efficiency of abiotic synthesis of complex organic compounds, such as nucleotides and dipeptides, from nucleosides and simple amino acids deposited on grains was evaluated in a quantitative manner; (2) the role of each energy source in this synthesis was assessed; (3) the role of mineral surface on abiogenic synthesis was evaluated.

All young Solar system objects are subjected to energetic processing by photons and ions. As a result, the chemical and physical properties of the materials composing these objects will change significantly over time. Energetic processing of organic compounds into more complex species can be driven by a significantly enhanced UV-field in starforming regions, high-energy particle bombardment, and UV-radiation from the T-Tauri phase in stellar birth; at the early stage of evolution and at the present, UV-radiation of different wavelengths, protons of the Solar wind, and flares can drive this process. The surviving remnants of the population of planetesimals which filled the inner early Solar system during the late stages of collapsing solar nebula had a tremendous total surface, and the reactions of complex organic compounds could have taken place on the solid surface of those bodies. It is therefore important to test experimentally how far the process of chemical evolution could have taken place on the surface of space bodies under the action of energy sources available during that period and how it could take place at the present time.

Meteorites and other small bodies which orbit the Sun are presently exposed to high-energy particles in quantities large enough to produce a measurable effect of abiogenic synthesis within a few million years. At the present time, the Sun is the largest source of protons which impinge upon meteorites or interplanetary dust grains (Sandstrom, 1965). Protons emanating from the Sun are divided into two principal groups: solar-wind protons, with energies of a few keV and fluxes of roughly 1016 particles/cm2 per year, and solar- flare protons, with energies of more than 0.5 MeV and average fluxes of 1011 particles/cm2 per year. The third group of protons is those which have their source within the galaxy but outside the Solar system; these have average energies of about 1 GeV and average fluxes of about 108 particles/cm2 per year. These particles do not normally enter into chemical reactions themselves, but they can produce at least ten secondary protons which could react chemically. Estimated fluxes of energy for ice processing environments are given in Colangeli et al. (2004).

The chemical evolution of the simple molecules would continue on the surface of these bodies under the action of the energy sources available in open space. Simple oligopeptides can be formed on solid material not only by proton radiation as described here but also by vacuum UVlight, heat, and γ -radiation (Simakov et al., 1996). Thus, it can be assumed that the chemical evolution would have taken place during the early stage of the

Solar system origin and reached the stage of polymerization before the end of planet accretion. Such complex biomolecules as oligopeptides would have been brought onto the primordial Earth with comets, carbonaceous chondrites, and interplanetary dust particles (Chyba and Sagan, 1992). The next stage of evolution would be the origin of the first protoliving organisms.

Chondritic material would have been a common component of the inner Solar system shortly after its formation, and the biologically useful products of clay mineral-organic matter interactions would have also widespread and delivered to planetary surfaces through the accretion of carbonaceous asteroids. A large amount of meteoritic material has undoubtedly showered the Earth through time. The organic compounds which have been delivered to the primordial Earth would have had a very complex structure and reached the second stage of chemical evolution-polymerization.

Meteorites, being remnants of asteroids, may have imported prebiotic elements to the early Earth which, when mixed with liquid water, would have enabled the synthesis of all molecules thought to be necessary for the origin of life.

2. EXPERIMENTS

In our experiments the solid mixtures of biologically significant compounds (nucleosides and amino acids) were exposed to either vacuum UV-photons or ionizing radiation (Kuzicheva et al., 1996; Kuzicheva and Simakov, 1999; Simakov et al., 2002). Both irradiation and photolysis may destroy molecules as well as allow the synthesis of new and more complex ones. Two types of reactions were investigated here: (1) the abiogenic synthesis of nucleotides from mixtures of nucleoside + inorganic phosphate; (2) abiogenic synthesis of dipeptides from mixtures of simple amino acids. The reaction mixture in the form of a solid film contains (1) nucleoside and dihydrogen phosphate; (2) two different amino acids. Seven different nucleosides (thymidine, cytidine, uracil, adenosine or deoxyadenosine, guanosine or deoxyguanosine) and four mixtures of aromatic (tyrosine or triptophan) and aliphatic (glycine or alanine) amino acids were investigated. Mixtures were irradiated as solid films with different sources of energy: (1) VUV-light of 145 nm; (2) high-energy protons (2–6 MeV); (3) the action of all spectra of the open space energy sources during the entire time of flight (327 h; an outstanding container was installed on the surface of biosputnik).the

The thin solid films, which simulated the surface of such solar system bodies as comets, carbonaceous chondrites, and interplanetary dust particles, were irradiated with highenergy protons, which are the predominant components in cosmic ray, in order to verify the possibility of forming complex amino acids together with the formation oligopeptides.

2.1 Abiogenic synthesis of nucleotides

Nucleosides, nucleotides, phosphate, and methanol were used as received from commercial available sources. All compounds were of the highest purity commercially available. The investigated compounds were irradiated as solid thin films prepared by air drying an aqueous solution (300 mL) of an equimolar mixture of adenosine with inorganic phosphate (NaH2PO4). The area of the film was ~1 cm2. The films contained 1.3 mM nucleoside and 1.3 mM NaH2PO4. The H2PO4 – ion was chosen as a source of phosphate based on the results of Kuzicheva et al. (1993), which showed that the ability to phosphorylate decreases in the order H2PO4 –>HPO4 2– PO4 3– but that the process takes place in all cases. When the effect of a substrate was being studied, an aqueous solution of the reagents was added to 3 mg of mineral component and the suspension was dried as above. The following minerals were used as a substrate: kaolinite (Al2)(Si2)O5(OH)4, montmorillonite (Mx (Al2–xMgx)(Si4)O10(OH)2×nH2O), and ash from the Tyatya volcano. Homoionic Na+ montmorillonite was prepared by titration (Banin et al., 1985). The exchange capacity (CEC) of the cations Na+-montmorillonite used in this study was 92 meq/100 g clay

2.2 Abiogenic synthesis of dipeptides

The investigated mixtures of amino acids were irradiated as solid thin films, which were prepared by air drying an aliquot of aqueous or acetic solution of single amino acids, dipeptides or a mixture of different amino acids on a specially cut microscopic cover-glass (area ~1 cm2) and set in a sample holder for irradiation. Such amino acids as glycine, alanine, tyrosine and tryptophan and dipeptides as glycyl-glycine (Gly-Gly), tyrosyl-glycine (TyrGly), glycyl-tyrosine (Gly-Tyr), tyrosyl-tyrosine (Tyr-Tyr), tryptophyl-tryptophan (Trp-Trp), glycyl-tryptophan (GlyTrp), tryptophyl-glycine (Trp-Gly), alanyl-tryptophan (AlaTrp), tryptophyl-alanine (Trp-Ala) and alanyl-alanine (AlaAla) were investigated.

2.3 Irradiation

A lamp with a barrier discharge in Kr (145 nm), a MgF2 window, and a radiation power of 10 mW was used as a source of VUV radiation. The quantum efficiency was about 1015 quanta/cm2 per second. The irradiation time was 1–9 h. The total doses, the incident energy per area, and the absorbed doses were virtually the same—1.5–14×107 J/m2.

A low-pressure mercury lamp LB-30 (254 nm) with a radiation capacity of 6 mW/cm2 or 8×1014 quants/cm2 per second being absorbed by the sample and measured by a radiometer (UVX Digital Radiometer, USA) was used as a source of UV-radiation.

The films were irradiated with protons from a cyclotron U-120 of the Institute of Physics, St. Petersburg University. The beam of protons, with an initial energy of 6.75 MeV, was directed into a dispersion chamber with diameter about 1 m and then dissipated on a gold target with a thickness of 11.2 mg/cm2. The irradiating samples were arranged in dissipated beam of protons under angle 6°. The energy of the protons acting on the reaction mixture was 6.6 MeV. In order to avoid a heat influence on the sample, the maximal flow did not exceed 3×107 protons/cm2 per second. Beam intensities were monitored with a semiconductor detector which has a small diaphragm. The doses were 2×109– 5×1011 protons/cm2.

2.4 Space experiments

The system is an outstanding container (OC), which is a large pan with a deployable lid fixed to the exterior of a Russian BION satellite. Closed and sealed during launch and reentry operations, the system is opened upon command during the Earth orbit. Before flight and after integration of the experiments, the system was flushed with nitrogen in order to protect the samples from oxygen and humidity. The spacecraft was launched on December 24, 1996 and landed back on Earth on January 07. The duration of the flight was 13.5 days with a lid-opening time of 10 days. The orbit was not synchronous; it had a 225 km perigee altitude and a 401 km apogee altitude. The spacecraft eccentricity was 0.01239 and its inclination was 62.8°. The lid-bearing experiments were not thermally controlled during the flight. The temperature in the cavities oscillated between −31°C and +26°C. The lid was programmed to close automatically at temperatures below −35°C. The UV flux was measured with a 260–400 nm spectral response UV-sensor (G36 14, Hamamatsu). After landing, vacuum was maintained in BIOPAN until the instrument was vented with nitrogen. The samples were recovered 4 days after landing. During pre- and post flight operations, the experiments were protected against light and humidity.

The reaction mixture in the form of a solid film contained 1.3 mM of nucleoside and 1.3 mM of NaH2PO4. Four different mixture samples, consisting of equimolar quantities of nucleoside (thymidine, cytidine, adenosine, or deoxyadenosine) and inorganic phosphate, were investigated.

Mixtures were prepared by air drying of aqueous solutions on a piece of glass. The area of the film was ~1 cm2. The glasses were installed on the surface of sputnik in OC, on Plate #3. Seven samples of each mixture investigated were exposed to the action of all spectra of the open space energy sources during the entire flight. One control sample of the each mixture was placed inside and sheltered from temperature variations, high-energy charged particles, and UV- and ionizing-radiation.

3. ANALYSIS

3.1 Nucleotides

Following the irradiation of the solid films, the resulting products were dissolved in water and analyzed on a Varian- 5000 high performance liquid chromatograph (HPLC). The sample volume was 10 mL. A reverse phase column (MZRP 18; 4×250 mm; particle size 10 mm) was used, and a buffer solution of 0.05 M KH2PO4 containing 5% of methanol was employed as the mobile phase. The elution rate was 1 mL/min, and the pressure in the column was approximately 160 atm. This method shows a good separation of different nucleotides. The products were identified by coinjection with authentic samples, and yields are expressed as a mole % relative to reactants.

3.2 Dipeptides

Following irradiation, the resulting products were dissolved in water and analyzed with Varian-5000 HPLC. A reverse phase column (MZ-RP 18; 4×250 mm; 10 mm) was used, and a 40% solution of methanol in 0.005 M ammonium acetate was employed. The elution rate was 1 ml/min, the pressure in the column was approximately 160 atm, and the chromatographic sample volume was 10 mL. For each run a solution of known concentration was used for calibration. Product yields were measured from elution profiles monitored at 275 nm.

4. RESULTS AND DISCUSSION

4.1 Polymerization of amino acids

The reactions of the amino acids in solid films was the primary objective of this investigation—primarily, the abiogenic synthesis of dipeptides from mixtures of simple amino acids. Four mixtures of aromatic (tyrosine or tryptophan) and aliphatic (glycine or alanine) amino acids were investigated. These mixtures were prepared by air drying aqueous solutions and irradiating these with different sources of energy: (1) VUV-light of 145 nm; (2) UV-light of 254 nm; (3) high energy protons (2–6 MeV).

Films containing a mixture of amino acids have been shown to yield various oligopeptides after being exposed to VUV-radiation and protons (Simakov et al., 1996, 1997a). Such dipeptides as Trp-Trp, Trp-Gly, Gly-Trp, Ala-Trp, Trp-Ala, Tyr-Tyr, Tyr-Gly, and Gly-Tyr were detected as products of irradiation with summary yields of ~2% in the case of VUV-irradiation (Table 1) and ~2.5% for proton irradiation (Table 2).

When thin solid films of single amino acids—tyrosine or tryptophan—were irradiated with protons, dipeptides, such as tryptophyl-tryptophan and tyrosil-tyrosine, were found. The yields of dipeptides were 2.43% for Trp-Trp and 2.38% for Tyr-Tyr under a maximal dose of irradiation (5×1011 protons/cm2). In the case of a mixture consisting of aromatic and aliphatic amino acids (tyrosine + glycine), the yields of synthetic reactions did not essential differ from those for single amino acid. Dipeptides Gly-Tyr and TyrGly were formed with the equal probability, together with Tyr-Tyr and, possibly, Gly-Gly. This last dipeptide was not found with the method of analysis used in the present investigation. The control experiments without irradiation did not yield any trace of oligopeptides.

These values suggest that dipeptide formation does not depend on the type of side chain residue. The super-excited state that is produced during the interaction of high-energy particles with amino acids may play a major role in these processes. Free radicals are formed due to the collision of protons with molecules, and these free radicals could react with each other.

The initial amino acids are able to undergo processes of destruction as well as the process of abiogenic synthesis of olygopeptides. Of the amino acids tested, glycine is the most resistant to decomposition by UV-photons. The rate of decomposition for aromatic amino acids (tyrosine, tryptophan and phenylalanine) is significantly accelerated under these conditions, mostly due to the strong UV-absorption of the aromatic group (Orzechowska et al., 2007).

Table 1. The abiogenic synthesis of dipeptides under action of VUV-radiation (dose 3 ×106 J/m2)

Irradiated mixture	Products	Yields (%)
Gly + Trp	Gly-Trp	0.39
	Trp-Gly	0.37
	Trp-Trp	0.48
Gly + Tyr	Gly-Tyr	0.36
	Tyr-Gly	0.37
	Tyr-Tyr	0.52
Ala + Trp	Ala-Trp	0.45
	Trp-Ala	0.45
	Trp-Trp	0.51
Trp	Trp-Trp	1.87
Tyr	Tyr-Tyr	1.92

Table 2. The abiogenic synthesis of dipeptides under action of protons (6.6 MeV).

Irradiated mixture	Products	Yield (%) at doses protons/cm^2		
		2×10^9	8×10^9	500×10^9
Trp	Trp-Trp	0.47	0.98	2.43
Tyr	Tyr-Tyr	0.54	1.08	2.38
Trp + Gly	Gly-Trp	0.09	0.31	0.71
	Tyr-Gly	0.10	0.31	0.67
	Tyr-Tyr	0.16	0.33	0.62

Polymerization is an essential step in prebiological evolution, and the results reported here show that this process probably could take place even at an early stage of the Solar system formation, before planet accretion, on the surface of small bodies

4.2 Synthesis of nucleotides

VUV-irradiation of the mixture of nucleoside and inorganic phosphate resulted in the production of the natural monophosphates of the corresponding nucleosides. The main products were nucleoside-5′ -monophosphates (5’ NMP) and some small amount of by-products (2′ - and 3′ - monophosphates, 2′ 3′ - and 3′ 5′ -cyclomonophosphates)

(Table 3). The yields of products were small, within a few percentages; however, the effectiveness of the abiogenic synthesis with VUV irradiation was higher than that with UVradiation (λ = 254 nm) and heat. When the investigated films were irradiated with protons, the full mixture of nucleotides was also found (Table 4). The maximal summary yield was 9.43%. The main product was 5′ AMP under the Ado, adenosine; Guo, guanosine; dAdo, deoxyadenosine; dGuo, deoxyguanosine; Urd, uridine; Cyt, cytidine; dThy, deoxythymidine. driving action of VUV

Table 3. The abiogenic synthesis of nucleotides under action of VUV-radiation (dose: 3×106 J/m2).

Nucleoside	Nucleotide's yield (%)					
	5′	2′	3′	2′3′c	3′5′c	Sum
Ado	0.36	0.21	0.20	0.20	*traces*	0.97
Guo	0.30	0.21	0.22	0.27	0.05	1.05
dAdo	0.13	—	0.15	—	0.02	0.30
dGuo	0.23	—	0.13	—	*traces*	0.36
Urd	2.80	1.21	1.07	2.13	0.56	7.77
Cyd	3.75	1.32	1.67	2.46	0.86	9.46
dThd	2.59	—	2.00	—	0.84	5.43

Table 4. The abiogenic synthesis of nucleotides under space conditions

Nucleoside	Nucleotide yields (%)					
	5′	2′	3′	2′3′c	3′5′c	Sum
Ado	3.28	0.82	0.71	1.12	0.01	5.94
dAdo	1.87	—	0.48	—	*traces*	2.35
Cyd	2.68	0.61	0.55	0.94	*traces*	4.78
dThd	1.70	—	0.36	—	*traces*	2.06

The dominant formation of 5′ -adenosinemonophosphate (5′ -AMP) indicates that the hydroxyl group in the 5′ - position of the carbohydrate residue is most reactive under these conditions. A similar regularity was also observed under the integral influence of extraterrestrial energy sources on dry films of nucleoside and inorganic phosphate

The solid films made of mixtures of different nucleosides and inorganic phosphate were exposed to space conditions during the space flight experiment onboard the BION-11 satellite. The abiogenic synthesis of the full set of the natural nucleotides was observed, with yields of up to 6%.

After returning back to Earth the experimental films were dissolved in water and analyzed by HPLC. The abiogenic synthesis of nucleotides was observed. Such nucleotides as 5′ -, 3′ -, and 2′ -monophosphates of adenosine, deoxyadenosine, cytidine, and thymidine were detected together with cyclomonophosphates of these nucleosides

In the case of purines and pyrimidines, the main products of the reactions were 5′ -monophosphates of the corresponding nucleosides. The preference to 5′ -monophosphate formation is indicative of the more advantageous spatial position of the 5′ -hydroxyl group in the carbohydrate residue in comparison with the 2′ - and 3′ -hydroxyl groups. The summary nucleotides yields were: for adenosine, 5.8%; for cytidine, 4.7%; for deoxyadenosine, 2.3%; for thymidine; 2.0%. The yield of products was calculated in relation to the initial amount of nucleoside.

The yield of phosphorylated nucleoside is dependent on the kind of sugar moiety. Compounds which have ribose in its structure form 5′ -nucleotides twofold more frequently than those with deoxyribose. For example, the yield of 5′ AMP was 3.23%, while for that 5′ dAMP was 1.87%. This result is related to the higher reaction capability of the ribose in the ribonucleosides. Purine nucleosides react slightly more easily than the pyrimidine ones. Thus, the yield of 3′ AMP was 0.71% while that of 3′ CMP was 0.55%. In general, however, it can be concluded that the formation of nucleotides in the space environment does not depend significantly on the nature of the nucleic acid base.

The 2′ , 3′ -cyclomonophosphates were formed on the solid films under open space energy sources as well (if it was possible based on the structure of nucleosides). It is interesting to note that the yield of 2′ , 3′ -cyclomonophosphates is more than those of 2′ - and 3′ - monophosphates. There have been previous reports of an interaction between all three nucleotides under VUVradiation (Kuzicheva et al., 1996; Simakov et al., 1997b) and γ -radiation (Kuzicheva et al., 1993). This possibility can be extended to the open space condition as well.

It is possible that the temperature variations (which were about 130◦C) on the satellite surface during its rotation also contribute noticeably to abiogenic synthesis. Despite the fact that UV-radiation is the most abundant energy source in space, the heat used in the ground experiments (Kuzicheva et al., 1992), is the most soft and plausible factor for the abiogenic synthesis of biochemical compounds. The radiation level was as much as 6 Gy (600 rad) per day and up to 80 Gy (8000 rad) per 15-day mission (depending on solar activity), and solar electromagnetic radiation was on the order of 40 solar constant hours per mission. The total dose of solar light (including UV) radiation during the pace mission was 1.83 (109 J/m2 at a UV-photon flow of 1.03×1021 photons/cm2.

Both the destruction of the initial compounds as well as the synthesis of products were observed. The ribose residue was more resistant to the action of destructive influences than the deoxyribose residue. This was observed for both the purine as well as the pyrimidine nucleotides. This is confirmed by earlier received data for adenosine and deoxyadenosine (Kuzicheva et al., 1996). On the other hand, the double ring of purine nucleotides is more stable than the pyrimidine ring. During the flight time, 66% of the cytidine was destroyed, which is the maximal value of decomposition of all nucleosides. Adenosine was the most stable nucleotide in the open space environment, with 54% of the initial nucleoside surviving the flight time. Reactions involving the decomposition of the heterocyclic rings, cleavage of the N-glycoside bond, and destruction of the sugar moiety were observed in the investigated mixtures. About 20% of each nucleoside was decomposed with a cleavage of the glycoside bond and the elimination of a free nucleic acid base. In such reactions, VUV-radiation at wavelengths below 200 nm has the greatest effect (Kuzicheva et al., 1996). UV-light at wavelengths of about 254 nm could be observed to be a factor in the destruction of the heterocyclic rings. Broad spectrum UV-light from the Sun is the principal source of energy for the destruction of organic molecules. In the ground experiments, quanta yields of nucleosides as a result of photochemical destruction have been found. For cytidine, these are 4×10–3 for VUV-radiation and 4×10–5 for UV (254 nm)-radiation.

The 5′ -nucleotides are the dominant products of the nucleoside reaction with inorganic phosphate with all sources of energy, both in the lab (Simakov et al., 2002) and in flight (Kuzicheva and Simakov, 1999). In addition to nucleotide formation, the destruction of the initial adenosine, such as depurination (base release) and decomposition of the base moiety, also may take place.

The results of our laboratory experiments and experiments in open space show that many kinds of energy were effective in the chemical evolution on the surface of asteroids. Hydrated minerals are found within a chondrite matrix of meteorites together with significant amounts of organic matter. Clays are likely to have formed when water was presented on the parent meteoritic bodies. In order to verify the existence of a relationship between the abiogenic synthesis of nucleotides and the inorganic components of the meteorites, possible abiogenic reactions in the presence of different clay minerals (montmorillonite, kaolinite) and the basaltic sample (Tjatja's volcanic ash) were investigated in the presence of open space energy sources for modeling of different exobiological environments on the surface of small space bodies (Simakov and Kuzicheva, 2005). Clay minerals may have concentrated organic compounds, thereby also promoting polymerization reactions. An adsorption/binding of nucleic acid components by clay crystals could change the electron distribution and/or the conformation of the molecules.

The irradiation of dry adenosine and inorganic phosphate films with VUV in the presence of mineral substrates (kaolinite, montmorillonite, volcanic ash) produced 5′ -, 2′ -, and 3′ -adenosine phosphates along with cyclic products, such as 2′ , 3′ - and 3′ , 5′ -cycloadenosinemonophosphates. Under the maximal used irradiation dose (4×105 J/m2), the total yields of nucleotides were 5.96%, 5.55%, and 3.88% in the presence of montmorillonite, Tjatja's ash, and kaolinite, respectively. In the absence of the mineral substrate, the yield was 2.12%. The results clearly demonstrate that, depending on the mineral substrate used, increases in the yield of nucleotides can vary by 1.8–2.8 times. Kaolinite and montmorillonite are clay minerals, whereas the Tjatja's ash is compositionally very similar to terrestrial and extraterrestrial basalts (e.g. Martian soil).

The whole spectrum of natural nucleotides is present in the reaction products. The maximal yield of the main product, 5′ AMP, was 3.48% after irradiation in the presence of montmorillonite. The yield increased with increasing irradiation dose in the presence or absence of the mineral substrate. The second major reaction product was 2′ , 3′ -cycloadenosinemonophosphate (2′ , 3′ cAMP). The maximal yields were 0.33% without the mineral substrate and 0.89% with montmorillonite; i.e., the mineral component increases the nucleotide yield as in the case of 5′ AMP synthesis. It is interesting to note the yield of 2′ , 3′ cAMP is greater than those of the 2′ - and 3′ -monophosphates, respectively.

The yields from nucleoside decay are 11% (Tjatja), 13% (kaolinite), 15% (without the mineral), and 18% (montmorillonite). Here we can see a shielding effect of the volcanic ash and catalytic effect of the montmorillonite. The clay facilitates the destruction of the initial adenosine.

In the case of using the mineral substrate, a maximal influence on the reaction yield is typical of the montmorillonite. This clay gave the highest yields, both in terms of nucleotide formation and adenosine destruction. The mineral is composed of units made up of two silica tetra hedral sheets and an alumina octahedral sheet. There are isomorphic substitutions in both tetrahedral and octahedral sheets, which result in charge deficiency. This charge de- ficiency is counterbalanced by the metal ions, called interlayer cations or exchangeable metal ions, held between the layers. In contemporary clays these are mainly Na+ and Ca2+. The hydration or dissolving of these metal ions result in the expansion of the layers (increase of the interlayer distance), thereby permitting the entrance of organic molecules into the interlayer. The isomorphic substitutions are much less in kaolinite than in montmorillonite, resulting in fewer numbers of interlayer cations and, therefore, less expansion and a shorter interlayer distance.

This structure can adsorb not only inorganic ions, but also organic material. For example, 1 g of this mineral can absorb up to 5×1010 mol of adenine (El Amri et al., 2002). It is precisely this property that makes montmorillonite different from kaolinite, another clay mineral used in our experiment. Kaolinite possesses layers consisting of a 1:1 pair of octahedral and tetrahedral sheets. The shorter distance between the layers prevents any accumulation of organic molecules into the kaolinite. Consequently, nucleoside molecules can be adsorbed only by its surface. The specific surface area of both clays is also different: montmorillonite has 78 m2/g and kaolinite, 8 m2/g. The higher catalytic activity of montmorillonite, relative to kaolinite, has also been established by several other researchers studying the synthesis of abiogenic BSCs such as amino acids (Yushkin, 1999). Ash from the Tjatja volcano represents an erupted silicate product with an extremely porous texture and various oxides of different elements. The abiogenic synthesis of many organic compounds occurs precisely at the surface of ash particles in the ash-gas volcanic channels (Markhinin, 1980; Basiuk and Navarro-Gonzalez, 1996). On the other hand, this material could serve as an analogue of Martian soil, and ash from Hawaiian volcanoes is often used for such purpose (ten Kate et al., 2006).

When adenosine associates with the mineral substrate, its behavior depends on the properties of this material. Ash particles are almost nontransparent to VUV-radiation, whereas clay minerals are sufficiently permeable to VUV radiation (Boillot et al., 2002). The influence of mineral substrates on the abiogenic synthesis may depend on the following mechanisms: (1) the protective function of the mineral, which prevents the decay of both the initial components and newly formed reaction products; the screening effect of clays has been studied earlier (Bonner et al., 1984), and it has been found that the clays provided a significant protection against photolysis; (2) catalytic properties of the mineral in heterogeneous photochemical phosphorylation reactions (Otroshchenko et al., 2002); (3) the presence of residual water molecules, which can generate radicals under the influence of VUV radiation, in the interlayer space of clays; (4) the dependence of the reactivity of both adenosine and inorganic phosphate molecules on their absorption by mineral substrates. The results obtained suggest that the action of the VUV-radiation on films of BSCs makes them more complex and produces important natural compounds. One can assume that the abiogenic synthesis of relatively complex compounds can proceed under conditions of open space. The discovery that meteoritic organic compounds may be trapped and protected within a clay mineral matrix has implications for our understanding of prebiotic molecular evolution in the early Solar system (Pearson et al., 2002). Clay minerals may have concentrated organic compounds, thereby also promoting polymerization reactions. This idea was first proposed by Bernal (1949). Organic molecules can then enter the interlayer region and interact with the negatively charged basal surfaces of montmorillonite. The interaction between the adsorbed molecules and the clay surface may be via simple electrostatic attraction, van der Waals forces, the formation of hydrogen bonding, or a combination of any two or three of the above. Adsorption of the compounds in the interlayer brings the molecules into close proximity of each other. Furthermore, it may orient them into a favorable position, thereby enabling intermolecular interactions. An adsorption/binding of nucleic acids components by clay crystals could change the electron distribution and/or the conformation of the molecules. The remnant water molecules in the clay sheets could also influence the course of the reaction. In addition, an immobilization of phosphate on clay also could play an important role in our reaction (Goldberg and Sposito, 1985). Chondritic material would have been a common component of the inner Solar system shortly after its formation and the biologically useful products of clay mineral-organic matter interactions would have also been widespread and ultimately delivered to planetary surfaces through the accretion of carbonaceous asteroids. The mineral components selected for substrates can serve as a model for the surface of minor bodies in the Solar.

5. EXOBIOLOGICAL IMPLICATIONS

The problem of the origins of life is a very intriguing question of contemporary science. Geological data indicate that complex photosynthetic organisms possibly related to cyanobacteria existed 3.5 billion years (Gyr) ago (Schopf, 1993). There is also evidence of biologically mediated carbon isotope fractionation in the 3.85 Gyr old Isua metasediments, which indicate that the photoautotrophy biochemical processes must have existed at that time (Mojzsis et al., 1996).

All Solar system objects, such as planets, satellites, rings, comets, asteroids, meteorites, and interplanetary dust particles (IDPs), are subjected to energetic processing by different kinds of open space energy. The study of the role of charged particles, UV-radiation of different wavelengths, and other energy sources in the abiogenic synthesis of biologically significant compounds (BSCs) is closely related to exobiological investigations in the Earth's orbit related to chemical evolution and the issue of the origin of life on Earth and throughout the Universe

According to recent views, the majority of biologically significant compounds may have been introduced to the primeval Earth by micrometeorites, meteorites, and comets (Chyba and Sagan, 1992). These cosmic objects contain large quantities of complex organic compounds, including carbohydrates, amino acids, and heterocyclic bases of nucleic acids (Cottin et al., 1999; Cooper et al., 2001; Botta and Bada, 2002). Evidently, the chemical evolu tion proceeds under conditions of open space, which serves as a huge chemical reactor. "Simulated space ice conditions" experiments have shown the synthesis of simple BSC in the form of the precursors of amino acids (Bernstein et al., 2002; Munos Caro et al., 2002) and pyrimidine bases (uracil, cytosine and thymine) of the nucleic acids (Kobayashi et al., 2004). The present investigation dealt with the subsequent reaction of the nucleic acid components to nucleotides, which are the main components of RNA and DNA. Abiogenic synthesis of nucleotides has been reported to take place under conditions of space flight (Kuzicheva and Simakov, 1999) in the presence of different types of energy-proton beams (Simakov et al., 2002), VUV- (Kuzicheva et al., 1996; Simakov et al., 1997b), UVand γ -radiation (Kuzicheva et al., 1993).

In contrast, just before that period the surface of Earth may have been melted by high intensive comet and meteoritic bombardment, and any structures related to life may have been killed off by such catastrophic events (Sleep et al., 1989). Life would therefore arise rapidly on Earth once conditions became clement, and this process could have been very rapid, almost instantaneous on the geological time scale. Only 300 million years (from the end of the late-heavy bombardment, 4–3.8 billion years ago, up to the first well-evolved living organisms at 3.5 billion years ago) appear to be left for terrestrial chemical evolution to lead to primitive, yet fully developed organisms. There is a very short time interval for producing the biochemical compounds in the atmosphere and on the surface of planet, but this type of matter evolution had to occur before the biological one.

It can be suggested that the chemical evolution which had proceeded at the early stages of the Solar system origin would lead to the appearance of very complex organic compounds. The compounds synthesized under these conditions could have survived inside comets and meteorites, the last witnesses of the formation of our Solar system, and have been delivered to early Earth at the stage of the "late heavy bombardment" and later. The comets, carbonaceous chondrites, and other primitive bodies of the Solar system, including interplanetary and interstellar dust particles, are responsible for essentially all carbon compounds and water currently present in our planet. Therefore, these highly complex organic molecules have been proposed as the starting material for early biological activity on Earth and on other planets in the Universe (Oro et al., 1992; Flynn, 1996).

Anders (1989) estimated the flux of intact organic matter reaching the contemporary Earth in IDPs. Our planet is currently accreting ~3.2×105 kg/year of intact organics. An additional source of exogenous organics on early Earth would have been the terrestrial accretion of dust as the Solar system passed through interstellar clouds. Greenberg (2002) that during its first 7×108 years, Earth should have passed through approximately four to five such clouds, thereby accreting organic molecules during each approximately 6×105 year passage at a rate 106–107 kg/year.

REFERENCES

1. Anders, E., Pre-biotic organic matter from comets and asteroids, Nature, 342, 255, 1989.
2. Banin, A., J. G. Lawless, J. Mazzurco, F. M. Church, L. Margulies, and J. B. Orenberg, pH profile of the adsorption of nucleotides onto montmorillonite, Orig. Life Evol. Biosphere, 15, 89–101, 1985.
3. Basiuk, V. and R. Navarro-Gonzalez, Possible role of volcanic ash-gas clouds in the Earth's prebiotic chemistry, Orig. Life Evol. Biosphere, 26, 173–194, 1996.
4. Bernal, J. D., The physical basis of life, Proc. Roy. Soc. Lond. A, 357, 537–558, 1949.
5. Bernstein, M. P., J. P. Dworkin, S. A. Sandford, G. W. Cooper, and L. J. Allamandola, Racemic amino acids from the ultraviolet photolysis of interstellar ice analogues, Nature, 416, 401–403, 2002.
6. Boillot, F., A. Chabin, C. Bure, M. Venet, F. Belsky, M. Bertrand-Urbaniak, F. Delmas, A. Brack, and B. Barbier, The Perseus exobiology mission on Mir: Behaviour of amino acids and peptides in Earth orbit, Orig. Life Evol. Biosphere, 32, 359–385, 2002.
7. Bonner, W. A., H. Hall, and Y. Liang, Racemization of amino acids on clays, Orig. Life Evol. Biosphere, 15, 103–114, 1984.
8. Botta, O. and J. L. Bada, Extraterrestrial organic compounds in meteorites, Surv. Geophys., 23, 411–467, 2002.
9. Bradley, J. P., Chemically anomalous, preaccretionally irradiated grains in interplanetary dust from comets, Science, 265, 925, 1994.
10. Chyba, C. F. and C. Sagan, Endogenous production, exogenous delivery, and impact-shock synthesis of organic molecules: An inventory for the origins of life, Nature, 355, 125–132, 1992.
11. Colangeli, L., J. R. Brucato, A. Bar-Nun, R. L. Hudson, and M. H. Moore, Laboratory experiments on cometary materials, in Comets II, edited by M. C. Festou, H. U. Keller, and H. A. Weaver, pp. 695–717, University of Arizona Press, Tucson, AZ, 2004.
12. Cooper, G. W., N. Kimmich, W. Belisle, J. Sarinana, K. Brabham, and L. Garrel, Carbonaceous meteorites as a source of sugar-related organic compounds for the early Earth, Nature, 414, 879–883, 2001.
13. Cottin, H., M. C. Gazeau, and F. Raulin, Cometary organic chemistry: a review from observation, numerical and experimental simulation, Planet. Space Sci., 47, 1141–1162, 1999.
14. Cronin, J. R., S. Pizzarello, and D. P. Cruickshank, Organic matter in carbonaceous chondrites, planetary satellites, asteroids and comets, in Meteorites and the Early Solar System, edited by J. R. Kerridge and M. S. Mattews, pp. 819–857, University of Arizona Press, Tucson, 1988.
15. El Amri, A. G., M.-H. Baron, and M. C. Maurel, Adenine in mineral samples: development of a methodology based on surface enhanced Raman and Raman microscopy for picomole detection, in Proceedings of the 2nd European Workshop on Exo/Astrobiology (Graz, 2002), edited by Lacoste, H., pp. 449–450, ESA Publication Division, Noordwijk, The Netherlands, 2002.
16. Flinn, G. J., The delivery of organic matter from asteroids and comets to the early surface of Mars, Earth, Moon and Planets, 72, 469, 1996.

17. Goldberg, S. and G. Sposito, On the mechanism of specific phosphate adsorption by hydroxylated mineral surfaces: a review, Commun. Soil Sci. Plant. Anal., 16, 801–821, 1985.
18. Greenberg, J. M., Cosmic dust and our origins, Surface Science, 500, 793–822, 2002.
19. Kobayashi, K., T. Kasamatsu, T. Kaneko, J. Koike, T. Oshima, T. Saito, T. Yamamoto, and H. Yanagawa, Formation of amino acid precursors in cometary ice environments by cosmic radiation, Adv. Space Res., 16(2), 21–26, 1995.
20. Kobayashi, K., T. Kaneko, T. Saito, and T. Oshima, Amino acid formation in gas mixtures by high energy particle irradiation, Orig. Life Evol. Biosphere, 28, 155–165, 1998.
21. Kobayashi, K., Y. Takano, H. Masuda, H. Tonishi, T. Kaneko, H. Hashimoto, and T. Saito, Possible cometary organic compounds as sources of planetary biospheres, Adv. Space Res., 33, 1277–1281, 2004.
22. Kuzicheva, E. A. and M. B. Simakov, Abiogenic synthesis of nucleotides in conditions of space flight of the biosputnik "BION-11", Adv. Space Res., 23, 387–391, 1999.
23. Kuzicheva, E. A., V. F. Chernova, and O. S. Veselkina, Abiogenic synthesis of adenine nucleotides in presence of lunar soil, Zhurn. Evol. Biokhimii i Fisiologii, 28, 281, 1992 (in Russian).
24. Kuzicheva, E. A., V. F. Zhukova, and M. B. Simakov, Role of UV and γ-radiation in the abiogenic synthesis of nucleotides in solid state, Zhurn. Evol. Biokhimii i Fisiologii, 29, 339–343, 1993 (in Russian).
25. Kuzicheva, E. A., M. B. Simakov, I. L. Malko, N. Ya. Dodonova, and N. B. Gontareva, Role of vacuum ultraviolet (VUV) radiation in abiogenic synthesis of adenine nucleotides, Adv. Space Res., 18, 65–68, 1996.
26. Markhinin, E. K., Volcanoes and Life, Mysl, Moscow, 1980 (in Russian).
27. McDonald, G. D., L. J. Whited, C. Deruiter, B. N. Khare, A. Patnaik, and C. Sagan, Production and chemical analysis of cometary ice tholins, Icarus, 122, 107–117, 1996.
28. Mojzsis, S. J., G. Arrhenius, K. D. McKeegan, T. M. Harrison, A. P. Nutman, and C. R. Friend, Evidence for life on Earth before 3,800 million years ago, Nature, 384, 55–59, 1996.
29. Munoz Caro, G. M., U. J. Meierhenrich, W. A. Schutte, B. Barbier, A. Arcones Segovia, H. Rosenbauer, W. H.-P. Thiemann, A. Brack, and J. M. Greenberg, Amino acids from ultraviolet irradiation of interstellar ice analogues, Nature, 416, 403–406, 2002.
30. Orzechowska, G. E., J. D. Goguen, P. V. Johnson, A. Tsapin, and I. Kanik, Ultraviolet photolysis of amino acids in a 100 K water ice matrix: Application to the outer Solar System bodies, Icarus, 187, 584–591, 2007.
31. Otroshchenko, V. A., V. A. Alekseev, and V. K. Ryabchuk, Nonequilibrium processes of organic substances synthesis in interstellar gasdust clouds, in Astrobiology in Russia, Proceedings of the International Workshop, edited by Simakov, M. B. and A. K. Pavlov, pp. 63–92, 23–29 March 2002, St. Petersburg, 2002.
32. Oro, J., T. Mills, and A. Lazcano, Comets and the formation of biochemical compounds on the primitive Earth—a review, Origins Life Evol. Biosphere, 21, 267, 1992.
33. Pearson, V. K., M. A. Sephton, A. T. Kearsley, P. A. Bland, I. A. Franchi, and I. Gilmor, Clay mineral—organic matter relationships in the early Solar system, Meteor. Planet. Sci., 37, 1829–1833, 2002.
34. Sandstrom, A. D., Cosmic Ray Physics, North-Holland Publ. Co., 1965.
35. Schopf, J.W., Microfossils of the early Archean Apex chert: new evidence of the antiquity of life, Science, 260, 640–644, 1993.
36. Simakov, M. B. and E. A. Kuzicheva, Abiogenic photochemical synthesis on surface of meteorites and other small space bodies, Adv. Space Res., 36, 190–194, 2005.
37. Simakov, M. B., E. A. Kuzicheva, I. L. Mal'ko, and N. Ya Dodonova, Abiogenic synthesis of oligopeptides in solid state under action of vacuum ultraviolet light (100–200 nm), Adv. Space Res., 18, 61–64, 1996.

38. Simakov, M. B., E. A. Kuzicheva, N. Y. Dodonova, and A. E. Antropov, Formation of oligopeptides on the surface of small bodies in solar system by cosmic radiation, Adv. Space Res., 19, 1063–1066, 1997a.
39. Simakov, M. B., E. A. Kuzicheva, and I. L. Malko, Abiogenic synthesis of pyrimidine nucleotides in solid state by vacuum ultraviolet radiation, Adv. Space Res., 19, 1059–1062, 1997b.
40. Simakov, M. B., E. A. Kuzicheva, A. E. Antropov, and N. Ya. Dodonova, Abiogenic synthesis of nucleotides on the surface of small space bodies with high energy particles, Adv. Space Res., 30, 1489–1494, 2002.
41. Sleep, N. H., K. J. Jahnle, J. F. Kasting, and H. J. Morowitz, Annihilation of ecosystems by large asteroids impacts on the early Earth, Nature, 342, 139–142, 1989.
42. Stocks, P. G. and A. W. Schwartz, Nitrogen-heterocyclic compounds in meteorites: significance and mechanisms of formation, Geochim. Cosmochim. Acta, 45, 563–566, 1981.
43. Strazzulla, G. and R. E. Johnson, Irradiation effects on comets and cometary debris, in Comets in the Post-Halley Era, edited by Newburn Jr., R., M. Neugebauer, and J. Rahe, pp. 243–275, Kluwer, Dordrecht, 1991.
44. Takahashi, J., T. Hosokawa, H. Masuda, T. Kaneko, K. Kobayashi, T. Saito, and U. Utsumi, Abiotic synthesis of amino acids by X-ray irradiation of simple inorganic gases, Appl. Phys. Lett., 74, 877–879, 1999.
45. Takahashi, J., H. Masuda, T. Kaneko, K. Kobayashi, T. Saito, and T. Hosokawa, Photochemical abiotic synthesis of amino-acid precursors from simulated planetary atmospheres by vacuum ultraviolet light, J. Appl. Phys., 98, 024907, 2005.
46. ten Kate, I. L., J. R. C. Garry, Z. Peeters, B. Foing, and P. Ehrenfreund, The effects of Martian near surface conditions on the photochemistry of amino acids, Planet. Space Sci., 54, 296–302, 2006.
47. Yushkin, N. P., Hydrocarbon crystallization of life, Earth Sci. Frontiers, 6, 71–78, 1999.

CHAPTER 2

Origin of High Orbital Eccentricity and Inclination of Asteroids

Makiko Nagasawa, Shigeru Ida, and Hidekazu Tanaka

Department of Earth and Planetary Sciences, Tokyo Institute of Technology, Meguro-ku, Tokyo 152-8551, Japan

ABSTRACT

We investigate the origin of the high orbital eccentricity and inclination of present asteroids. The relative velocity between asteroids corresponding to the observed high eccentricity and inclination is ~5 km/sec, which is much larger than their surface escape velocity. Thus collisions result in disruption rather than accretion. It is important to clarify when and how their eccentricity and inclination were pumped up, in order to know the environment of asteroidal disruption. The origin of high eccentricity and especially that of inclination have not been well understood yet. We propose sweeping secular resonances due to the gravitational potential change associated with the depletion of the protoplanetary nebula as the excitation mechanism of eccentricity and inclination. We calculated magnitudes of excited eccentricity and inclination through three-dimensional orbital integration. We found that both eccentricity and inclination are excited enough if the protoplanetary nebula is depleted from inside to outside in the region between 5 AU to 10 AU on a timescale $\gtrsim 5 \times 10^5$ years. We found that the secular resonances sweep the asteroid belt after the nebula gas is removed from there. Then the asteroidal motion is not affected by gas drag significantly. High velocity collisions between asteroids would occur in a gas-free environment.

1. INTRODUCTION

There are two features of the present asteroid belt that cannot be accounted for by planetary perturbations in the present configuration: (1) both eccentricity (e) and inclination (i) of asteroids are too high; (2) the observationally estimated total mass in the asteroid belt (a few × 1024 g) is only 0.1% of solid component mass inferred from the minimum mass solar nebula model (Weidenschilling, 1977; Hayashi, 1981).

To consider the collisional evolution of asteroids and production of meteorites, it is important to know when and how their eccentricities and inclinations are excited. In the present paper, we focus on the problem of large orbital excitation, though the second problem should be closely related with the first one.

The random velocities of present asteroids defined by Vran = (e2 + i 2)1/2Vkep, where Vkep is Keplerian velocity at individual semi-major axes, are shown in the Fig. 1. Inclination (i) is given in radian in this paper. Relative velocities between asteroids are similar to the random velocities. The root mean square of random velocities is about 4.8 km/s. Above the dotted-broken line, orbits of asteroids with i = 0 cross the Martian orbit (1.52 AU) and above the dashed line they cross the Jovian orbit (5.20 AU). Asteroids above these lines closely approach the planets and can be strongly scattered. Arrows show the locations of strong resonances. The inclinations of asteroids are excited at the location of the secular resonance called ν16, and their eccentricities are excited at the location of the secular resonance called ν6. These secular resonances define the inner edge of the main asteroid belt. The locations of secular resonances shown in the Fig. 1 by arrows are those for i = 0, e = 0. The other arrows show the locations of mean-motion resonances. The orbits of asteroids can be largely affected by planetary perturbations only at the narrow regions around resonances, however, e and i (or Vran) of asteroids are very high in the entire belt. Vran and accordingly relative velocity between asteroids are much larger than their surface escape velocities given by 1×(m/1025 g)1/3 km/s, where m is mass of asteroid with the density of 3 g/cm3. Gravity of asteroids accelerates collision velocity up to the surface escape velocity even if Vran = 0. If Vran > Vesc, the collision velocities are more than a few times Vesc, so that the collision products cannot be bound by their gravity (e.g., Ohtsuki et al., 1993). The collision would rather result in disruption (e.g., Greenberg et al., 1978).

The high eccentricities and inclinations of asteroids cannot result from gravitational scatterings between asteroids. The random velocity acquired by mutual scattering does not exceed Vesc, because collision cross section is larger than effective gravitational scattering cross section for Vran > Vesc (e.g., Safronov, 1969). Even for the largest asteroid, Ceres, its escape velocity is only ~1 km/sec. It is impossible for mutual gravity of asteroids to excite Vran up to the present observed value ~5 km/sec. Wetherill (1989) considered a model that Earth-sized planetesimals were scattered into the asteroid belt and ejected from there after pumping up e and i of asteroids. Earth-sized planetesimals with mass M can pump up Vran of asteroids up to their surface escape velocity 10 × (M/M⊕)1/3 km/s. Using statistical methods, Wetherill (1992) showed that after the encounter of asteroidal embryos, remained collisional fragments have a velocity distribution like that of the present asteroids. Petit et al. (2001) considered the effect of planetary embryos on the asteroids starting from the stage when the runaway growth of terrestrial planets are ended (Chambers and Wetherill, 1998). They showed that the asteroids stably stay in the asteroid belt with these embryos before Jovian formation. When Jupiter is formed, the orbits of embryos are highly excited by Jovian perturbation and start to affect objects in an entire region of asteroid belt. Most of the asteroids are ejected from asteroid belt (more than 99%) by the gravitational stirring by the embryos. The eccentricities and inclinations of remaining asteroids are excited to the observed level. Their model may need optimal timing of formations of terrestrial planets, asteroids, and Jupiter, although we cannot rule out these possibilities of planetary embryos.

Similarly, the high eccentricities and inclinations are not caused by collisional disruptions. Except for a small fraction of fragments, ejected velocities of fragments are significantly smaller than the impact velocities (e.g., Fujiwara and Tsukamoto, 1980). Because collisions are a dissipative process, they damp the eccentricities and inclinations rather than excite them, on average. Actually, members of an asteroid family do not disperse so widely in semi-major axis, eccentricity and inclination

As suggested in Fig. 1, the secular resonances strongly affect the orbits of asteroids. At present, the resonances are discretely located and pump up e and i in the discrete narrow regions. However, as explained below, secular resonances migrate over the entire asteroid belt when protoplanetary nebula was being depleted. The sweeping of secular resonances is inevitable, because the depletion of the nebula after formation of Jupiter and Saturn is inevitable as explained below.

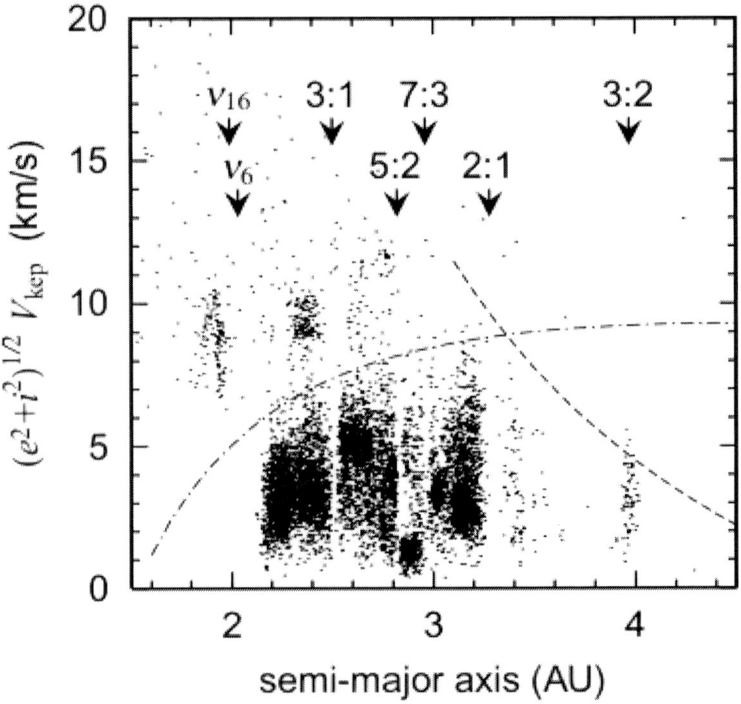

Figure 1. The random velocities ((e2 + i 2)1/2Vkep) versus semi-major axes of the present asteroids. Above the dotted line, orbits of asteroids with i = 0 cross Jovian orbit, while they cross Martian orbit above the dotted-broken line. The root mean square is about 4.8 km/s. Arrows show the locations of important mean motion resonances with Jovian orbit and secular resonances. The locations of secular resonances are shown for i = 0, e = 0.

Secular resonances occur when the proper frequency associated with the longitude of perihelion or that of the ascending node of an object coincides with one of the eigenfrequencies of planetary (perturber's) precessions. Coincidence of the longitude of perihelion causes excitation of eccentricity, and that of ascending node causes excitation of inclination. When the gravitational potential of the system (nebula potential) changes, secular resonances move, pumping up the eccentricity and the inclination of field objects. This mechanism is called sweeping secular resonances. The importance of the sweeping secular resonances was first suggested by Ward et al. (1976). Jovian and Saturnian gravity causes the secular resonances in the asteroid belt. According to the standard model (e.g.,

Safronov, 1969; Hayashi et al., 1985), solid cores of Jupiter and Saturn accrete in the protoplanetary disk. When a solid core becomes large enough (~10 M⊕), the pressure gradient no longer supports its atmosphere against gravity of the core and gas accretion onto the core starts, which results in formation of gas giants (Mizuno, 1980; Bodenheimer and Pollack, 1986). This means that the nebula gas with at least Jovian mass certainly existed when Jupiter and Saturn accreted at present locations. When the protoplanetary nebula existed, the nebula potential significantly altered locations of the secular resonances because the protoplanetary nebula has more than 10 Jovian masses even in the case of the minimum mass nebula model. This nebula was depleted afterward. There is no nebula gas in the present Solar system. The observations show that the nebula is depleted before a central star becomes a main sequence star (e.g., Strom et al., 1993; Zuckerman et al., 1995). As the nebula was depleted, the secular resonances moved from their initial locations to the present ones.

Heppenheimer (1980) studied the pumping-up of the eccentricities of the asteroids due to the sweeping secular resonances through two-dimensional linear calculations. He found that the eccentricities of asteroids could be pumped up to the present magnitude in the entire region of asteroid belt. Ward (1981) developed techniques to calculate time evolution of secular resonances in a time-dependent axis symmetric potential. Heppenheimer (1980) and Ward (1981) concluded that characteristic timescales of nebula depletion must be shorter than few ×104 years, in order to avoid excess eccentricities and inclinations of the orbit of terrestrial planets. Including the effect of gas drag, Lecar and Franklin (1997) studied the effect of the sweeping secular resonances in the nebula gas through two-dimensional numerical orbital integration, using a uniform nebula depletion model where nebula surface density decreases with the depletion factor independent of locations in the nebula similar to Heppenheimer (1980) and Ward (1981). They also concluded that the depletion time of ~104–105 years is preferable. If depletion timescale is longer, the asteroids (with size

In this paper, we show how random velocities of asteroids are excited during the depletion of the protoplanetary nebula, according to Nagasawa et al. (2000). In Section 2, we explain models we used. In Section 3 we show the results. In Section 4 we give conclusions. The random velocities of asteroids are pumped up to observed magnitude if non-uniform nebula depletion with reasonable timescale (>~ 5 × 105 years) is considered.

2. MODELS AND PROCEDURES

We consider a system that consists of Jupiter, Saturn, massless asteroids, and the nebula gas, and simulate the orbital evolution of asteroids. We use the numerical techniques of Nagasawa et al. (2000) for orbital integration. In our simulations, we initially put 38 asteroids (10–100 km sized) on nearly circular, co-planar orbits (e, i = 0.01) between 2.05 AU and 3.9 AU with equal interval of 0.05 AU. Initial conditions of Jupiter and Saturn are given by velocities and positions at J2000.0 with the present masses. As for the nebula model, we adopt the minimum mass solar nebula model by Hayashi (1981), in which the distribution of gas density is given by

$$\rho_H = 1.4 \times 10^{-9} \left(\frac{r}{1\ \mathrm{AU}}\right)^{-11/4} \times \exp\left\{-\left(\frac{z}{0.047(r/1\ \mathrm{AU})^{5/4}}\right)^2\right\} \mathrm{g/cm^3} \quad (a < 36\ \mathrm{AU}), \tag{1}$$

where (r,z) are heliocentric cylindrical coordinates. We assume the nebula has a gap with half width 1.233 AU near the orbit of Jupiter (e.g., Lin and Papaloizou, 1993). Nagasawa et al. (2000) showed that non-uniform type of nebula depletion is needed to excite inclinations. We consider the nebula is depleted from inside to outside as

$$\rho(r, z) = \begin{cases} 0 & \text{for } r < r_{\mathrm{edge}}(t) \\ \rho_H(r, z) & \text{for } r > r_{\mathrm{edge}}(t), \end{cases} \tag{2}$$

with

$$\dot{r}_{\mathrm{edge}} > 0, \tag{3}$$

where migration speed of the nebula edge, $\dot{r}_{edge}$, is given by a constant parameter. When redge = 5 – 10 AU, secular resonances sweep the asteroid belt. Hence the model we adopt is essentially equivalent to the model that a gap around Jupiter expands with time (Takeuchi et al., 1996). When the nebula is depleted from outside or depleted uniformly, the resonances of inclinations do not sweep in the asteroid belt region (Nagasawa et al., 2000). We also include the effect of hydrodynamic gas drag according to Adachi et al. (1976). As will be shown later, the gas drag does not damp the excited eccentricities and inclinations of asteroids, since secular resonances pass through the asteroid belt and the eccentricities and inclinations are excited after the nebula gas was cleared in the asteroid belt (after redge became larger than outer edge of the asteroid belt). In each simulation, orbits are integrated at least for 106 years.

3. RESULTS OF NUMERICAL SIMULATIONS

Figure 2 shows an example of the orbital evolution of an asteroid with initial a = 2.35 AU. The edge of the nebula moves at a constant velocity (r˙edge = 5 × 10−6 AU/year). When the secular resonances pass though, the eccentricities and inclinations of asteroids are pumped up. The eccentricities are pumped up three times (t = 6 × 105, 1.8 × 106, and 3.4 × 106 years). First two passages are both due to the ν5 resonance and last passage is due to the ν6 resonance. The inclinations are pumped up four times (t = 5 × 105, t = 6 × 105, 1.8 × 106, and 3.4 × 106 years). These are due to the ν16, ν15, ν15, and ν16 resonance, respectively. First two passages are difficult to distinguish in this figure. At the passage at t = 1.8 × 106 years, e and i are excited more strongly than the other passages. After the nebula depletion, mean values of i and e attain 0.73 and 0.48, respectively. Figure 3 shows the locations of the secular resonances versus the location of the nebula edge. The lines are calculated according to the Nagasawa et al. (2000). Both eccentricity and inclination resonances sweep occur in the asteroid belt. The resonances pass where the gas was depleted. The ν5, ν15, and ν16 resonances migrate near Jovian orbit from inside to outside first. These resonances follow behind the nebula edge because of the local change of gravitational potential near the edge. Next, they return and pass through the asteroid belt again from outside to inside, and finally settles down to the present location. The ν6 resonances occurs after the nebula edge pass though the Saturnian orbit. The ν16 resonance pass a = 2.35 AU when the nebula edge is at redge = 2.5 AU and redge = 16.7 AU. For the edge velocity r˙edge = 5 × 10−6 AU/year, these passage is at 6 × 105 and 3.3 × 106 years (see Fig. 2). The results of our simulations show good agreement with the analytical estimates.

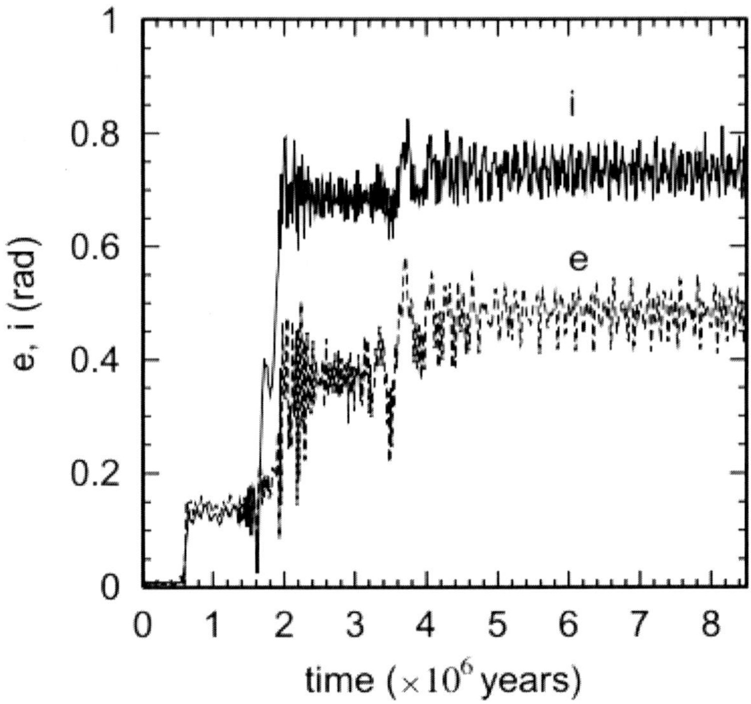

Figure 2. Orbital evolution of an asteroid with initial a = 2.35 AU, e,i = 0. The edge velocity r˙edge = 5 × 10−6 AU/year

Figure 4 shows snapshots of inclination versus semimajor axis of test particles. The edge of the nebula moves at a constant velocity (r˙edge = 5 × 10−6 AU/year). Horizontal bars show the distance from perihelia (a(1 − e)) to aphelia (a(1+e)). With the horizontal bars, we can also see the excitation of eccentricities. The orbits of particles cross each other when these bars overlap in semi-major axis. The panel of t = 0 shows the initial state (e = i = 0.01). At t = 6.5×105 years, the first secular resonances begin to pass the asteroid belt from inside to outside. Inside 2.6 AU, the inclinations and eccentricities of particles are excited by the resonance passage. In this area, random velocities (Vran = (e2 + i 2)1/2Vkep) are 3–5 km/s, which already exceeds surface escape velocities (~1 km/s) of 1000 km-sized particles. The collision cross sections between particles are larger than scattering cross sections with this speed. The nebula edge passes through a location of a at 2×105(a/1 AU) years. We found that the resonances pass after the nebula edge passes through (see Fig. 3), i.e., the orbits of asteroids are exited after the nebula is depleted around those asteroids (right upper panel of Fig. 4). Therefore, the gas drag is not effective, and the semi-major axes of particles are not damped by gas drag except at narrow regions of low-order mean motion resonances. Both secular resonances of inclination and eccentricity pass through the asteroid belt three times as seen from Fig. 2. At t = 1.3 × 106 years, the first resonances have passed the asteroid belt. Some particles acquire large enough eccentricity for orbit crossing with Jupiter and are ejected from the asteroid belt by Jovian perturbations. The random velocities of most particles are excited to ~5 km/s.

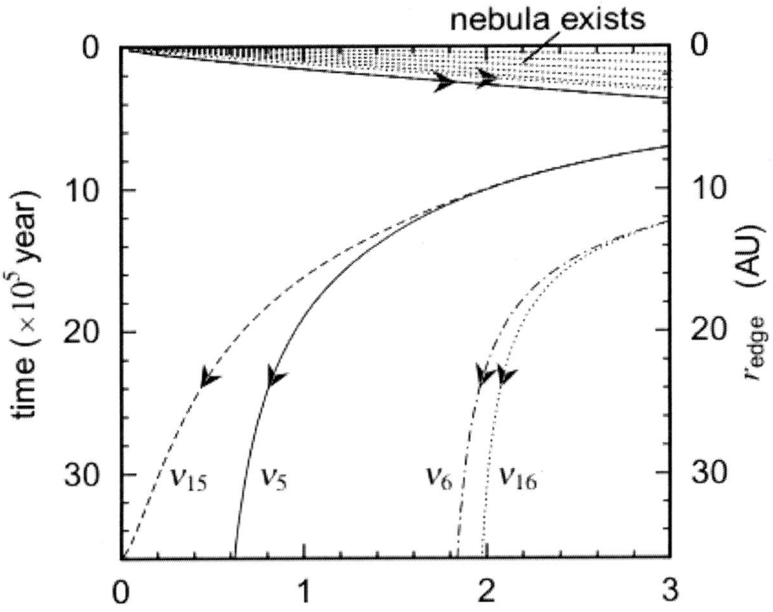

Fig.ure 3. Locations of the secular resonances ν5, ν6, ν15, and ν16 as functions of the time. The migration speed of nebula edge is r˙edge = 1 × 10−5 AU/year. The relations between the position of nebula edge (redge) and the locations of the secular resonances do not depend on the migration speed of nebula edge. The shadowed aria shows the region where the nebula is remained.

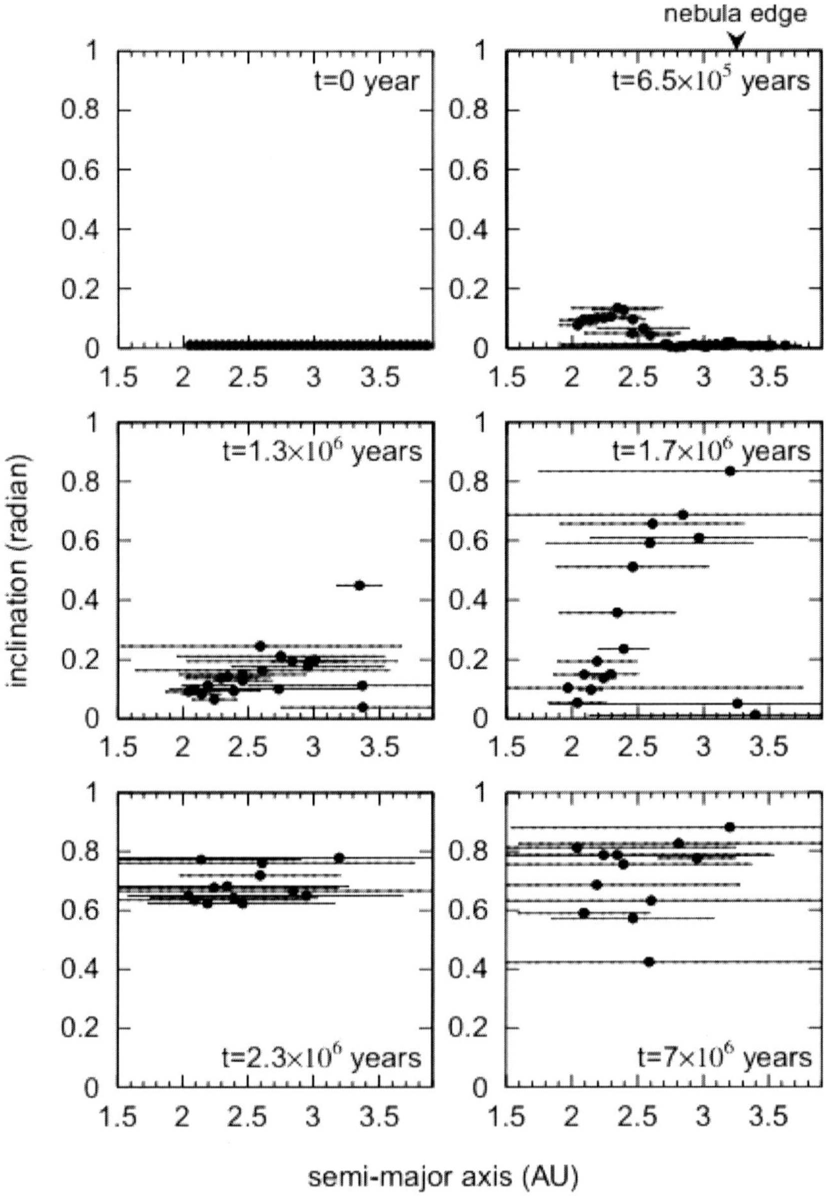

Figure 4. The evolution of inclinations vs. semi-major axis in the cases that the velocity of nebula edge is 5 × 10−6 AU/year. Horizontal bars show radial excursion due to eccentricities, more precisely, the distance from perihelia to aphelia.

At t = 1.7 × 106 years, the second resonances are passing through the asteroid belt from outside to inside. The second resonances pass over at t = 2.3 × 106 years. The last resonances pass though the asteroid belt from outside to inside at about t = 4×106 years. The particles obtain large eccentricities and inclinations. The random velocities are excited up to 20 km/s. Since the resonances can be either constructive or destructive when eccentricities and inclinations are not zero, final random velocities are spread out to some degree. Although the random velocities of all particles stay at high levels in this simulation, they would be diffused by later mutual collisions

We performed the simulation in the cases of r˙edge = 4 × 10−4, 2 × 10−5, and 5 × 10−6 AU/year. The compiled results of the final eccentricities and inclinations are shown in Fig. 5. The random velocities are excited to larger values when the nebula edge migrates more slowly. Nagasawa et al. (2000) analytically showed that the excitation level is roughly proportional to (r˙edge)−1/2. The observed random velocities are explained in the cases that r˙edge is slower than 10−5 AU/year. The excitation magnitudes of inclinations depend on the discrepancies between the nebula mid-plane and the planets. In this paper, we fix the nebula mid-plane on the current ecliptic plane. If the nebula mid-plane exactly coincides with the invariant plane of Jupiter and Saturn, the excitation effect is the minimum. In that case, the required r˙edge to excite the inclinations of asteroids up to the observational level is slower by a factor 10 than in the case that the nebula mid-plane coincides with the elliptic plane.

4. CONCLUSIONS

We have investigated the evolution of eccentricities and inclinations of asteroids caused by the sweeping secular resonances associated with depletion of the protoplanetary neb ula. The observed high random velocities (orbital eccentricities and inclinations) of asteroids cannot be explained by planetary perturbations in the present configuration. Sweeping secular resonances can be responsible for the high random velocity. We performed orbital calculations of a system that consists of Sun, Jupiter, Saturn, test particles (corresponding to asteroids), and nebula gas, according to Nagasawa et al. (2000). In order to pump up orbital inclination, the nebula must be depleted non-uniformly, e.g., from inside to outside. We found that random velocities are pumped up high enough to be comparable to the observed magnitude in the entire asteroid belt if the nebula edge migrates from 5 AU to 10 AU on a timescale longer than 5×105 years in the case of the inside-out depletion of nebula. In the case of r˙edge

= 5 × 10−6 AU/year, random velocities are excited up to 20 km/s mostly during when the nebula edge is moving in 5–10 AU.

We found that in this inside-out depletion model, the secular resonances pass through the asteroid belt after nebula gas has already been depleted there. As a result, the particles remain in the asteroid belt without damping of eccentricities and inclinations due to gas drag even if the nebula depletion timescale is as long as the observationally inferred one. Therefore, high velocity collisions between asteroids would have occurred in gas-free environment. The deficiency of total mass in the asteroid belt is still unsolved problem in our mechanism of sweeping secular resonances. Although a timing of erosion is not matter in our model, the mass depletion problem must be solved consistently with observed properties of taxonomic asteroidal families

ACKNOWLEDGMENTS

We acknowledge K. Nakazawa and H. Emori for fruitful discussions. We thank J. Lissauer and S. J. Weidenschilling for careful reading. We used the parallel computer (Silicon Graphics Origin 2000) of Earthquake Information Center of Earthquake Research Institute, the University of Tokyo.

REFERENCES

1. Adachi, I., C. Hayashi, and K. Nakazawa, The gas drag effect on the elliptical motion of a solid body in the primordial solar nebula, Prog. Theor. Phys., 56, 1756–1771, 1976.
2. Bodenheimer, P. and J. B. Pollack, Calculations of the accretion and evolution of giant planets: The effects of solid cores, Icarus, 67, 391–408, 1986.
3. Chambers, J. E. and G. W. Wetherill, Making the Terrestrial planets: N-body integrations of planetary embryos in three dimensions, Icarus, 136, 304–327, 1998.
4. Fujiwara, A. and A. Tsukamoto, Experimental study on the velocity of fragments in collisional breakup, Icarus, 44, 142–153, 1980.
5. Greenberg, R., W. K. Hartmann, C. R. Chapman, and J. F. Wacker, Planetesimals to planets—Numerical simulation of collisional evolution, Icarus, 35, 1–26, 1978.
6. Hayashi, C., Structure of the solar nebula, growth and decay of magnetic fields and effects of magnetic and turbulent viscosities on the nebula, Prog. Theor. Phys. Suppl., 70, 35–53, 1981.
7. Hayashi, C., K. Nakazawa, and Y. Nakagawa, Formation of the solar system, in Protostars and Planets II, edited by D. C. Black and M. S. Mathews, pp. 1100–1153, Univ. of Arizona Press, Tucson, 1985.
8. Heppenheimer, T. A., Secular resonances and the origin of eccentricities of Mars and the asteroids, Icarus, 41, 76–88, 1980.

9. Ida, S. and D. N. C. Lin, Long Term Gas Drag Effect on the Structure of the Asteroid Belt and its Implications for the Solar Nebula, AJ, 112, 1239–1246, 1996.
10. Lecar, M. and F. Franklin, The Solar Nebula, Secular Resonances, Gas Drag, and the Asteroid Belt, Icarus, 129, 134–146, 1997.
11. Lin, D. N. C. and J. C. B. Papaloizou, On the tidal interaction between protostellar disks and companions, in Protostars and Planets III, edited by E. H. Levy and J. I. Lunine, pp. 749–835, Univ. of Arizona Press, Tucson, 1993.
12. Mizuno, H., Formation of the Giant Planets, Prog. Theor. Phys. Suppl., 64, 544–557, 1980.
13. Nagasawa, M., H. Tanaka, and S. Ida, Orbital Evolution of Asteroids during Depletion of the Solar Nebula, AJ, 119, 1480–1497, 2000.
14. Ohtsuki, K., S. Ida, Y. Nakagawa, and K. Nakazawa, Planetary accretion in the solar gravitational field, in Protostars and Planets III, edited by E. H. Levy and J. I. Lunine, pp. 1089–1107, Univ. of Arizona Press, Tucson, 1993.
15. Petit, J. M., A. Morbidelli, and J. Chambers, The primordial excitation and clearing of the asteroid belt, 2001 (preprint).
16. Safronov, V. S., Evolution of Protoplanetary Cloud and Formation of the Earth and Planets, Nauka, Moscow, NASA Tech. Trans. F-677 1969.
17. Strom, S. E., S. Edwards, and M. F. Skrutskie, Evolutionary timescales for circumstellar disks associated with intermediate- and solar-type stars, in Protostars and Planets III, edited by E. H. Levy and J. I. Lunine, pp. 837–866, Univ. of Arizona Press, Tucson, 1993.
18. Takeuchi, T., S. M. Miyama, and D. N. C. Lin, Gap Formation in Protoplanetary Disks, ApJ, 460, 832–847, 1996.
19. Ward, W. R., Solar nebula dispersal and the stability of the planetary system. I—Scanning secular resonance theory, Icarus, 47, 234–264, 1981.
20. Ward, W. R., G. Colombo, and F. A. Franklin, Secular resonance, solar spin down, and the orbit of Mercury, Icarus, 28, 441–452, 1976.
21. Weidenschilling, S. J., The distribution of mass in the planetary system and solar nebula, Astrophys. Space Sci., 51, 153–158, 1977.
22. Wetherill, G. W., Origin of the asteroid belt, in Asteroids II, edited by R. P. Binzel, T. Gehrels, and M. S. Matthews, pp. 661–680, Univ. of Arizona Press, Tucson, 1989.
23. Wetherill, G. W., An alternative model for the Formation of asteroids, Icarus, 100, 307–325, 1992.
24. Zuckerman, B., T. Forveille, and J. H. Kastner, Inhibition of Giant Planet Formation by Rapid Gas Depletion around Young Stars, Nature, 373, 494–496, 1995.

CHAPTER 3

Incidental Catch and New Distributional Records of Asteroids by Bottom Trawling Activities from Southeast Coast of India

K. Chamundeeswari*, S. Saranya, S. Shanker, D. Varadharajan, S. Rajagopal and T. Balasubramanian

Faculty of Marine Sciences, Centre of Advanced Study in Marine Biology, Annamalai University, Parangipettai-608 502, Tamil Nadu, India

ABSTRACT

A study was undertaken for a period of one year between January 2011 and December 2011 from Parangipettai coastal waters (Mudasalodai and Annankoil) southeast coast of India. Present study discusses the distribution and systematic position of asteroids from these study areas. There are 8 species viz., Luidia maculate (15.6%), Astropecten indicus (54.9%), A. hemprichi (6.9%), Stellaster equestris (21.7%) and Anthenea pentagonula (0.1%), Protoreaster linki (0.2%), Pentacerasterr regulus (0.1%) and P. affinis (0.2%) was observed. 4 species viz., Luidia maculata, Astropecten indicus, A. hemprichi, Stellaster equestris are commonly available in both stations, where as P. lincki and P. regulus observed only from station II and Anthenea pentagonula is newly recorded from station I. diversity indices of Shannon (H'), Simpson (1-D), Evenness, Mergalef's species richness and cluster analysis, MDS also derived based on the number of observed species. Species diversity, abundance, richness and evenness are higher in station I than the station II. The present study concluded that station I has higher resources of asteroids (89.1%) than station II (10.9%) and this study also provides additional information about diversity ofasteroids.

Keywords: Abundance; Trash fish; A. pentagonula; Diversity; Shallow waters; Intertidal

INTRODUCTION

Asteroidea is one of the five main classes in the phylum Echinodermata. Representatives of them are commonly called as sea stars or star fishes are marine pentaradiate with the bilateral ancestors comprising of 1890 extant species under 36 families [1] and about 180 species of asteroids are prevailing in Indian waters including Andaman and Nicobar Islands [2]. Sea stars exists in ecosystems of sea grass, coral, rocky substratum etc. from shallow to deepest region in the ocean which shows different feeding modes like predators of small invertebrates (paxillosids), detritivores (valvatidans), planktivores (brisingids) and coral polyps eaters [3]. They play an important role in ecology as an indicator of fishing intensity, biomarkers for pollution [4]. They act as key species due to their predatory nature and they are documented as determining factors in patterns of distribution, abundance and diversity within low intertidal habitats [5,6]. They widely used in the developmental and experimental studies and used as model organism in the area of community structure and feeding ecology [7,8]. The recent threat for sea stars is habitat destruction and ocean acidification. Habitat destruction is the major threat to the sea star species those has restricted distribution. Sea stars live on the rocky substrates are very susceptible to increase of sand and silt deposition. Coastal development such as jetties and breakwaters which alter the shoreline and changing water currents will cause deposition of sand and silt. Deterioration of sea grass habitat by pollution will affect the some herbivorous sea stars. For specific algal feeding sea stars, changes in water quality or nutrients may susceptible to those feed on particular algae. Collection of sea stars for aquarium keeping also one of the threats for species decline for e.g. Patiriella vivipaara, Smilasterias tasmaniae and Marginaster littoralis are restricted to Tasmania and there are considered as threatened species. Ocean acidification is another serious threat to sea stars because they required calcium carbonate in the form of high magnesium calcite for construction of skeleton. Magnesium calcite is more soluble and sensitive to ocean acidification than the other calcite forms [9]. Studies on the biology or other related work of sea stars are scarce from Indian waters. Some lucid contribution on taxonomy was done by [10-25]. The knowledge on asteroids occurrence is poorly known from these coastal areas, hence, the present survey was undertaken to know abundance and distribution of asteroids through incidental catch along the Parangipettai coastal waters, southeast coast of India.

MATERIALS AND METHODS

Description of the study area

Samples were collected from Parangipettai specifically at Mudasalodai (Station I, Lat.11o 28'N; Long 79° 46'E) and Annankoil (Station II, Lat.11°29'N; Long.79°46'E) fish landing centre during the study period from January 2011 to December 2011. Station I & II (Figure 1) is a small coastal fishing village 15km south of Periyakuppam and it is merely closer to the marine biological station of Annamalai University. In station I, about 82 small trawlers and 15 non-motorized catamarans are engaged in regular fishing activity. Shrimp trawl net, fish trawl net, crab net are the major gears being operated. Fish, shrimp, crab and squid are the main target fishing. It is carried out at the depths range of 5-35m. Sea bottom is muddy with enormous silt, received from the Vellar River. In station II, about 200 FRP motorized boat and 130 small motorized (Lambadi) are engaged in fishing. Gillnets (Kanangatha valai-58mm and Katuvalai-166-118mm) used for capturing commercial fishes like Mackerel, sardine, squid and shrimps etc.

Sample collection

Collected samples were brought to the laboratory where samples cleaned with water to remove adhered mud and other particles then identified them based on the morphology, spines in the abactinal (including supero-marginals, disc, radial arms, paxillae) actinal (infero-marginals, adambulacral, adoral)with the help of appropriate literature [10]. Data were collected fortnightly, pooled seasonally and this was repeated throughout the study period.

Statistical software used

The univariate methods such as Diversity indices of Shannon (H index), species richness indices of Mergalef index (d), Simpson (1-D) and evenness index were used for estimation of relative abundances of different species of sea stars for four seasons. The multivariate methods of classification (Bray-Curtis cluster analysis) and ordination (nonmetric multidimensional scaling ordinations {MDS}) was analyses by using software PRIMER version6 for assessing the relative importance in terms of abundance.

RESULTS

Species composition

Totally 806 individuals of 8 species such as Luidia maculata, Astropecten indicus, A. hemprichi, Stellaster equestris, Anthenea pentagonula, Protoreaster lincki, Pentaceraster regulus and P. affinis (plate I) from 6 genera under 2 orders viz., Paxillosida and Valvatida were noticed at station I and II (Table 1). Species composition was varied among the observed study area. Maximum of 426 individuals were observed for A. indicus at station I and minimum of 1 for A. pentagonula station I.

Percentage contribution

The highest percentage of 55 was given by A. indicus during the survey, followed by 21.7% for S. equestris and 15.6% for L. maculata. The lowest of 0.1%was set by both P. regulus and A. pentagonula respectively (Figures 2 and 3).

Taxonomic description of species

Luidia maculate Muller & Troschel, 1842

Synonyms: Luidia maculate and Luidia varia

Material examined: 10 specimens from station I and 4 from station II were examined.

Description: Rays 7 much elongated and cylindrical; R/r ranged from 90-100/8.0-8.2mm and arm breadth (br) 8.0-8.2mm. Disc flat, margins raised with the distinct rows of paxillae. No prominent supero-marginal plates. Abactinal plates with distinct paxillae arranged in 5-6 transverse rows arranged along the arms. The disc arrangement confused centrally. Paxillae are well spaced and square or rectangular at proximal and rounded, consisted at distal. Single series of actinal plates between infero-marginals and adambulacrals. Stalked pedicellariae found enormously on oral side.

Habitat: Sandy bottoms in bay, lagoon or in the deep reef, benthic, bentho pelagic and coastal.

Color: Creamy-white bars and blotches interspersed with brown bands on the aboral and creamy on the oral side.

Geographical distribution: Commonly distributed in the Indo- Pacific region.

Remarks: Alternative bold color patterns of dark and whitish plates on aboral are very clear in all the specimens. We observed species with 7 arms only. No marked madreporite. Stalked bivalve pedicellariae are numerous along the adambulacrals. All the specimens were with lost and degenerating arms. Dissection of adult specimen showed presence of bifurcated ampullae. Tube feet pointed. Clark HES et al. [26] reported that L. varia is the synonymy of L. maculata. This species is differing from L. magnifica by having 9-10 arms.

Astropecten indicus Doderlein, 1888

Common name: Fringed starfish

Synonyms: Astropecten indicus

Material examined: 12 specimens from station I and 5 from station II were observed.

Description: Arms 5, R/r ranged between 32-40/10-13 mm and br: 9-12 mm. No anus instead involuted cone like structure at disc centre. Paxillae regularly arranged along the dorso-lateral arm with single row of dark stripes between it. Supero-marginal plates mostly 22, well developed and without spines but bearing alternatively arranged simple granules throughout. Infero-marginal plates with single series of spines and plates were extended to the adambulacrals. Two actinal plates were observed at the interradial region. Inter actinal region with spines. Ambulacral groove bordered with two rows of spines and adambulacral spines larger than the furrow spines. Tube feet pointed arranged in a single row on either side of the ambulacral ridge.

Habitat: Subtidal sandy, detritus and muddy bottoms often burrowing into the sea bottom.

Color: Greyish or muddy color on the aboral and creamy on the oral.

Distribution: Tropical Indo-Pacific.

Remarks: They generally inhabit sandy and muddy substrate. Stomach was always filled with bivalves and gastropods. Relative lengths of arms are the unique character of this species. Superomarginal plates without spines but bears alternatively arranged two series of tiny tubercles towards arm tip. This species is available in a larger quantity throughout the year.

Astropecten hemprichi Muller & Troschel, 1842

Synonyms: Astropecten hemprichi, Astropecten articulates and Astropecten mauritianus.

Material examined: 6 from station I and 4 from station II were examined.

Description: Arms 5 four times longer than the central disc, R/r: 91-102/20-22mm and br: 18-20mm. Supero-marginal plates were numerous and no spines at the first four plates. About 15paxillae at each inter radius between the basal supero-marginal plates are distinct. Brown color alternative bands along the inter radius and edges of arms are clearly seen in juveniles. Single series of short spinelets found at the inner end of supero-marginal plates. Single series of enlarged spines found along the infero-marginal spines and also backed with numerous small spines. Two series of adambulacral spines and the inner are more elongated.

Habitat: Sandy bottom

Color: Brownish on aboral

Distribution: Aldabra, Eastern Africa and Madagaskar, Mauritius, Mozambique, Red sea, Seychelles, South Africa, Tanzania, Bay of Bengal, Arabian Sea, West Indian Ocean.

Remarks: Young ones are more fragile and distinguish from adult by bearing stripes along interradius. Sastry DRK [25] described that, specimens lake supero-marginal spines instead, short conspicuous and conical spines are observed.

Stellaster equestris Retzius, 1805

Synonyms: Asterias equestris, Stellaster childreni, Stellaster equestris childreni, Stellaster equestris, Stellaster incei and Stellaster belcheri.

Material examined: 6 from station I and 5 from station II were observed.

Description: Arm 5, shorter and narrower than disc. R/r: 60-65/29-33 mm and br: 8-11 mm. Center of the disc is slightly raised and arms flat. There is a small groove in each inter radius. Aboral plates are more or less round and are different sizes and covered by small granules. The median carinal plates are slightly elevated from the rest which carries 2-3 enlarged granules on each arm. Inter radial plates adjacent to the supero-marginal plates are slightly elevated. The first few infero-marginal plates are small and oval in shape, remaining plates are almost square shaped and flat. Each infero-marginal plates bear a peculiar type of single series of spines. Actinal plates are small, rounded fewer in number and gives smooth texture. The adambulacral armature consists of two rows of spines. The outer row consists of single spine at the centre, a pedicellariae at the proximal and an enlarged granule at the distal end. The second row consists of 6 short flat spines of which the middle ones are the largest.

Remarks: This species is more distinguished from other species by having unique polygonal plates on oral and aboral side. Clark AM [10] stated that, juvenile specimen of Goniodiscaster have smooth, even granulation like Stellaster.

Habitat: Sandy and muddy environments

Color: Purple on aboral purplish with creamy white on oral side

Anthenea pentagonula Lamarck, 1816

Synonyms: Asterias pentagonula, Anthenea pentagonula, Anthnea rudis and Anthenea tuberculata.

Material examined: Only one specimen from station I.

Description: Arm 5 shorter and broad at base. Disc is wider. R/r: 75/45 mm, br: 43 mm. Body pentagonal, edges are stretched by prominent and vertically placed well developed supero-marginal plates. Disc larger, skin is thick which bears spinelets and granules. The carinals are slightly raised. Wider interradials, all supero and inferomarginal plates are stacked with each other and engraved with thick granules. Numerous large bivalved pedicellariae scattered on oral side are interesting characters of this species.

Habitat: Sandy and muddy environments.

Color: Reddish on pentagonal aboral and the rest of the body with pale color.

Geographical distribution: Gulf of Mannar, Andaman and Nicobar Islands and tropical Indo-west Pacific regions.

Remarks: Remarkable point is that few bristles like a structure may be of parasites exposed on the aboral surface of this species.

Protoreaster lincki Blain ville, 1830

Common name: African Red-Knob sea star; Red Knob; Red Spine Star; Crimson Knobbed Star Fish; African Sea Star.

Synonyms: Asterias lincki, Oreaster linckii, Pentaceros lincki, Pentaceros muricatus, Oreaster muricatus, Oreaster reinhartdi.

Material examined: Only two specimens were observed from station II.

Description: Arm 5, elongated from the disc and triangular, R/r: 120-114/35-39 and br: 38-42 mm. disc raised upwardly, plates reticulated, angular inter radius, single central spine near anus, single spine on each primary plate and 3 on secondary plates. Spines are stout with pointed tips. Carinal spines are stout and shorter than primary one. Either lateral side towards arm tip of adult specimen has 3 lateral spines. Bivalve pedicellariae are more scattered on oral side.

Habitat: Coral and sea grass environment

Color: Pinkish on grey background on aboral and fully pinkish on oral

Geographical distribution: Southeast Arabia, Ceylon, Bay of Bengal, East Indies, Australia, western and central Indian Ocean.

Remarks: One of the dissected specimens had shown disc type tube feet and it was pinkish from the podia and bifurcated ampullae whitish. Clark AM [10] mentioned that, Pentaceraster alveolatus superficially resembles Protoreaster lincki in the development of abactinal and supeo-marginal spines but differs in having infero-marginal spines.

Pentaceraster regulus Muller & Troschel, 1842

Synonyms: Pentaceros regulus, Pentaceros australis

Material examined: Three specimens were observed from Station II.

Description: Arm 5, relatively short & narrow, disc enlarged, R/r: 70-90/35-45 mm and br: 25-39 mm, dorsal armament found only on the disc. Elevations and spines form regular longitudinal series. Dorso lateral elevations or spines developed along the arm rarely few in number. No inter radial supero-marginal spines. Valvate type of pedicellariae distributed on the actinal surface only. Tube feet disc type. Ventro-lateral plates 3 rows and distinct.

Habitat: Sandy and muddy environments.

Color: Maroon on light or pale background.

Geographical distribution: Bay of Bengal, East Indies, North Australia, Philippine, China, south Japan and South Pacific Islands and Lakshadweep to western Pacific Ocean.

Remarks: Clark AM [10] mentioned, P. regulus and P. gracilis has similar appearance. Karuppaiyan M [27] stated that this species shows remarkable color patterns like bluish, greyish, yellowish.

Pentaceraster affinis Muller & Troschel, 1842

Synonyms: Pentaceros affinis and Pentaceraster affinis.

Material examined: Two specimens were observed from station I.

Description: Arm 5, R/r: 95-110/ 48-55 mm and br: 50-53 mm. This species is looks similar like P. regulus, but it is differ by its dorso lateral armament on the entire aboral surface. Abactinal plate more reticulated, small pedicellariae were scattered over on it. Porous are contained approximately 50 pores per each area. Single tubercleon each supero-marginals, spinelets were grouped on infero-marginals. Primary, carinal and latero-ventral tubercles stouter and were relatively smaller. Armsmore or less elongated. Inter marginal plate present between inter radial infero and supero-marginal plates. Abactinal armaments also spinulated.

Habitat: Sandy

Color: Maroon patches on grey background

Geographical distribution: Tropical Indo-west Pacific

Remarks: This species has similarity with P. mammillatus by having aboral reticulated patterns and tubercles.

Diversity indices

Season wise contribution: The maximum of 275 individuals were accounted during pre monsoon and minimum of 103 during summer at station I and it was 31 during pre monsoon and 15 during monsoon at station II. The Shannon index (H') showed maximum value of 1.262 during summer and minimum 0.9294 during monsoon at station I and it was 1.585 during post monsoon & 1.288 on pre monsoon at station II. The higher value for Simpson (1-D) index was 0.6821 on summer and minimum of 0.4815 on monsoon for station I and 0.7744 during post monsoon & 0.7055 pre monsoon for station II. Evenness (eH/S) showed maximum of 0.716 at pre monsoon and minimum of 0.5667 at post monsoon from station I and 0.993 during monsoon and 0.8136 during pre monsoon at station II. Likewise, the higher value of Margalef index for station I & II was0.863 (summer), 1.553 (post monsoon) and minimum 0.5341 (pre monsoon) & 0.8736 during pre monsoon respectively (Figures 4 and 5).

Cluster analysis: At station I, the cluster analysis showed only one group was formed by L. maculata and S. equestris at the level of 69% similarity (Figure 6). Three groups formed at station II (Figure 7). The first group is formed by linking L. maculata and S. equestris at the highest similarity level of 85%, the second group is formed by A. indicus and A. hemprichi at 82% and third group by P. lincki and P. regulus at the level of 67%.Similarity level for both the stations (Figure 8) showed three groups. First group formed by L. maculata and S. equestris at 82% level followed by two groups such as A. pentagonula, P. regulus, P. lincki and P. affinis formed at the level 66%.

MDS: There are three clusters formed at the stress level 0 for station I. Top right showed one cluster formed by P. affinis, bottom left showed one cluster formed by A. pentagonula and another cluster formed at the mid-left of bottom. Four clusters formed from station II at stress level 0. Two cluster from each right and left of top and bottom respectively. By overall, only three clusters formed at the stress level 0.05 viz., top right and middle left of bottom and right side (Figure 9,10 and 11).

DISCUSSION

Sea stars from the study areas were represented by 8 species and four of them were found abundantly in both stations. The family Luidiidae included only one species L. maculata. The family Astropectinidae included A. indicus and A. hemprichi. Family Goniasteridae included S. equestris and Oreasteridae includes A. pentagonula, P. lincki, P. regulus and P. affinis. In the total percentage composition, A. indicus alone gives foremost contribution and least was set by A. pentagonula and P. regulus. Sea star A. pentagonula was reported for the first time from Mudasalodai. Karuppaiyan M [27] reported that this species is restricted to Gulf of Mannar. Many environmental factors control the abundance and distribution of these organisms but one vital reason is abundant food supply. A. indicus was abundantly caught in both the stations reveals that, this species predominantly found in the shallow regions. Wells FE et al. [28] have stated that, the species of Astropecten are found throughout the world in the intertidal and subtidal sandy environments. Clark HES et al. [26] also affirmed that, the large family Astropectenidiae has a worldwide distribution and is known from shallow subtitdal to greater depths. Sea star P. regulus and A. pentagonula is a shallow water sandy habitant, occasionally caught by trawl nets hence it's caught was minimum in both stations. The maximum occurrence was due to spawning during summer followed by catching of more juveniles during pre monsoon by bottom trawl from station I and catching of organisms by gillnets operation at station II. The minimum occurrence was due to fishing holidays at station I and due to monsoon at station II.

Species diversity is a simple and useful measure of a biological system [29]. The quantification of biodiversity is fundamental to the identification of changes that may take place to our understanding of their possible consequences and it can be measured of many levels and in different ways [30,31]. In the present study, the maximum diversity indices were noticed during summer at station I and minimum during monsoon. The same result was obtained to station II but the values were shifted to post monsoon and pre monsoon respectively. The Shannon and Simpson diversity was correlated with each other for station I. The low diversity obtained during the survey was monsoonal flood. Species evenness was more even on pre monsoon and lesser on post monsoon for station I, it was higher on monsoon & lower on post monsoon for station II. The evenness for the macro benthos depends on the many environmental factors like habitat, food availability, migration etc. controls the distribution of sea stars. Mostly, the paxillosidans are found common on the shallow sandy and muddy areas but valvatididans found little deeper and sandy and reef environments. There is a general consideration that, the species are more or less equal to evenness when the benthic community undisturbed [32,33]. Richness of species also found higher during summer at station I & on post monsoon at station II.

The taxonomic diversity is sum of biodiversity which is seen to be natural extension of a form of Simpson index, incorporating taxonomic (or phylogenetic) information since it mixes taxonomic relatedness with the evenness properties of the abundance distribution. The value of this univariate index appears not to be influence by sampling effort. The diversity measures derived in this study is based on the observation of number of sample only. The MDS 2 or 3D map of samples reflects the similarities of their biological communities, for example the species has clustered much closer have similar community structure and species are far apart have different level of abundance. The goodness of fit is measured by the stress value; an ideal representation having zero stress. Relative stress values increase with increasing number of entities and decreasing dimensions. The MDS ordinations derived in this present study revealed that, the same grouping of stations as in the cluster analysis. The stress values found in all the MDS configurations from the two stations were at 0 levels indicating good representation of the interrelationship between the different groups of sea stars in both stations.

Asteroids are a unique group of marine invertebrates which has more confusion in identifying species also this group is puzzle for zoologists and paleontologists. Very few works of it has been done so for in India and other neighboring countries, some most spectacular works were done by James DB and Sastry DRK [14-25], [34] from Indian waters. A remarkable work on the global trend of asteroid diversity by Mah CH et al. [1] provides updated information on it. From this study, we conclude that, the Parangipettai coastal water has 8 different species of sea stars, amongst, 4 of them were common to both the stations and station I has high species abundance. Based on the number of species, the diversity is more in the station II even though this study provides some information on asteroids still, more study is needto understand the deeper water sea stars. This study provides sufficient taxonomical information about asteroids. This study also implies that this is the additional work for the existing work. The untargeted asteroid resources have to be utilized for human needs with sustainable management.

ACKNOWLEDGEMENTS

Authors are thankful to the authorities of Annamalai University for providing necessary facilities and to the Ministry of Earth Sciences, New Delhi, India for the financial assistance.

REFERENCES

1. Mah CH, Blake DB (2012) Global Diversity and Phylogeny of the Asteroidea (Echinodermata). PLoS one 7: e35644.
2. Venkataraman K, Mohideen W (2005) Coastal and marine biodiversity of India. I J Mar Sci 34: 57-75.
3. Cintra-Buenrostro CE, Reyes-Bonilla H, Herrero-Perezrul MD (2005) Oceanographic conditions and diversity of sea stars (Echinodermata: Asteroidea) in the Gulf of California, Mexico. Rev Biol Trop 53: 245-261.
4. Everaarts JM (1995) DNA integrity as a biomarker of marine pollution: Strand breaks in seastar (Asterias rubens) and dab (Limanda limanda). Mar Pollut Bull 31: 431-438.
5. Lubchenco J, Menge BA (1978) Community Development and Persistence in a Low Rocky Intertidal Zone. Ecol Monogr 59: 67-94.
6. Menge BA (1982) Effects of feeding on the environment: Asteroidea. In: Echinoderm Nutrition, M Jangoux, JM Lawrence (eds.). Balkema, Netherlands. pp 521-551.
7. Menge BA, Daley BA, Lubchenco J, Sanford E, Dahlhoff E, et al. (1999) Top-Down and Bottom-Up Regulation of New Zealand Rocky Intertidal Communities. Ecol Monogr 69: 297-330.
8. Ortiz M, Jesse S, Stotz W, Wolff M (2003) Feeding behavior of the asteroid Meyenaster gelatinosus in response to changes in abundance of the scallop Argopecten purpuratus in northern Chile. Arch Hydrobiol 157: 213-225.
9. UNEP (2010) UNEP Emerging Issues: Environmental Consequences of Ocean Acidification: A Threat to Food Security.
10. Clark AM, Rowe FWE (1971) Monograph of shallow water Indo-west Pacific Echinoderms. Trustees of the British Museum (Natural History), London vii. p 238.
11. Sladen WP (1882) The Asteroidea of HMS Challenger Expedition, Part I - Pterasteridae. J Linn Soc Zool 16: 186-246.
12. Sladen WP (1883) The Asteroidea of HMS Challenger Expedition, Part II – Astropectinidae. J Linn Soc Zool 17: 214-269.
13. Sladen WP (1889) On the Asteroidea of the Mergui Archipelago, Collected for the Trustees of the Indian Museum, Calcutta by Dr. John Anderson, FRS Superintendent of the Museum. J Linn Soc 21: 319-331.
14. James DB (1997) Notes on the family Goniasteridae (Echinodermata: Asteroidea) from the Indian seas. J Mar Biol Ass India 38: 133-138.

15. James DB (1969) Catalogue of Echinoderms in the reference collections of the Central Marine Fisheries Research Institute, Mandapam. Bull Cent Mar Fish Res Inst 7: 51-62.
16. James DB (1972) Note on the development of the asteroid Asterina burtoni Gray. J Mar Biol Ass India 14: 883-884.
17. James DB (1973) Studies on Indian Echinoderms 5. New and little known Starfishes from the Indian Seas. J Mar Biol Ass India 15: 556-559.
18. James DB (1983) Research on Indian Echinoderms-A review. J Mar Biol Ass India 25: 91-108.
19. James DB (1986) Zoogeography of the shallow-water Echinoderms of Indian seas. In: Recent Advances in Marine Biology, PSBR James (ed). Today and Tomorrow printers and publishers, New Delhi. pp 569-591.
20. James DB (1998) Echinodermata, B. In: Alfred JRB (Ed.) Faunal diversity in India. Zoological Survey of India Culcutta, I-VIII. Publisher - Envis Centre. pp 398-403.
21. Sastry DRK (1995) Asteroidea, Ophiuroidea and holothuroidea (Echinodermata). Zool. Surv. India. Estuarine Ecosystem Series. Part 2: Hugli Matla Estuary. 327-228.
22. Sastry DRK (1998) Echinodermata. Zool. Sur. Of India. Culcutta Fauna series 3: Fauna of West Bengal. Part 10: 463-489.
23. Sastry DRK (1999) Echinodermata of Great Nicobar Island, Bay of Bengal. J Andaman Sci Asso 15: 91-93.
24. Sastry DRK (1999a) New Records of Echinodermata from Andaman Islands. J Andaman Sci Ass 15: 17-20.
25. Sastry DRK (2001) Echinodermata. Zool. Surv. India. Estuarine Ecosystem Series 4, Fauna of Godavari Estuary. 83-84.
26. Clark HES, McKnight DG (2001) The marine fauna of New Zealand: Echinodermata: Asteroidea (Sea-stars). NIWA Biodiversity Memoir. p 196.
27. Karuppaiyan M (2007) Diversity of Echinoderms along the southeast coast of India. Ph.D., Thesis. Faculty of Marine Sciences, Annamalai University, India. p 266.
28. Wells FE, Lalli CM (2003) Astropecten sumbawanus (Echinodermata: Asteroidea) in Withnell Bay, northwestern Australia. In: The Marine Flora and Fauna of Dampier, Western Australia Western Australian Museum, Perth, Wells, F. E., D. I. Walker & D. S. Jones (eds.). pp 209-216.
29. Redding DW, Hartmann K, Mimoto A, Bokal D, DeVos M, et al. (2008) Evolutionarily distinctive species often capture more phylogenetic diversity than expected. J Theor Biol 251: 606-615.
30. Harper JL, Hawksworth DL (1994) Biodiversity: measurement and estimation. Philos Trans R Soc Lond B Biol Sci 345: 5-12.
31. May RM (1994) Biological Diversity: Differences between Land and Sea. Philos Trans R Soc B Biol Sci 43: 105-111.
32. Varadharajan D, Soundarapandian P, Gunalan B, Babu R (2010) Seasonal Abundance of Macro Benthic Composition and Diversity along the South East Coast of India. European J Appl Sci 2: 1-5.
33. Manoharan J, Varadharajan D, Thilagavathi B, Priyadharsini S (2011) Biodiversity and abundance of benthos along the South East Coast of India . Advan Appl Sci Res 2: 554-562.
34. Radhadrishnan Y, Ganapati PN (1969) Fauna of Kakinada Bay. Bull Nat Inst Sci India 38: 689-699.

CHAPTER 4

Towards Understanding The Dynamical Evolution of Asteroid 25143 Itokawa: Constraints from Sample Analysis

Harold C Connolly Jr[1,2,3,4*], Dante S Lauretta[4] , Kevin J Walsh[5] , Shogo Tachibana[6] and William F Bottke Jr[5]

[1]*Earth and Environmental Sciences, The Graduate Center of CUNY, 365 5th Ave, New York, New York 10016, USA.*

[2] *Department Physical Sciences, Kingsborough Community College of CUNY, 2001 Oriental Blvd, Brooklyn, NY 100235, USA.*

[3] *Department Earth and Planetary Sciences, AMNH, Central Park West, New York, NY 10024, USA.*

[4] *Lunar and Planetary Laboratory, University of Arizona, Tucson, AZ 85721, USA.*

[5] *Southwest Research Institute, 1050 Walnut St, Suite 300, Boulder, CO 80302, USA.*

[6] *Department of Natural History Sciences, Hokkaido University, N10W8, Sapporo 060-0810, Japan.*

ABSTRACT

The data from the analysis of samples returned by Hayabusa from asteroid 25143 Itokawa are used to constrain the preaccretion history, the geological activity that occurred after accretion, and the dynamical history of the asteroid from the main belt to near-Earth space. We synthesize existing data to pose hypotheses to be tested by dynamical modeling and the analyses of future samples returned by Hayabusa 2 and OSIRIS-REx. Specifically, we argue that the Yarkosky-O'Keefe-Radzievskii-Paddack (YORP) effect may be responsible for producing geologically high-energy environments on Itokawa and other asteroids that process regolith and essentially affect regolith gardening.

Keywords: Asteroids Hayabusa Itokawa Cosmic-ray exposure ages Chondrites

INTRODUCTION

Numerous remote-sensing techniques coupled with dynamical simulations have provided important constraints on the characteristics and potential origins of near-Earth asteroids (NEAs). While these studies have provided valuable insight, we argue that the only method for validating the conclusions from such work is through analysis of samples returned from these objects. In addition, a key method for understanding the geological and dynamical evolution of asteroids in general, and NEAs in particular, is to develop hypotheses that are testable through the analysis of returned samples. The Hayabusa mission returned samples in the form of numerous small grains of regolith from the surface of asteroid 25143 Itokawa. Research on these returned particles has been a binding force for the sample analysis, remote sensing, and dynamic-modeling communities (Keller and Berger 2014; Langenhorst et al. 2014; Thompson et al. 2014).

We contend that there are two major lines of evidence generated from the analysis of Hayabusa samples that can be used to test hypotheses on the dynamical evolution of Itokawa in the recent geological past: (1) the cosmic-ray exposure (CRE) ages of the returned grains and (2) the overall physical characteristics of these grains. We present a synthesis of existing data from the analysis of Hayabusa samples and chondritic meteorites. We pose new hypotheses and a scenario for the evolution of NEAs that can be tested by future dynamical modeling. We pose a scenario that links seemingly unrelated data, CRE ages of returned regolith grains from Itokawa and those of chondrites to the physical shape of the returned Itokawa grains, and pose processes that affected the dynamical evolution of NEAs with the consequences of producing rounded grains on Itokawa that have young CRE ages and thus affecting regolith gardening.

Cosmic-ray exposure ages, tidal disruptions, and YORP?

CRE ages, either of meteorites or of returned sample from Itokawa, into a context related to the dynamical evolution of asteroids, specifically NEAs, have been a considerable challenge. CRE ages represent two kinds of processes. First, the penetration depth of cosmic rays into rock is on the order of a meter, so meteorites that show damage from isotropic directions, which is often referred to as 4π exposure, are assumed to have been free-floating bodies in space smaller than a few meters in diameter prior to reaching Earth. Here, the CRE ages record the time of liberation from their parent bodies (i.e., asteroids) to when they were delivered to Earth.

Damage from a more uniform direction, called 2π exposure, tells us that the sample in question was close to or on the surface of a larger body. Several analyzed grains returned from the asteroid Itokawa have these kinds of CRE ages. If the ages can be correctly interpreted in the context of a process whereby they were exposed near the surface of Itokawa, they can potentially provide us with clues to the dynamical evolution of Itokawa itself.

The classical interpretation for what CRE ages of meteorites record is a catastrophic disruption event through a major collision within the main belt that delivered the meteoroids to resonances and then the Earth (Eugster et al. 2006; Herzog 2005). If this was the main mechanism, however, we should see a plethora of young CRE ages among all meteorite groups, for our purposes chondrites in particular, which is not observed (Bottke et al. 2006).

The 4π CRE ages for different types of chondrites span a wide range (Eugster et al. 2006; Herzog 2005). For the highly lithified ordinary chondrites (OC), most of their ages range between approximately 0.5 and 80 Myr. Interesting peaks in the data are found at 6 to 10 Myr for H chondrites, approximately 40 Myr for L chondrites, and approximately 15 Myr for LL chondrites.

The CRE age distribution of carbonaceous chondrites (CC) is fairly similar to those of ordinary chondrites. The least friable of the CC, the CV, CK, and CO chondrites, have average CRE ages that fall around approximately 13, 23, and 22 Myr, respectively. For the intermediate-strength CR chondrites, a cluster exists around an age of approximately 8 Myr. CRE ages for CH chondrites fall within the range of CRs. CB chondrites have a calculated age of approximately 26 Myr. Finally, the CI and CM chondrites, some of the most friable chondrites, show the youngest CRE ages, with averages of approximately 1.8 and 2.8 Myr, respectively.

The 4π CRE age data for OC and CC demonstrate that some chondrite samples were delivered to Earth in a relatively short time span of a Myr or less, inconsistent with their origin from catastrophic disruption within the main belt (Bottke et al. 2006). As has been discussed in the literature, in addition to catastrophic disruptions, the Yarkovsky effect plays a major role in the orbital evolution of asteroids and meteoroids from the main-belt to near-Earth space (Bottke et al. 2006, for a recent review). The Yarkovsky effect results from the reemission of absorbed solar radiation as thermal radiation, which produces a steady drift in the semi-major axis of the asteroid's orbit. Slow Yarkovsky transportation in the main belt is hypothesized to be the mechanism that explains why meteorites have so few young ages. Thus, the CRE ages of chondrites are thought to describe the time between (1) their origin as small meter-sized meteoroids in the main belt, (2) their subsequent evolution by Yarkovsky thermal forces into a main-belt escape hatch, and (3) their delivery onto planet-crossing orbits that can eventually take them to Earth (see Bottke et al. 2006 for a recent review) or NEA production. As we discuss below, it can also be argued that at least some chondrites delivered to Earth may have come from larger precursors that shed mass impact events or other processes while on Earth-crossing orbits.

To set the stage for our scenario, it is important to review the evidence that links chondrites to asteroids and what we know about the age of the surface of asteroids as constrained from Hayabusa samples. The parent asteroid spectral type for ordinary chondrites was hypothesized to be the S-type asteroids (Binzel et al. 2001; McCoy et al. 2001; Binzel et al. 2010; Nesvorny et al. 2010; Nakamura et al. 2011). The Hayabusa samples provide a key data point for referencing an S-type asteroid to ordinary chondrites, supporting the above hypothesis. Particles returned from the surface of Itokawa are LL chondrite material, ranging from metamorphic type 4 to 6 (Nakamura et al. 2011; Tsuchiyama et al. 2013). In addition, we now have data from samples returned from Itokawa that constrains the age of the surface of that asteroid. The 2π CRE ages for Itokawa samples taken from the surface show a young age of approximately 1 Myr or less, (Meier et al. 2013, 2014), with an upper limit of approximately 8 Myr (Nagao et al. 2011). We note that these data are within the lower range, if not even younger, than the 4π CRE age for ordinary chondrites, indicating that the surface of Itokawa is as young if not even younger than many ordinary chondrites recovered on Earth.

The parent asteroid spectral type for carbonaceous chondrites is hypothesized to be the C-type asteroids (Bus and Binzel 2002; Clark et al. 2011). However, we have no unequivocal evidence that links C-type asteroids to a specific type of carbonaceous chondrite. Therefore, we have no unequivocal evidence of CRE age for material from the surface of C-type asteroids. The goal of the Hayabusa 2 and OSIRIS-REx missions is to return samples from 1999 JU3 and Bennu, respectfully, and provide ground truth on these asteroids through sample analysis.

Explaining the very young CRE age of sample returned from Itokawa requires that we think of all processes that may have affected the asteroids dynamical evolution. We pose the following major questions to explore: (1) Why are the CRE ages for particles found on the surface of Itokawa young? (2) Is there a relationship between the ages of the Itokawa particles and a process that affected and/or affects the evolution of this asteroid? (3) Finally, can we apply what we have learned from the materials returned by Hayabusa to make some wide-scale predictions on the range of CRE ages determined for chondritic materials that are recovered on Earth?

Exploring the questions posed above: it is important to note that the 2π CRE ages of samples from the Moon are on the order of 100 Myr to many hundreds of Myr (Heiken et al. 1991), by far longer than the ages found for grains returned from Itokawa. Thus, the surface of Itokawa is a much more dynamic place than the Moon, and different processes may be shaping the surface. The standard party-line answer to question 1 could be regolith gardening, and that remains a valid hypothesis; at least in part, gardening might be responsible for the age of Itokawa regolith. The question is what processes produce gardening on small bodies like asteroids? The overall morphology of Itokawa is a strong statement that the asteroid has been dynamically disrupted. The issue is what were the affects of this disruption on regolith or regolith production and how does the process that produced the disruption relate to the questions posed above? To explore question 1 and begin to investigate question 2 above, it is important to explore processes other than catastrophic disruption events and the Yarkovsky effect that affect the dynamical orbital evolution of NEAs: the Yarkosky-O'Keefe-Radzievskii-Paddack (YORP) effect and tidal disruptions during planetary close encounters.

The YORP effect is a result of the same physical phenomenon that produces the Yarkovsky effect, namely the reflection/absorption of sunlight and subsequent reemission as thermal radiation. YORP can create a torque that modifies the rotation of small asteroids so they spin up or down, while also affecting their obliquity. In some cases, the YORP effect can add enough rotational angular momentum to change the asteroid shape, with particles, rocks, boulders, etc. moving across the asteroid surface in response to the resulting centrifugal forces (Walsh et al. 2008).

The other dynamical process that can affect the physical properties of NEAs is tidal disruptions from a close flyby to a terrestrial planet. The tidal forces of a planet acting across the asteroid passing within a few planetary radii of a planet like Earth or Venus can entirely disrupt or distort the asteroid (Richardson et al. 1998). Nesvorny et al. (2010) argued that tidal effects could disturb the surface of a NEA in a number of ways: (1) The interior structure of a rubble-pile NEA might find a new equilibrium by rearranging its components; this could result in landslides or displacement of near-surface material, etc. (2) Tidal stresses applied to a fractured interior could produce seismic shaking, allowing fresher material to reach the surface. (3) The tidal torques could add or subtract enough rotational angular momentum to cause it to shed or rearrange its surface regolith. We argue that all of these processes could have contributed to producing new regolith on the surface of Itokawa and thus explain the young CRE ages of returned sample.

Nesvorny et al. (2010) also explored how tidal forces might resurface NEAs in the context of space weathering. It has been found that many fresh-looking, non-space-weathered ordinary-chondrite-like bodies called Q-type asteroids have orbits similar to asteroids that would be most likely to undergo tidal encounters. By tracking asteroids during close encounters with Earth and Venus, and then comparing these results to the orbital distribution of space-weathered S-types and Q-types in the NEA population, Nesvorny et al. (2010) were able to calculate the timescale of resetting by space weathering for the asteroid population. The order of their timescale, approximately 1 Myr at 1 AU for objects making typical passes of 5 planetary radii near Earth or Venus, represents the interval during which a fresh Q-type affected by space weathering will remain a Q-type; at least the dominant signature from the asteroid surface is from materials that produce Q-type spectra.

To explore question 2 from above, we hypothesize that the physical processes we have discussed that affect near-Earth objects, such as YORP or tidal encounters, may provide an explanation for the short CRE ages of the grains found on Itokawa. One end-member possibility is that the CRE age data suggest that the last major breakup and reaccretion experienced by Itokawa occurred approximately 8 Myr (upper limit), with subsequent processing occurring at potentially approximately 1.5 Myr or <0.5 Myr (or at all ages). During these breakups, either as a result of YORP or tidal disruptions, or a combination of these processes, Itokawa would have had its surface reshaped and it is highly likely that it lost mass (Richardson et al. 1998; Walsh et al. 2008; Binzel et al. 2010; Nesvorny et al. 2010).

Within our discussion, we have set the stage for a scenario that can provide an answer to question 3 posed above: extending our reasoning to meteorites, we hypothesize that YORP and tidal disruption had the net effect of causing NEAs to shed mass, with perhaps some of this lost mass arriving to the surface of Earth (Richardson et al. 1998; Walsh and Richardson 2006). Thus, we hypothesis that chondrites (or other meteorites) that have young 4π CRE ages can be explained not only from an impact event, but as having been the result of mass loss from their parent asteroid(s) due to YORP spin-up or tidal disruption. A range of CRE ages within chondrites or chondrite type could be reflecting whole-scale restructuring of their parent asteroids due to YORP or tidal disruption, potentially mixing rock as cobbles, boulders, etc., to produce a combination of young (fresh even) and older regolith. We can extend this scenario to grains returned from Itokawa being the product of mixing due to YORP or tidal disruption as discussed above.

Geologic high-energy environments on Itokawa: abrasion of grains and YORP effects?

Based on their overall general shape, grains returned by Hayabusa from Itokawa have been divided into two main groups: angular and rounded (Tsuchiyama et al. 2011, Tsuchiyama et al. 2013). The presence of angular grains on the surface of Itokawa was not unexpected. However, the presence of rounded grains is a surprise. The presence of rounded rock or minerals in any geological setting on a planet indicates a geologically high-energy environment - one that is or was actively moving material resulting in mechanical weathering, which must be more than a one-time event in order to turn angular grains into rounded ones. Mechanical weathering of rocks or minerals is a differential process that in part depends on characteristics such as composition of the material and in the case of rocks the level of their lithification. Such geologically high-energy environments on planetary surfaces are produced by wind- or water-dominated geological settings, even ice-dominated such as glacial movement. However, on an airless body such as Itokawa, weathering of material on surfaces cannot be the product of movement of any kind of fluid, particularly on the surface of bodies. Whatever mechanism or combination of mechanisms produced the rounded grains, the young CRE ages of the returned grains indicate that they were formed or exposed near the surface relatively recently and science must explain their existence. We thus pose two questions directly related to the rounding of grains on airless bodies such as Itokawa: (1) What geologically high-energy environments exist or existed that could physically weather rock to produce rounded grains on the surface of an asteroid? (2) Does the overall physical shape and characteristics of the Itokawa particles constrain the dynamical evolution of Itokawa and, by extension, other asteroids?

Exploring the questions posed above: a geological high-energy process and one known to have occurred on asteroids is macro- or micro-impacts, which without question is a powerful process for weathering of materials and the production of regolith. Tsuchiyama et al. (2011, 2013) hypothesized that the geologically high-energy environment needed to produce the physical weathering of rounded Itokawa grains was the byproduct of shaking of the asteroid due to impacts. We envision that what is implied by this hypothesis is that grains repeatedly migrate within the regolith due to repeated impacts. Whereas we cannot completely refute this hypothesis through quantitative argument, qualitatively the number and size of impacts on the surface of such small asteroids as Itokawa that could produce an environment energetic enough to physically weather particles seems unlikely. In addition, as discussed above, Itokawa's surface is not littered with impact craters (Michel et al. 2009) adding additional qualitative support to our argument. To date, however, no quantitative modeling has been performed to test the impact hypothesis, and thus, we cannot nor do we completely rule it out as a contributor to grain production through mechanical weathering.

To explain the physical weathering of rounded grains on Itokawa, we hypothesize that the dominant mechanism is from instabilities in the rotational axis and, most importantly, YORP effects. As was argued by Walsh et al. (2008), YORP can play a significant role in evolving the regolith on the surface of small asteroids. Whereas we also cannot discount the potential importance of tidal forces affecting asteroids, and certainly it may have produced angular fragments of rocks and regolith, YORP occurs steadily over the lifetime of the asteroid, compared to the whole-scale stochastic restructuring of an asteroids physical shape from tidal forces alone. The spinning-up of an asteroid due to YORP would produce an environment where material is colliding frequently and thus physically weathering, both on the spin-up and in the reaccretion. Once an asteroid is reformed by YORP, we speculate that the surface of the object may have new slopes; thus, regolith that exists and new regolith that forms could be migrating downslope, as may be true for asteroid Bennu (Lauretta et al. 2014), which could have the net effect of weathering small grains to become rounded. For Itokawa, it is not clear if grains could be migrating downslope to the Muses Sea. Miyamoto et al. (2007) argued that grain migration might have been due to gravity forcing landslides or grain convection on the asteroid surface. They also discussed the potential for grain movement due to tidal forces. We speculate that the roundness of the grains may have been largely related to grain collisions resulting from YORP-driven processes, which may also have caused major mass-wasting processes such as landslides (Miyamoto et al. 2007). Tidal forces, however, are most likely not responsible for multiple reshaping of regolith grain from angular to rounded. If small, dust-sized grains are to be rounded, it may require multiple or continuous restructuring of the asteroid, and that we argue is consistent with YORP.

It is not clear that Itokawa went through or could experience YORP and some debate exists within the literature (Kitazato et al. 2007; Breiter et al. 2009; Lowry et al. 2014; Mazrouei et al. 2014; Scheeres et al. 2007). We contend that the existence of rounded grains is additional evidence that Itokawa experienced YORP. Indeed, some of the boulders on the surface also appear rounded, which may provide additional evidence for a geological high-energy environment created by YORP.

As discussed above, all particles returned by Hayabusa thus far investigated have relatively young CRE ages. We have discussed a potential mechanism that could have produced rounded grains. It would be remiss of us if we did not also apply our reasoning to the production of angular grains. Angular regolith grains could have been produced from a number of mechanisms: (1) impacts, (2) YORP-driven processes, (3) tidal forces, and/or (4) thermal fatigue (Murdoch et al. 2012; Delbo et al. 2014). Constraining which one (or combination of processes) was the dominant mechanism may require a data set that carefully compares the CRE ages of rounded grains with those that are angular. We predict that there should be a higher abundance of rounded grains with slightly older CRE ages than angular grains due to the fact that time is needed to produce the physically weathered rounded grain from an angular one.

Future missions and predictions on the nature of the returned sample

In the future, Hayabusa 2 and the OSIRIS-REx missions will return samples from two NEAs, 1999 JU3 and Bennu that are hypothesized to be very different chemically from Itokawa and certainly, as asteroids, have different overall shapes. We predict, however, that the CRE ages of the returned materials will be young, closely matching that of CI and CM chondrites (Clark et al. 2011), which are proposed as the closest analogs to type C asteroids. We further predict that some of the particles returned will have experienced weathering to produce rounding, and we further speculate that, based on our reasoning above, because of the overall shape of 1999 JU3 and Bennu, the proportion of rounded grains to angular will be higher than material returned from Itokawa. Material may be continuously migrating from polar to the equatorial regions on these asteroids with the overall shape of these objects, at least for Bennu, having been shaped in part by YORP (Nolan et al. 2013). It is one of the goals of these missions to test the hypotheses put forth in this paper through the analysis of the returned samples and for modelers to make additional predictions on the affects of YORP and tidal disruptions that will be tested by the returned samples.

SUMMARY

We have proposed that the relatively young CRE ages of samples returned from Itokawa may be reflecting either YORP or tidal disruption. Given the fact that rounded grains returned by Hayabusa require a geological high-energy environment for their production and that YORP has the potential for repeatedly affecting the surfaces of asteroids over time whereas tidal disruption does not, we propose that YORP was a major process that helped produce and shape the regolith on Itokawa (Figure 1). Furthermore, we extend our reasoning to predict that mass loss from NEAs that occurs during YORP is responsible, at least in part, for delivering chondrites with young CRE ages to Earth.

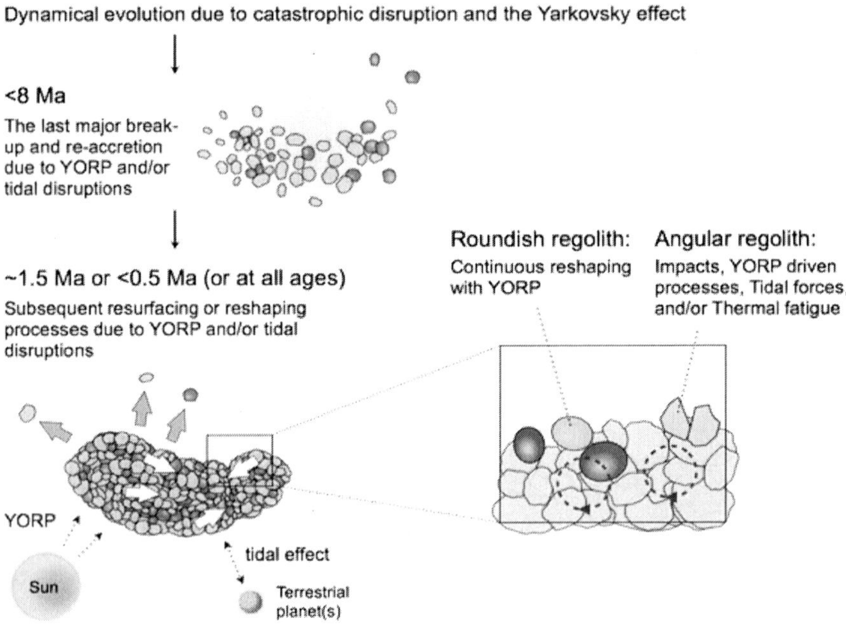

Figure 1. A schematic drawing summarizing the potential evolution of the surface of itokawa.

ACKNOWLEDGEMENTS

We thank two anonymous reviewers for their helpful comments and Dr. Toru Yada for his editorial handling of the paper and for his helpful comments. This work was supported by NASA contract NNM10AA11C (D. S. Lauretta, PI; H. C. Connolly Jr. Co-I, W. F. Bottke Jr., Co-I, K. J. Walsh, Co-I).

COMPETING INTERESTS

The authors declare that they have no competing interests.

AUTHORS' CONTRIBUTIONS

HCCJr, DSL, KJW, ST, and WFBJr contributed to the interpretation and preparation of this manuscript. All authors read and approved the final manuscript.

REFERENCES

1. Binzel RP, Rivkin AS, Bus SJ, Sunshine JM, Burbine TH (2001) MUSES-C target asteroid (25143) 1998 SF36: a reddened ordinary chondrite. Meteorit Planet Sci 36:1167–1172
2. Binzel RP, Morbidelli A, Merouane S, DeMeo FE, Birlan M, Vernazza P, Thomas CA, Rivkin AS, Bus SJ, Tokunaga AT (2010) Earth encounters as the origin of fresh surfaces on near-Earth asteroids. Nature 463:331–334
3. Bottke WF Jr, Vokrouhlicky D, Rubincam DP, Nesvorny D (2006) The Yarkovsky and YORP effects: implications for asteroid dynamics. Annu Rev Earth Planet Sci 34:157–191
4. Breiter S, Bartczak P, Czekaj M, Oczujda B, Vokrouhlicky D (2009) The YORP effect on 25143 Itokawa. Astronomy Astrophysics 507:1072–1081
5. Bus SJ, Binzel RP (2002) Phase II of the Small Main-belt Asteroid Spectroscopy Survey: a feature-based taxonomy. Icarus 158:146–177
6. Clark BE, Binzel RP, Howell ES, Cloutis EA, Ockert-Bell M, Christensen P, Barucci MA, DeMeo F, Lauretta DS, Connolly HC Jr, Soderberg A, Hergenrother C, Lim L, Emery J, Mueller M (2011) Asteroid (101955) 1999 RQ36: spectroscopy from 0.4 to 2.4 μm and meteorite analogs. Icarus 216:462–475
7. Delbo M, Libourel G, Wilkerson J, Murdoch N, Michel P, Ramesh KT, Ganino C, Verati C, Marchi S (2014) Thermal fatigue as the origin of regolith on small asteroids. Nature 508:233–236
8. Eugster O, Herzog GF, Marti K, Caffee MW (2006) Irradiation records, cosmic-ray exposure ages, and transfer times of meteorites. In: Lauretta DS, McSween HY (eds) Meteorites and the early solar system II. University of Arizona Press, TucsonGoogle Scholar
9. Heiken GH, Vaniman DT, French BM (1991) Lunar source book: a user's guide to the Moon. Cambridge University Press, CambridgeGoogle Scholar
10. Herzog G (2005) Cosmic-ray exposure ages of meteorites. In: Davis AM (ed) Treatise on Geochemistry. 1 Elsevier, San DiegoGoogle Scholar
11. Keller L, Berger E (2014) A transmission electron microscope study of Itokawa regolith grains. Earth Planet Space 66:71
12. Kitazato K, Abe M, Ishiguro M, Ip W-H (2007) 25143 Itokawa: direct detection of the current decelerating spin state due to YORP effect. Astronomy Astrophysics 472:L5–L8
13. Langenhorst F, Harries D, Pollok P, van Aken A (2014) Mineralogy and defect microstructure of an olivine-dominated Itokawa dust particle: evidence for shock metamorphism, collisional gragmentation, and LL chondrite origin. Earth Planets Space 66:118
14. Lauretta DS, Bartels AE, Barucci MA, Bierhaus EB, Binzel RP, Bottke WF, Campins H, Chesley SR, Clark BC, Clark BE, Cloutis EA, Connolly HC, Crombie MK, Delbo M, Dworkin JP, Emery JP, Glavin DP, Hamilton VE, Hergenrother CW, Johnson CL, Keller LP, Michel P, Nolan MC, Sandford SA, Scheeres DJ, Simon AA, Sutter BM, Vokrouhlicky D, and Walsh KJ (2014) The OSIRIS-REx target asteroid 101955

Bennu: constraints on its physical, geological, and dynamical nature from astronomical observations. Meteorit. Planet Sci. In press.

15. Lowry SC, Weissman PR, Duddy SR, Rozitis B, Fitzsimmons A, Green SF, Hicks MD, Snodgrass C, Wolters SD, Chesley SR, Pittichová J, van Oers P (2014) The internal structure of asteroid (25143) Itokawa as revealed by detection of YORP spin-up. Astronomy Astrophysics 562:A48
16. Mazrouei S, Daly MG, Barnouin OS, Ernst CM, DeSouza I (2014) Block distributions on Itokawa. Icarus 229:170–180
17. McCoy TJ, Burbine TH, McFadden LA, Starr RD, Gaffey MJ, Nittler MJ, Evans LG, Izenberg N, Lucey P, Trombka JI, Bell JF, Clark BE, Clark PE, Squyres SW, Chapman CR, Boynton WB, Veverka J (2001) The composition of 433 Eros: a mineralogical-chemical synthesis. Meteorit Planet Sci 36:1661–1672
18. Meier MMM, Alwmark C, Bajt S, Botter U, Busemann H, Gilmour J, Heitmann U, Hubers H-W, Marone F, Pavlov S, Schade U, Spring N, Stampanoni M, Terfelt F, Weber I (2013) Determing a precise He, Ne, cosmic-ray exposure age for grains from Itokawa. Paper presented at the HAYABUSA 2013: Symposium of Solar System Materials. Abstr# 1017–1145. JAXA Sagamihara Campus, Japan, 16–18 October 2013
19. Meier MMM, Alwmark C, Bajt S, Botter U, Busemann H, Fujiya W, Gilmour J, Heitmann U, Happe P, Huber H–W., Marone F, Ott U, Pavlov S, Schade U, Spring N, Stampanoni M, and Weber I (2014) A precise cosmic-ray exposure age for an olivine grain from the surface of near-earth asteroid (25143) Itokawa. 45th Lunar. Planet Sci Conf Abstr#1247.
20. Michel P, O'Brian DP, Abe S, Hirata N (2009) Itokawa's cratering record as observed by Hayabusa: implications for its age and collisional history. Icarus 20:503–513
21. Miyamoto H, Yano H, Scheeres DJ, Abe S, Barnouin-Jha O, Cheng AF, Demura H, Gaskell RW, Hirata N, Ishiguro M, Mickikami T, Nakamura AM, Nakamura R, Saito J, Sasaki S (2007) Regolith migration and sorting on asteroid Itokawa. Science 316:1011–1014
22. Murdoch N, Delbo M, Libourel, G, Ganino C, Michel P, & Verati C. (2012, September). Regolith formation on asteroids via thermal fatigue. In: European Planetary Science Congress 2012, held 23–28 September, 2012 in Madrid, Spain. http://meetings.copernicus.org/epsc2012, id. EPSC2012-581 Vol. 1, p. 581.
23. Nagao K, Okazaki R, Nakamura T, Miura YN, Osawa T, Bajo K, Matsuda S, Ebihara M, Ireland TR, Kitajima F, Naraoka H, Noguchi T, Tsuchiyama A, Yurimoto H, Zolensky ME, Uesugi M, Shirai K, Abe M, Yada T, Ishibashi Y, Fujimura A, Mukai T, Ueno M, Okada T, Yoshikawa M, Kawaguchi J (2011) Irradiation history of Itokawa regolith material deduced from noble gases in the Hayabusa samples. Science 333:1128–1131
24. Nakamura T, Noguchi T, Tanaka M, Zolensky ME, Kimura M, Tsuchiyama A, Kawaguchi J (2011) Itokawa dust particles: a direct link between S-type asteroids and ordinary chondrites. Science 333:1113–1116
25. Nesvorny D, Bottke WF, Vokrouhlicky D, Chapman CR, Rafkin S (2010) Do planetary encounters reset surfaces of near Earth asteroids? Icarus 209:510–519
26. Nolan MC, Magri C, Howell ES, Benner LAM, Giorgini JD, Hergenrother CW, Hudson RS, Lauretta DS, Margot J-L, Ostro SJ, Scheeres DJ (2013) Shape model and surface properties of the OSIRIS-REx target Asteroid (101955) Bennu from radar and lightcurve observations. Icarus 226:629–640
27. Richardson DC, Bottke WF, Love SG (1998) Tidal distortion and disruption of Earth-crossing asteroids. Icarus 134:47–76
28. Scheeres DJ, Abe M, Yoshikawa M, Nakamura R, Gaskell RW, Abell PA (2007) The effect of YORP on Itokawa. Icarus 188:425–429
29. Thompson MS, Christoffersen R, Zega TJ, Keller LP (2014) Microchemical and structural evidence for space weathering in soils from asteroid Itokawa. Earth Planets Space 66:89

30. Tsuchiyama A, Uesugi M, Matsushima T, Michikami T, Kadono T, Nakamura T, Uesugi K, Nakano T, Sandford SC, Noguchi R, Matsumoto T, Matsuno J, Nagano T, Imai Y, Takeuchi A, Suzuki Y, Ogami T, Katagiri J, Ebihara M, Ireland TR, Kitajima H, Zolensky ME, Mukai T, Abe M, Yada T, Fujimura A, Yoshikawa M, Kawaguchi J (2011) Three-dimensional structure of Hayabusa samples: origin and evolution of Itokawa regolith. Science 333:1125–1128
31. Tsuchiyama A, Nakamura T, Yurimoto H, Ebihara M, Noguchi T, Nagao K, Naraoka H, Kitajima F, Ireland TR, Sandford SA, Zolensky ME, Abe M, Okada T, Yada T, Ishibashi Y, Karouji Y, Uesugi M, HAYAYBUSA sample preliminary examination team members (2013) Outline of Hayabusa sample preliminary examination results. Paper presented at the HAYABUSA 2013: Symposium of Solar System Materials. Abstr# 1016–1100. JAXA Sagamihara Campus, Japan, 16–18 October 2013
32. Walsh KJ, Richardson DC (2006) Binary near-Earth asteroid formation: rubble pile model of tidal disruptions. Icarus 180:201–216
33. Walsh KJ, Richardson DC, Michel P (2008) Rotational breakup as the origin of small binary asteroids. Nature 454:188–190

CHAPTER 5

The Key Role of Meteorites in The Formation of Relevant Prebiotic Molecules in a Formamide/Water Environment

Luca Rotelli[1], Josep M.Trigo-Rodríguez[2], Carles E. Moyano-Cambero[2], EleonoraCarota[1], Lorenzo Botta[1], Ernesto Di Mauro[1] & Raffaele Saladino[1]

[1] *Biological and Ecological Department (DEB), University of Tuscia, 01100 Viterbo, Italy.*

[2] *Institute of Space Sciences (CSIC-IEEC), Meteorites, Minor Bodies and Planetary Sciences Group, Campus UAB Bellaterra, Carrer de Can Magrans, s/n 08193 Cerdanyola del Vallés, Barcelona, Spain*

ABSTRACT

We show that carbonaceous chondrite meteorites actively and selectively catalyze the formation of relevant prebiotic molecules from formamide in aqueous media. Specific catalytic behaviours are observed, depending on the origin and composition of the chondrites and on the type of water present in the system (activity: thermal > seawater > pure). We report the one-pot synthesis of all the natural nucleobases, of aminoacids and of eight carboxylic acids (forming, from pyruvic acid to citric acid, a continuous series encompassing a large part of the extant Krebs cycle). These data shape a general prebiotic scenario consisting of carbonaceous meteorites acting as catalysts and of a volcanic-like environment providing heat, thermal waters and formamide. This scenario also applies to the other solar system locations that experienced rich delivery of carbonaceous materials, and whose physical-chemical conditions could have allowed chemical evolution.

INTRODUCTION

It is assumed that meteorites played a role in the origin of life, behaving as carriers of organic compounds during the Heavy Bombardment period on Earth[1,2], before the earliest appearance of living organisms[3]. Carbonaceous chondrites are ancient space objects containing minerals condensed in the Solar Nebula or inherited from the delivery from nearby stars, partially modified by secondary processing histories[4]. They derive from asteroids that did not undergo planetary differentiation[5] or from outer regions of the protoplanetary disk, and are characterized by reactive minerals, high water content and organic molecules, as amino acids, in the range of parts-per-billion (ppb)[6,7]. Theoretical and experimental models suggest that a part of the organic materials present in carbonaceous chondrites could have been synthesized before the parent body accretion and then evolve under hydrothermal alteration after accretion inside the parent body[8] at temperature ranging from 100 °C to 150 °C due to radioactive phenomena[9,10]. It has been envisioned that water alteration in carbonaceous asteroids occurred under quite static conditions because they were "cooked" slowly and with restricted water availability (e.g. from bounded water in phyllosilicates, hydroxides or other minerals)[11].

Current evidence suggests a significant and heterogeneous water content in the parent bodies of carbonaceous chondrites (0–25 wt. % H_2O)[8]. Carbonaceous chondrites exhibit signs of static aqueous alteration that generated characteristic minerals in the fine-grained matrix, where reactions could take place as a function of the increasing complexity[12].

Catalytic properties of meteorites in prebiotic chemistry were observed related to thermal processes of neat formamide (NH_2CHO)[13,14]. Meteorites perform as catalysts better than terrestrial minerals, as exemplified by the one-pot formation of ribo- and 2'-deoxyribo nucleosides upon irradiation of NH_2CHO with high-energy proton beams, mimicking the effect of solar wind[15]. Comments on the prebiotic relevance of NH_2CHO are in Supplementary text SI # 1. Results on the use of HCN in presence of copper and UV light and involving a meteorite in the synthesis of some prebiotically relevant specific intermediates have been reported[16].

The role of water in prebiotic chemistry is controversial, due to its nucleophile character inducing solvolysis reactions. NH_2CHO and HCN are degraded in water, which might have limited their efficacy as prebiotic precursors17. In spite of water-caused instability and of possible dilution-related problems (potentially solved as in ref. 18), at one point molecular evolution shifted to water-based processes: extant genotypes and phenotypes are now profoundly adapted to and controlled by water.

Nucleosides are efficiently phosphorylated by reaction in NH_2CHO or in water with phosphate minerals19 in conditions allowing spontaneous oligomerization, ligation and rudimentary ribozyme activities20 and references therein. Is there a positive role for water in these initial prebiotic synthetic processes? The chemical rationale behind this question relies on the hypothesis that, due to its complete solubility in water, NH_2CHO might have provided, in the presence of carbonaceous chondrites, a smooth transition towards the aqueous chemistry onto which extant terrestrial life is based21.

RESULTS

We show that carbonaceous chondrites catalyze the synthesis of natural nucleobases, carboxylic acids, and amino acids from mixtures of NH_2CHO and water at 140 °C. Two general scenarios were analyzed: thermal water (TW) and seawater (SW), both tested in the presence of NH_2CHO and of one of six meteorites of the carbonaceous chondrite type: ALH 84028, EET 92042, MIL 05024, LAR 04318, GRO 95551, and GRO 95566. References, inorganic and organic composition, and cosmo-origin data of meteorites are detailed in Supplementary text SI # 2. NH_2CHO was heated in the presence of meteorite materials and in the presence or in the absence of water, and the products were analyzed by GC-MS. Meteorites were treated before the use to remove possible organic contaminants. In two selected cases, ALH 84028 and EET 92042, they were used also as untreated samples. The water samples were carefully filtered before the use to remove any possible microbial and organic contamination. After the treatment, the water samples did not release detectable trace of organic substances. A similar result was also observed after heating ALH 84028 and water samples (distilled water DW, TW and SW) at 140 °C, further confirming that meteorite and water samples do not release detectable endogenous organics under the applied experimental conditions. The chromatographic profiles of reactions are in SI #6. The NH_2COH/water ratio (40% water) was selected in accordance with results previously obtained in the thermal condensation of NH_2COH with iron-copper sulfur minerals22, in order to compare the catalytic performance of meteorites relative to terrestrial minerals. Materials and methods are detailed in Methods and in SI # 3. The reaction of NH_2CHO or NH_2CHO/DW mixture in the absence of meteorite material afforded purine as the only recovered product, besides to traces of formic acid23(Table 1, note a).

Table 1: Products of thermal condensation from NH_2CHO/water mixtures in the presence of ALH 84028, LAR 04318 and EET 92042 meteoritic components.

Product[a]			ALH 84028 (CV3)				LAR 04318 (CK4)			EET 92042 (CR2)		
			A	B	C	D	A	B	C	A	B	C
Acids (μg)	C2	Glycolic ac. 1	25,0 (25,1)[b]	10,3	9,8	----	----	----	----	9,8 (9,8)[b]	----	----
		Oxalic ac. 2	traces	5,2	9,0	1,3	125,0	95,0	60,3	----	----	----
	C3	Pyruvic ac. 3	2,5 (2,5)	1,8	----	----	25,0	25,1	----	68,4 (68,3)	66,8	46,3
		Lactic ac. 4	28,7 (28,6)	12,5	11,3	----	----	----	----	----	----	----
		Parabanic ac. 5	----	----	----	----	16,7	8,9	----	----	----	----
	C4	Malic ac. 6	----	----	----	----	6,0	4,4	----	----	----	----
		Succinic ac. 7	6,6 (6,6)	0,1	traces	----	4,9	2,3	2,5	----	----	----
		Oxaloacetic ac. 8	----	----	----	----	----	----	----	3,0 (3,0)	2,5	6,5
		Fumaric ac. 9	0,07 (0,06)	0,05	traces	----	----	----	----	----	----	----
	C5	Ketoglutaric ac. 10	0,08 (0,08)	0,07	traces	----	----	----	----	----	----	----
	C6	Citric ac. 11	4,8 (4,8)	3,3	1,3	6,9	16,4	6,0	3,2	5,6 (5,7)	3,4	2,1
	C16	Palmitic ac. 12	20,9 (20,8)	25,2	21,2	7,4	47,4	25,9	8,1	----	----	----
		Stearic ac. 13	37,7 (37,7)	14,5	9,0	7,9	83,1	42,6	33,1	----	----	----
Heterocycles (μg)	Uracil 14		9,7 (9,7)	3,6	2,5	----	14,4	0,1	----	----	----	----
	Adenine 15		1,5 (1,5)	2,4	1,3	----	----	0,5	----	3,3 (3,3)	1,4	1,1
	Guanine 16		1,4 (1,5)	1,2	1,2	----	----	----	----	3,1 (3,1)	2,9	2,9
	Hypoxanthine 17		3,7 (3,8)	1,3	----	----	----	----	----	5,1 (5,1)	5,3	2,7
	Isocytosine 18		12,6 (12,5)	0,9	0,1	----	----	----	----	----	----	----
	2,6-Diaminopurine 19		----	----	----	----	----	----	----	7,5 (7,5)	6,3	5,1
	4(3H)-pyrimidinone 20		----	----	----	----	----	----	----	4,8 (4,7)	1,7	1,6
	Uracil 5-carboxylic ac. 21		0,4 (0,4)	0,4	----	----	----	----	----		----	----
	2,4-diamino-6-hydroxypyrimidine 22		25,9 (25,9)	0,6	----	----	35,5	----	----	8,1 (8,1)	7,33	6,1
A.A. (μg)	Glycine 23		14,3 (14,4)	25,3	23,2	1,1	0,9	1,4	1,1	10,0 (9,9)	28,3	25,1
	Formyl glycine 24		48,7 (48,6)	3,4	3,3	----	traces	56,7	29,8	51 (51)	12,5	8,1
	Alanine 25		12,1 (12,1)	6,3	6,0	----	traces	2,2	1,4	7,1 (7,2)	9,8	9,0
Mix (μg)	Urea 26		0,9 (0,9)	0,5	----	1,0	1,1	0,5	----	----	----	----
	Guanidine 27		50,1 (50,1)	33,4	28,7	6,5	64,6	37,4	35,6	58,9 (58,9)	15,4	10,3

The samples of analyzed waters were: SW from the representative Mediterranean area of Montalto di Castro (Viterbo, Italy), and TW from the Bagnaccio volcanic spring area (Viterbo, Italy). The physical and chemical properties, and the ion composition of water samples are in SI # 4. The reaction performed with TW, SW and DW with NH_2CHO in the absence of meteorites only afforded purine.

ALH 84028

In the presence of ALH84028 and DW, NH_2CHO afforded a limited panel of compounds, including oxalic acid **2**, citric acid **11**, palmitic acid **12**, stearic acid **13**, glycine **23**, urea **26** and guanidine **27** (Fig. 1, Table 1, column D).

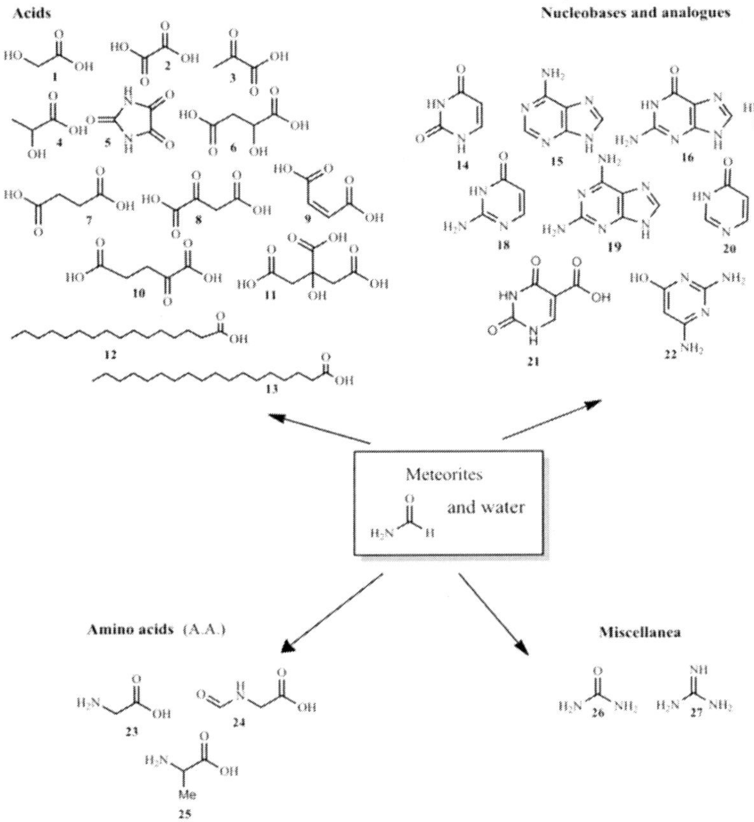

Figure 1: Prebiotic synthesis of biomolecules from meteorite and water in the presence of formamide.

Products of thermal condensation from NH_2CHO/water mixture in the presence of meteorites of carbonaceous chondrite sub-type. Experimental conditions: 1% meteorite, 59% NH_2CHO, 40% water, 140 °C, 24 h.

Better results were obtained in the presence of TW and SW, affording larger panels of products of biological relevance: carboxylic acids with increasing levels of structural complexity (from C2 to C16), as glycolic- **1**, oxalic- **2**, pyruvic- **3**, lactic- **4**, parabanic- **5**, malic- **6**, succinic- **7**, oxaloacetic- **8**, fumaric- **9**, ketoglutaric- **10**, citric- **11**, palmitic- **12**, and stearic- **13** acids; nucleobases and nucleobase analogues uracil **14**, adenine **15**, guanine **16**, hypoxanthine **17**, isocytosine **18** and 2,6-diamino purine **19**, purine and pyrimidine heterocycles 4 (3 H)-pyrimidinone **20**, uracil 5-carboxylic acid **21**, 2,4-diamino-6-hydroxypyrimidine **22**; amino acids glycine **23**, *N*-

formyl glycine **24**, alanine **25**. Urea **26** and guanidine **27** were also detected (Fig. 1, Table 1, columns B and C).

Carboxylic acids are key intermediates in energy metabolism and for the synthesis of metabolites. Compounds **3**, **6**, **7**, **8**, **9**, **10** and **11**, are seven of the ten components of the Krebs cycle (KC), one of the most ancient cellular metabolic pathway. The relevance of the KC in the early production of energy and organics was postulated, invoking its prebiotic role in the fixation of carbon oxides from the atmosphere of the primitive Earth24. Compounds **1** and **2** are components of the anabolic alternative pathway of the KC in plants, bacteria, protists, and fungi (the glyoxilate cycle), **32** of the Cori cycle and gluconeogenesis. Long-chained carboxylic acids **12–13** are constituents of lipids. Carboxylic acids can be formed from NH_2CHO by oligomerization to monoamino malonitrile (AMN) and diaminomalo nitrile (DAMN), followed by hydrolysis and successive redox processes25.

The complete set of RNA nucleobases was obtained (isocytosine **18**being bioisoster of cytosine), along with nucleobase analogues **17** and **19–21**, whose biological relevance was reviewed26. Reaction mechanisms for the formation of nucleobases from NH_2CHO were reported27.

The presence of glycine **23**, *N*-formylglycine **24** and alanine **25**, is probably due to a NH_2CHO-based Strecker-type synthesis (Strecker-cyanohydrin), in accordance with the recent theoretical observations on the key role of NH_2CHO in the Miller-Urey synthesis of amino acids28. Glycine is classically the most abundant amino acid found in meteorites. *N*-formylglycine **24** is produced from **23** by a formylation process that mimics the peptidase activity involving "*in situ*" generated carbodiimide (not isolated in this case), urea **26** and guanidine **27**29. A schematic representation of the reaction pathways leading to nucleobases, carboxylic acids and amino acids from NH_2CHO is in SI #8.

As a general trend, the reactions in TW or SW behaved as in the condensation of neat NH_2CHO, suggesting that the presence of the meteorite ALH 84028 effectively increased the reactivity of the system, thus balancing the expected inhibition due to water dilution (Table 1, columns B and C versus column A). The treated or untreated ALH 84028 meteorite materials performed as catalyst very similarly in the reaction with neat NH_2CHO (Table 1, column A). In some cases (oxalic acid, palmitic acid, adenine, glycine and alanine), the yield in the NH_2CHO/water systems was higher than that obtained in neat NH_2CHO (Table 1). The reaction in TW was always more efficient than in SW, affording a larger panel of products in higher yield (Table 1, columns B versus C).

LAR 04318, EET 92042, GRO 95551, GRO 95566 AND MIL 05024

The reaction was extended to other selected meteorites, LAR 04318, EET 92042, GRO 95551, GRO 95566 and MIL 05024, as reported in Tables 1 and 2. M/z values and peak abundances are in SI, Table C. Selected chromatograms and original m/z fragmentation spectra are in SI # 6 and SI # 7, respectively.

Table 2: Products of thermal condensation from NH_2CHO/water mixtures in the presence of GRO 95551, GRO 95566 and MIL 05024 meteorites.

Product			GRO 95551 (C-ung)			GRO 95566 (C2-ung)			MIL 05024 (CO3)		
			A	B	C	A	B	C	A	B	C
Acids (µg)	C2	Glycolic 1	—	—	—	—	—	—	0,5	0,5	—
		Oxalic 2	6,2	5,8	5,0	2,5	—	—	—	—	—
	C3	Pyruvic 3	8,3	10,9	5,9	46,9	10,9	11,0	16,7	13,6	12,4
		Lactic 4	6.5	2.9	2,3	6,8	9,4	2,2			
	C4	Succinic 7	3,0	2,5	1,4	2.9	—	—	12.9	11.5	8.3
		Oxaloacetic 8	—	—	—	1,4	1.5	—	—	—	—
		Fumaric 9	13.3	10.2	9.8	—	—	—	—	—	—
	C5	Ketoglutaric 10	6,8	1,2	—	—	—	—	—	—	—
		Citric 11	—	—	—	4,3	3,8	0,3	—	—	—
	C16	Palmitic 12	25,3	10,2	10,0	35,9	39,2	32,0	19,4	18.3	13.6
		Stearic 13	19,7	15,6	12,3	24,6	33,6	29,1	37,7	26,3	31,7
	Uracil 14		—	—	—	6,0	0,2	0,2	—	—	—
Heterocycles (µg)	Adenine 15		3,1	0,7	0,8	1,3	0,4	0,1	—	—	—
	Guanine 16		1.4	1.2	1,2	1,7	3,5	1,1	5,0	2,3	1,8
	Hypoxanthine 17		4,9	3.6	2,2	4,6	3,7	2.9	2,2	1,7	1,1
	Isocytosine 18		—	—	—	3,3	2,1	1,8	—	—	—
	2,6-Diaminopurine 19		7,1	3,6	2,1	8,3	—	—	—	—	—
	4 (3 H)-pyrimidinone 20		1,3	9,5	—	3,8	4,2	2,4	4.3	3,4	2,1
	Uracil 5-carboxylic ac. 21		5,2	—	—	13,6	—	—	—	—	—
	2,4-diamino-6-hydroxypyrimidine 22		32,5	2,7	2,8	12,4	8,9	6,6	—	—	—
A. A. (µg)	Glycine 23		3.3	2.2	2.0	5.9	3,3	2,2	3,7	3,1	2.8
	Formyl glycine 24		6.5	6.0	5.3	1,1	4,2	3.1	6,6	6.9	5,7
	Alanine 25		4,1	3,7	3,5	6,2	3,2	1,8	3,4	3,2	3,0
Mix (µg)	Urea 26		—	—	—	2,0	—	—	0,6	0,3	0,3
	Guanidine 27		85,3	55,9	24,7	2,8	—	—	6,6	5,9	—

Even though activity relationships cannot be established due to the overall complexity of the systems, general considerations can be derived from the analysis of the reaction pattern in terms of quality and quantity of synthesized products. Irrespective of the experimental conditions, carboxylic acids were isolated in higher variety and yield in TW relative to SW, in agreement with data obtained with ALH 84028. In both TW and SW, the yield was comparable with that obtained using neat NH_2CHO (Tables

1 and 2, columns B and C versus column A). Meteorites ALH 84028, LAR 04318 and GRO 95566 produced the highest number of derivatives (ten and eight carboxylic acids, respectively). Some of these derivatives were obtained with high selectivity, as malic acid (observed only with LAR 04318) and oxaloacetic acid (observed only with EET 92042 and GRO 95566). In terms of isolated compounds, LAR 04318 and GRO 95566 afforded the highest total amount of carboxylic acids (c.a. 210 μg/mL and 98 μg/mL, respectively), followed by ALH 84028, EET 92042 and MIL 05024, which showed a similar behavior (c.a. 70 μg/mL). A high quantitative efficacy in the synthesis of carboxylic acids did not necessarily coincide with a high variety of synthesized products (i. e., Table 1, ALH 84028 versus LAR 04318).

The yield of nucleobases was lower than that of carboxylic acids. In agreement with the general trend observed for carboxylic acids, nucleobases and nucleobase analogues were synthesized in higher yield in TW than in SW (Tables 1 and 2). ALH 84028 and GRO 9556 were the only meteorites that yielded the complete set of RNA nucleobases. Adenine and guanine were the most represented nucleobases, isocytosine being obtained only with ALH 84028 and GRO 95566. In terms of isolated compounds, ALH 84028, GRO 95566 and EET 92042 afforded the highest total amount of nucleobases and nucleobase analogues in NH_2CHO/TW (c.a. 9.5 μg/mL, respectively), followed by GRO 95566 (c.a. 5.5 μg/mL). Occasionally, nucleobases were synthesized in yield higher in NH_2CHO/TW than in neat NH_2CHO, as observed for guanine and adenine in the presence of GRO 95566 and ALH 84028, respectively. Glycine, *N*-formyl glycine and alanine were always synthesized, EET 92042 being the best catalyst, followed by ALH 84028 and MIL 05024. Again, treated or untreated EET 92042 materials performed as catalyst very similarly in the reaction of neat NH_2CHO.

CONCLUSIONS

A careful comparison of the relationships between the products obtained and the meteorite used as catalyst shows a complex scenario. For example, the EET 92042 (CR2) meteorite catalyzes the synthesis of a panel of products whose qualitative composition (namely, the panel of products and their relative abundance) is similar to the panel of its endogenous organics. At the contrary, ALH 84028 (CV3) catalyzes the formation of a panel of products richer than that observed in the original sample. Thus, the catalytic role played by meteorites in the transformation of NH_2CHO is tuned by the specific reaction environment, depending on the history and composition of the catalyst.

These data hint to the important role that chondrites could have played in an early Earth subjected to a strong projectile flux. They also indicate that, even after ablation of the volatile organic phases from the impactants during their fall, the mineral components that eventually reached the ground could have promoted the catalysis of organics in an environment that was more hydrous and oxidizing than the one they had in their progenitor asteroids30.

TW and, to lower extent, SW were more reactive than DW. The types of water we used largely differ for their inorganic composition (as detailed in SI, Tables A and B) and for their pH (pH 6.0 for TW versus pH 8.1 for SW). TW shows concentrations of alkaline earth metals Ca, Mg, Sr and Ba, largely higher than SW. In addition, transition metal ions Mn and Fe were detected only in TW. Transition metal ions very efficiently coordinate aliphatic amides, achieving their maximum coordination number31. Alkaline earth metals and transition metal ions coordinate the nucleophilic centers in NH_2CHO, to yield complexes with different geometries and stabilities, which are known to lose water in the presence of high energy sources to yield HCN32, one of the key intermediate in the synthesis of both nucleobases and carboxylic acids (SI #8). Different transition metal compounds have been recently reported as efficient catalysts for the prebiotic synthesis of a large panel of prebiotic molecules from NH_2CHO in geochemical scenarios33. Hydrothermal systems are accepted models of prebiotic environments, and several organics have been detected in abiotic hot springs34.

The selectivity of the reactions was tuned by the nature of each chondrite, probably as a function of the differently available amounts of minerals acting as catalysts. Glycine, *N*-formyl glycine and alanine were isolated in the highest yield in the presence of EET 92042 (classified as CR2 chondrite), followed by ALH 84028 (CV3) and MIL 05024 (CO3).

This order of reactivity is in agreement with the observed and theoretically calculated ability of carbonaceous chondrites to perform the Strecker-type synthesis, that is CR> CV3 (CK3)> CO335. This observation prompts us to clarify our interpretation of the data obtained, as follows: the syntheses of prebiotically relevant compounds was observed in a system composed of a reactive substrate (NH_2CHO), of catalysts (chondrite materials) in a typically Earth-like environment (heat + waters, thermal or otherwise). Thus, we frame our synthetic system in an earliest planet Earth scenario. Nevertheless, the very fact that the syntheses of certain compounds bona fide occurred according to mechanisms that were both theorized to be possible and observed to have occurred in meteorites during their non terrestrial history (i.e., Strecker-type synthesis of aminoacids, ref. 4), points that these processes are not exclusively terrestrial.

The CR chondrite group affords the highest yield of amino acids, greater by several orders of magnitude than any other subclass. This subclass is generally associated with alteration temperatures in the broad range of 0 °C to 240 °C36. In view of recent results, the catalytic role exhibited by the CR mineral components is not surprising as it is an extraordinarily pristine group, being identified as source of primitive organics37,38.

Deviations from this reaction pattern were observed for carboxylic acids and nucleobases. LAR 04318 (CK4), ALH 84028 (CV3), EET 92042 (CR2) and GRO 95566 (C2-ung) were found to be the most active catalysts for these compounds. The difference between the mechanism of a Strecker-type synthesis relative to that of the formation of carboxylic acids and nucleobases from NH_2CHO may account for this result (SI #8).

The complexity of the relationships between organics, water and early stages of aqueous alteration in pristine chondrites was described8. This study confirms the view that reactive minerals could have acted as catalysts promoting increasing organic complexity in chemical evolution. In particular, some secondary minerals were found to be the product of primordial aqueous alteration, during a short first stage of water release due to radiogenic heating39, and often the minerals formed exhibit clear features of static aqueous alteration with limited water availability12.

NH_2CHO/water solutions were recently shown to form a multipurpose reactive mixture in a totally different reaction system33. Unlike recent theoretical argumentation of the contrary40, the data described here and in ref. 33 show that NH_2CHO/water is a flexible and fertile prebiotic incubator.

These study highlights the role of water, of thermal processes and of one-carbon atom precursors (such as NH_2CHO) in the endogenous internal prebiotic chemistry of meteorites leading to the diverse organic compounds detected in chondrites, and opens a door to the understanding of the appearance of life in other planetary bodies that experienced an efficient delivery of carbonaceous chondrite materials, as Mars, Titan or Europa.

METHODS

Approximately 50 mg of the meteorite stone obtained after removal of the fusion crust were ground in an agate mortar and treated with NaOH 0.1 N (1.0 mL) and $CHCl_3$-MeOH mixture (3.0 mL; 2:1 v/v), followed by sulphuric acid 0.1 N (1.0 mL) and $CHCl_3$-MeOH mixture (3.0 mL; 2:1 v/v). ALH 84028 (CV3) and EET 92042 (CR2) were also used as untreated samples. The reactions were performed by heating freshly distilled NH_2COH (1.0 mL) at 140 °C for 24 hours in the presence of the appropriate meteorite sample (1.0% by weight relative to NH_2COH) and 40% in weight of distilled water

DW, thermal water TW or sea water SW. The water samples were carefully treated before the use to remove any possible microbial and organic contamination by filtration on 0.20 µm *Minisart Sartorius* (catalogue number 16534; Sterile-E0), followed by extractions with EtOAc (20 ml; ×3). Reactions with neat NH_2COH, with NH_2COH and water without the meteorite sample, or with meteorite (ALH 84028) and water (DW, TW and SW) without NH_2COH were also performed as references. At the end of the reaction the meteorite was recovered by centrifugation (6000 rpm, 10 min, Haereus Biofuge) and washed with MeOH. The excess NH_2COH and MeOH were then removed by distillation (40 °C, 4×10^{-4} barr). The crude product was analyzed by gas-chromatography associated to mass-spectrometry (GC-MS) after treatment with *N,N*-bis-trimethylsilyl trifluoroacetamide in pyridine (620 µL) at 60 °C for 4 h in the presence of betulinic acid acid [3β-hydroxy-2029-lupaene-oic acid] as internal standard (0.3 mg). Materials and methods are detailed in SI # 3.

ACKNOWLEDGEMENTS

JMTR acknowledges MICINN research projects AYA2011-26522 and AYA2015-67175-P funding support in the framework of which the current carbonaceous chondrites were chosen, characterized and requested to NASA. This work is supported by the Italian Space Agency (ASI) project "Esobiologia e Ambienti estremi: dalla chimica delle Molecola allaBiologia degli estremofili" Number 2014-026-R.0 (CUP: F 92I14000030005). C.E.M.C. contribution to this paper was done in the frame of a PhD. on Physics at the Autonomous University of Barcelona (UAB). The authors offer this contribution to the memory of pioneer Prof. Joan Oró, for his early support and encouragement to pursue research on carbonaceous chondrites.

REFERENCES

1. Chyba, C. F. The cometary contribution of the oceans of primitive Earth. Nature 330, 632–635 (1987).
2. 2.Oró, J. Comets and the formation of biochemical compounds on the primitive Earth. Nature 190, 389–390 (1961).
3. Mojzsis, S. J. et al. Evidence for life on Earth before 3,800 million years ago. Nature 384, 55–59 (1996).
4. Cronin, J. R., Pizzarello, S. & Cruikshank, D. P. Organic matter in carbonaceous chondrides, planetary satellites, asteroids and comets. In J. F. Kerridhe, M. S.

Matthews, editors Meteorites and the early Solar System (Tucson: University of Arizona Press) 819–857 (1988).
5. Callahan, M. P. et al. Carbonaceous meteorites contain a wide range of extraterrestrial nucleobases. Proc. Natl. Acad. Sci. USA108, 13995–13998 (2011).
6. Alexander, C. M. O.' D. et al. The provenances of asteroids and their contributions to the volatile inventories of the terrestrial planets. Science 337, 721–723 (2012).
7. Trigo-Rodríguez, J. M. & Martín-Torres, F. J. Clues on the importance of comets in the origin and evolution of the atmospheres of Titan and Earth. Planetary Space Sci. 60, 3–9 (2012).
8. Le Guillou, C. & Brearley, A. Relationships between organics, water and early stages of aqueous alteration in the pristine CR 3.0 chondrite MET 00426. Geochim. Cosmochim. Acta 131, 344–367 (2014).
9. Doyle, P. M. et al. Early aqueous activity on the ordinary and carbonaceous chondrite parent bodies recorded by fayalite. Nature Commun. 6, 7444 (2015).
10. Travis, B. J. & Schubert, G. Hydrothermal convection in carbonaceous chondrite parent bodies. Earth Planet. Sci. Lett.240, 234 (2005).
11. Zolensky, M. E., Krot, A. N. & Benedix, G. Record of low-temperature alteration in asteroids. Reviews Mineralogy Geochem. 68, 429–462, (2008).
12. Rubin, A., Trigo-Rodríguez, J. M., Huber, H. & Wasson, J. T.Progressive aqueous alteration of CM carbonaceous chondrites. Geochim. Cosmochim. Acta 71, 2361–2382 (2007).
13. Saladino, R., Botta, G., Delfino, M. & Di Mauro, E. Meteorites as catalysts for prebiotic chemistry. Chemistry-A European J. 19, 16916–16922 (2013).
14. Ferus, M. et al. High Energy chemistry of formamide: a simpler way for nucleobase formation. Proc. Natl. Acad. Sci. USA 112, 657–662 (2015).
15. Saladino, R. et al. Meteorite-catalyzed synthesis of nucleosides and other prebiotic compounds under proton irradiation. Proc. Natl. Acad. Sci. USA 112, 2746–2755 (2015).
16. Patel, Bhavesh H., Percivalle, Claudia, Ritson, Dougal J., Duffy, Colm D. & Sutherland, John D. Common origins of RNA, protein and lipid precursors in a cyanosulfidic protometabolism. Nature Chem. 7, 301–307 (2015).
17. Saladino, R., Crestini, C., Pino, S., Costanzo, G. & Di Mauro, E.Formamide and the origin of life. Phys. Life Rev. 9, 84–104 (2012).
18. Niether, D., Afanasenkau, D., Dhont, J. K. G. & Wiegand, S.Accumulation of formamide in hydrothermal pores to form prebiotic nucleobases. Proc. Natl. Acad. Sci. USA 113, 4272–4277 (2016).
19. Costanzo, G., Saladino, R., Crestini, C., Ciciriello, F. & Di Mauro, E.Nucleoside Phosphorylation by Phosphate Minerals. J. Biol. Chem.282, 16729–16735 (2007).
20. Šponer, J. E. et al. Emergence of the First Catalytic Oligonucleotides in a Formamide-Based Origin Scenario. A review. Chemistry-A European J. 22, 1–16 (2016).
21. Horneck, G. et al. AstRoMap European Astrobiology Roadmap. Astrobiology 15, 201–243 (2016).
22. Saladino, R., Neri, V., Crestini, C., Costanzo, G., Graciotti, M. & Di Mauro, E. Synthesis and degradation of nucleic acid components by formamide and iron sulfur minerals. J. Am. Chem. Soc. 130(46), 15512–15518 (2008).
23. Yamada, H. & Okamoto, T. A one-step synthesis of purine ring from formamide. Chem. Pharm. Bull. 20, 623–627 (1972).
24. Smith, E. & Morowitz, H. J. Universality in intermediary metabolism. Proc. Natl Acad. Sci. USA 101, 13168–13173 (2004).
25. Eschenmoser, A. On a Hypothetical Generational Relationship between HCN and Constituents of the Reductive Citric Acid Cycle. Chem. Biodivers. 4, 554–573 (2007).
26. Saladino, R., Botta, G., Pino, V., Costanzo, G. & Di Mauro, E.Materials for the onset. A story of necessity and chance. Frontiers Biosci. 18, 1275–1289 (2013).
27. Ferus, M. et al. On the Road from Formamide Ices to Nucleobases: IR-Spectroscopic Observation of a Direct Reaction between Cyano Radicals and Formamide in a High-Energy Impact Event. J. Am. Chem. Soc. 134, 20788–20796 (2012).

28. Saitta, A. M. & Saija, F. Miller experiments in atomistic computer simulations. Proc. Natl Acad. Sci. USA 111, 13768–13773 (2014).
29. Saladino, R., Botta, G., Pino, S., Costanzo, G. & Di Mauro, E.Genetics first or metabolism first? The formamide clue. Chem. Soc. Rev. 41, 5526–5565 (2012).
30. Trigo-Rodríguez, J. M. Aqueous alteration in chondritic asteroids and comets from the study of carbonaceous chondrites. In Planetary Materials. EMU Notes in Mineralogy 15, 67–87 (2015).
31. Zabicky, J. In The Chemistry of Amides, Patai S. Ed. Interscience Publishers John Wiley & Sons, London, 236 (1970).
32. Eizaguirre, A., Mó, O., Yanez, M. & Tortajada, J. Modelling peptide-metal dication interactions: Formamide-Ca2+ reactions in the Gas Phase. Org. Biomol. Chem. 10, 7552–61 (2012).
33. Saladino, R., Botta, G., Bizzarri, B. M., Di Mauro, E. & Garcia Ruiz, J. M. A Global Scale Scenario for Prebiotic Chemistry: Silica-Based Self-Assembled Mineral Structures and Formamide. Biochemistry55, 2806–2811 (2016).
34. Kompanichenko, V. N., Poturay, V. A. & Shlufman, K. V.Hydrothermal Systems of Kamchatka are Models of the Prebiotic Environment. Orig. Life Evol. Biosph. 45, 93–103 (2015).
35. Burton, A. S., Stern, J. C., Elsila, J. E., Glavin, D. P. & Dworkin, J. P.Understanding prebiotic chemistry through the analysis of extraterrestrial amino acids and nucleobases in meteorites. Chem. Soc. Rev. 41, 5459–5472 (2012).
36. Busemann, H., Alexander, C. M. O'. D. & Nittler, L. R.Characterization of insoluble organic matter in primitive meteorites by microRaman spectroscopy. Meteorit. Planet. Sci. 42, 1387–1416 (2007).
37. Marty, B. et al. Origins of volatile elements (H,C,N, noble gases)) on Earth and Mars in light of recent results from the ROSETTA cometary mission. Earth Planetary Sci. Lett. 441, 91–102 (2016).
38. Pizzarello, S. & Shock, E. The Organic Composition of Carbonaceous Meteorites: The Evolutionary Story Ahead of Biochemistry. Cold Spring Harb. Perspect. Biol. 2(3), a002105 (2010).
39. Fujiya, W., Sugiura, N., Sano, Y. & Hiyagon, H. Mn–Cr ages of dolomites in CI chondrites and the Tagish Lake ungrouped carbonaceous chondrite. Earth Planetary Sci. Lett. 362 130–142 (2013).
40. Bada, J. L., Chalmers, J. & Cleaves H. J. Is formamide a geochemically plausible prebiotic solvent? Phys. Chem. Chem. Phys. Accepted Manuscript (2016).

CHAPTER 6

Possible Duplicities of Five Asteroids

Isao Satō[1*], Hiromi Hamanowa[2], Hiroyuki Tomioka[3], Sadaharu Uehara[4]

[1]*Nihon University, Sakata, Yamagata, Japan*

[2]*Hamanowa Astronomical Observatory, Motomiya, Fukushima, Japan*

[3]*Hitachi, Ibaraki, Japan*

[4]*KEK, Tsukuba, Ibaraki, Japan*

ABSTRACT

Some evidences of possible duplicities of five asteroids are presented. A satellite of (279) Thule was convincingly detected by a stellar occultation on 2008 April 3 by Thule and also from follow-up light curve observations. The orbital period of the satellite is 3.007 or 6.014 days, and the minimum diameter is estimated to be 52 km. A satellite of (324) Bamberga was detected by secondary drops of the light curve in 2007. The rotation period of the primary is 1.22625 days, and the revolution period of the secondary is 3.00 or 6.00 days. Presumed contact duplicity of the main body of the L4 Trojan (624) Hektor was detected by a stellar occultation by Hektor on 2008 January 24. A possible satellite of (657) Gunlöd was suggested from a secondary occultation by Gunlöd on 2008 November 29. The minimum of the diameter is 7 km. A possible satellite of (739) Mandeville was suggested by stellar occultation observations on 1980 December 10.

Keywords: Asteroid, Occultation, Light Curve, Satellite, Binary

1. INTRODUCTION

Since the first discovery of Dactyl, a satellite of (243) Ida, by the Galileo spacecraft in 1993, several hundreds of satellites of asteroids have been detected by various means, such as asteroidal occultations, light curve observations, radar observations, and spacecraft encounters. Discovery of a satellite of an asteroid provides a total mass of the binary system from the third law of Kepler if its evolution period and semi-major axis of the orbit are obtained. We present some evidence of the duplicities of five asteroids detected in Japan through occultation observations.

2. SATELLITE OF (279) THULE

2.1. Stellar Occultation on 2008 April 3

The asteroid (279) Thule was discovered by J. Palisa in 1888. The semi-major axis of its orbit is 4.27 au, the eccentricity of the orbit is 0.01, and the revolution period is 8.83 years which is within the mean motion resonance of 4:3 with Jupiter. Asteroids with this resonance are part of a group called the Thule group. The other members of the Thule group known thus far are (185290) 2006 UB219 and (186024) 2001 QG207.Thule is a D-type asteroid [1] .

The occultation of the star TYC 497200415 by (279) Thule, the first observed stellar occultation by Thule, was observed on 2008 April 3 from four sites in Japan. The event was predicted by Satō (Figure 1). The position of the occulted star is α_{2000} = 13 h 31 m 32.3082 ± 0.020 s, δ_{2000} = −06°59'53". 701 ± 0.022, and the brightness of the star is m_v = 10.82. The assumed diameter of Thule is 127 km and the predicted extinction would be 3.7 mag for 8.6 seconds. The occultation track would go through Kanto and Chubu districts of Japan. The observation results are listed in Table 1. Three chords among four are timed by video with GPS clock, and one is visual.

Revealed occultation cross section is (126 ± 1 km) × (123 ± 7 km), the position angle of the ellipse is 80° ± 129° (Figure 2). Three chords by Yanagida Astronomical Observatory, Hideo Takashima, and Hideyoshi Karasaki are consistent with an ellipse of the primary, but the chord by Hiroyuki Tomioka is not coincident with it. This specific observation was made with GPS time-inserted video, hence the observation must be considered reliable. The extinction suggests a possible previously unknown satellite with diameter of 52 km. The following-up light curve observations of Thule was obtained after the occultation.

2.2. Light Curve Observations in 2008

A light curve observation was obtained at Hamanowa Observatory during 13 nights from 2008 April 8 to May 6. Figure 3 shows a total combined light curve. Four instances of secondary drops are seen in the light curve. The Light curve shows at rotation period of 7 h 26 m 24 ± 1 s, peak-to-peak amplitude of 0.081 mag with secondary drops of 0.1 mag. The result is coincident with the previous result of 7.44 ± 0.05 hours [2] . Figure 4 shows the light curves of four nights with secondary drops, 2008 April 6, 15, 30, and May 3. The secondary drops suggest a satellite with revolution period of 3.007 or 6.014 days. The assumed phase of the secondary drop at the time of the stellar occultation is coincident with the hypothesis that Tomioka's extinction occurred from a satellite with a diameter of 52 km, namely, it is assumed that the possible satellite was during the partial phase of a mutual phenomenon.

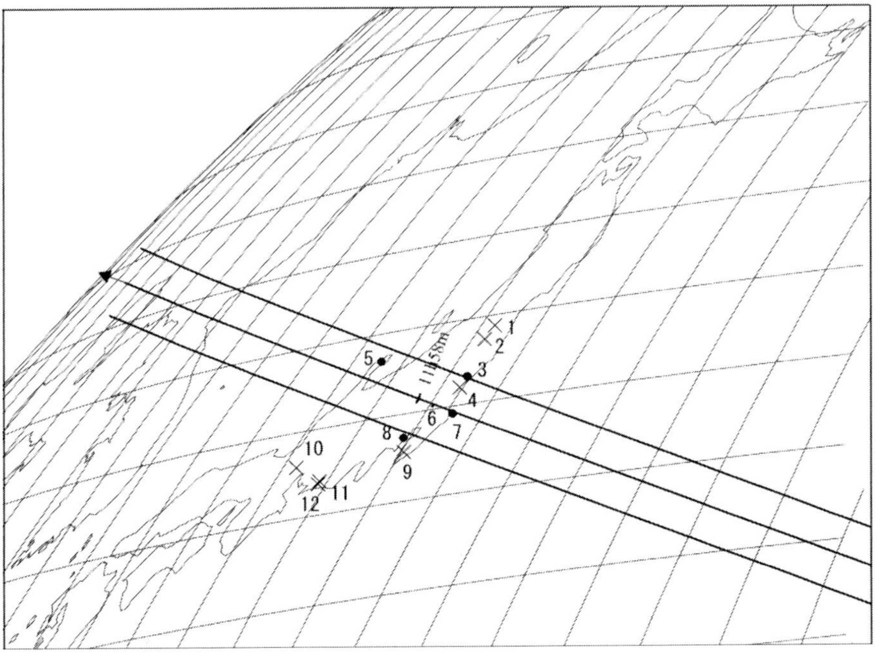

Figure 1. Occultation track and locations of observers of the occultation of TYC 497200415 by (279) Thule on 2008 April 3. The black circles (#3, 5, 7, 8) are successfully times sites and the crosses (#1, 2, 4, 9, 10, 11, 12) are places where no occultation was observed.

Table 1. Observationsdata of the occultation of TYC 497200415 by (279) Thule on 2008 April 3. Geodetic datum is WGS84.

No.	observer	location	longitude	latitude	height	telescope	method	phen	UTC
1	Hikaru Satō	Fukushima, Fukushima. Japan	E140°29'25.3"	N37°44'35.9"	90 m	16 cm N, F = 3.3,	GHS, TiVi video		no occultation
2	Hamanowa Observatory	Koriyama, Fukushima, Japan	E140°25'48.4"	N37°27'37.7"	260 m	40 cm N, F = 4.5	GHS .TiVi, video		no occultation
3	Kazuyuki Shinozaki	Takahagi. Ibaraki, Japan	E140°40'29.4"	N36°42'19.4"	40 m	35cm SC, F=11	GPS, visual		no occultation
4	Hiroyuki Tomioka	Hitachi, Ibaraki, Japan	E140°41'09"	N36°38'33"	33 m	30 cm C	GHS, TiVi, video	2D	11 h 57 m 57.90 s ± 0.03 s
								2R	11 h 58 m 01.37 s ± 0.03 s
5	Akira Tsuchikawa	Yanagida, Ishikawa, Japan	E137°08'13.4"	N37°20'04.1"	200 m	60 cm C	TEL, video	1D	11 h 58 m 02.50 s ± 0.03 s
								1R	11 h 58 m 10.09 s ± 0.03 s
6	Sadaharu Uehara	Tsukuba. Ibaraki, Japan	E140°07'04.2"	N36°05'11.2"	30 m	20 cm N, F = 4			cloudy
7	Hideo Takashima	Kashiwa, Chiba, Japan	E140°58'07"	N35°49'53"	28 m	20 cm S, F = 2	GHS, TiVi, video	1D	11 h 57 m 56.13 s ± 0.04 s
								1R	11 h 58 m 02.24 s ± 0.04 s
8	Hideyoshi Karasaki	Nerima, Tokyo, Japan	E139°47'00.3"	N35°44'25.4"	46 m	20.3 cm SC	TEL, TiVi, video	1D	11 h 57 m 58.6 s ± 0.6 s
								1R	11 h 57 m 59.9 s ± 0.6 s
9	Mamoru Urabe	Kamogawa, Chiba, Japan	E140°06'05"	N35°06'35"	8 m	30 cm N, F = 4	TEL, video		no occultation
10	Minoru, Owada	Hamamatsu, Shizuoka, Japan	E137°44'08.0"	N34°42'02.5"	8 m	10 cm R, F = 4	GHS, TiVi, video		no occultation
11	Hisashi Suzuki	Hamamatsu, Shizuoka, Japan	E137°42'48.4"	N34°45'53.8"	50 m	25 cm R, F = 5	GHS, TiVi, video		no occultation
12	Akira Arai	Inabe, Mie, Japan	E136°31'24.3"	N35°10'14.6"	187 m	35.5 cm SC, F = 11	GHS, TiVi, video		no occultation

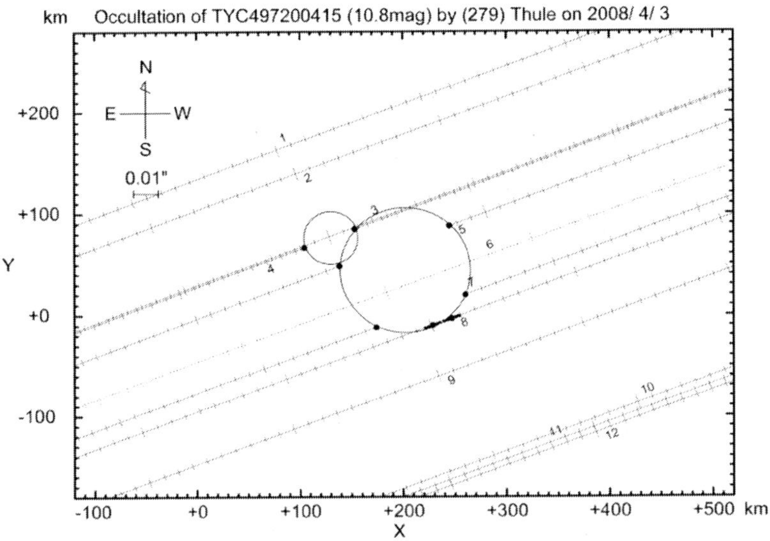

Figure 2. Result of the occultation of TYC 497200415 by (279) Thule on 2008 April 3. An ellipse of (126 ± 1 km) × (123 ± 7 km) is fitted for the primary, and a circle of a diameter of 52 km is fitted for the satellite.

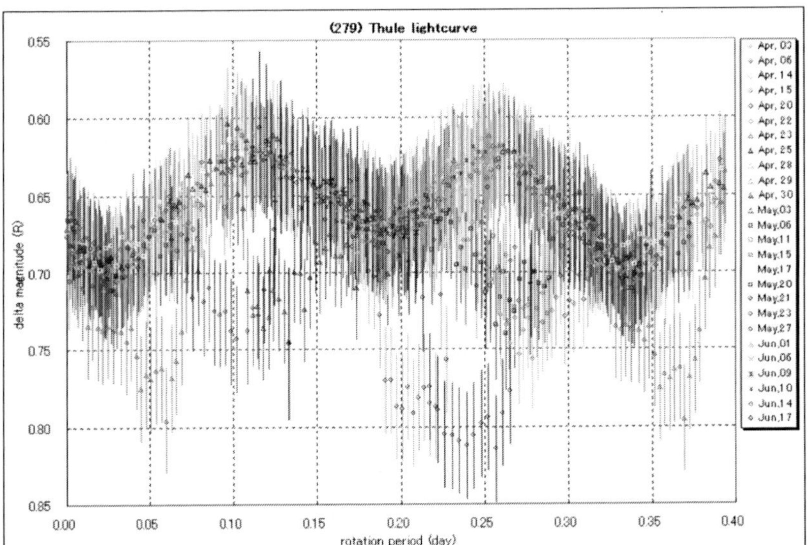

Figure 3. Composed light curve of (279) Thule from 2008 April 3 to May 6 at Hamanowa Astronomical Observatory. The main light curve is consistent with a period of 7.44 ± 0.05 hoursand a peak-to-peak amplitude of 0.081 mag. In addition to the main light curve, secondary drops are shown on four nights (April 6, 15, 30, and May 3).

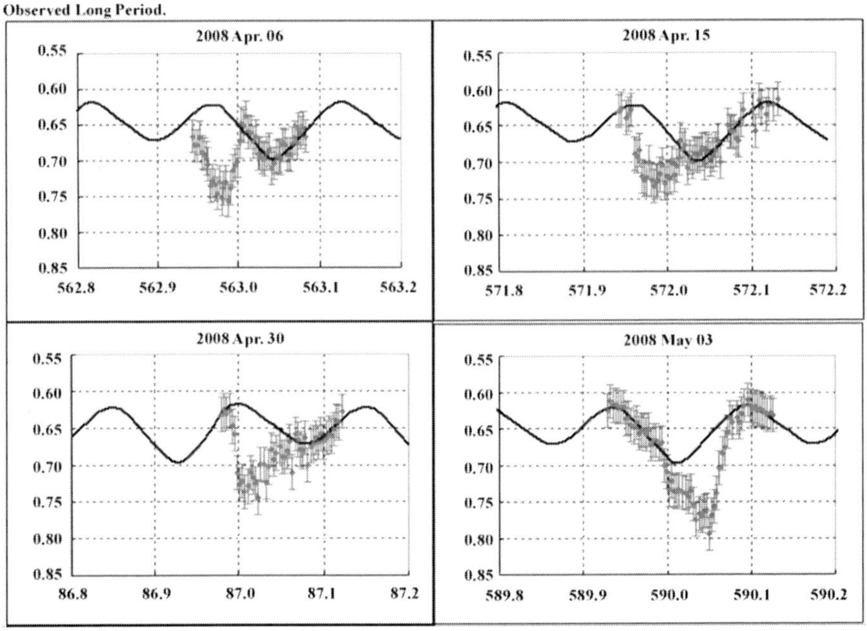

Figure 4. Detailed features of the light curves of (279) Thule on 2008 April 6, 15, 30, and May 3. Each figure shows a secondary drop on the main light curve. The period of the secondary drops is 3.007 or 6.014 days.

3. SATELLITE OF (324) BAMBERGA

(324) Bamberga was discovered by J. Palisa in 1892. The planet is a main belt asteroid with a revolution period of 4.4 years. The first stellar occultation by (324) Bamberga was observed from the USA, Japan and China on 1987 December 8. The occultation cross section showed a diameter of 227.6 ± 1.8 km [3] .

The second event, the occultation of HIP059807 (7.3 mag) by Bamberga, was observed from five sites in Australia on 2007 April 20 (Table 2, Figure 5). The occultation cross section shows an ellipse of (231 ± 4 km) × (211 ± 12 km), P = 106° ± 8°. Hamanowa Observatory performed light curve observations just after the second event from 2007 April 23 to May 22. As a result, the amplitude is 0.084 ± 0.0030 mag during the rotation period of 29 h 27 m 27 ± 3 s (Figure 6). Light curve observations were also obtained by [4] - [6] . The period is coincident with these results.

The light curve (Figure 6) also shows secondary drops on the three nights (2007 April 28, May 4 and 7). Expanded light curve from 2007 April 23 to May 8 (Figure 7) shows that the observations on the six nights (2007 April 23, 26, 27, 30, May 3, and 8) fit the shorter period (1.2274 days) sine curve but the observations on the three nights (2007 April 28, May 4 and 7) fit the long period (2.96 days) sine curve. This fact suggests an unknown satellite whose revolution period is 2.96 days.

From these two occultation cross sections and light curves, a 3-D model of Bamberga is reconstructed by the statistical inversion method [7] . The results are shown in Figure 8 and Figure 9. North and south pole is (λ, β) = (149° ± 11°, +68° ± 11°) or (329° ± 11°, −68° ± 11°), lengths of the three principal axes are (2a, 2b, 2c) = (246.3 ± 168.6 km) × (228.3 ± 9.5 km) × (220.4 ± 8.0 km). Constructed cross sections are shown in Figure 10 and Figure 11.

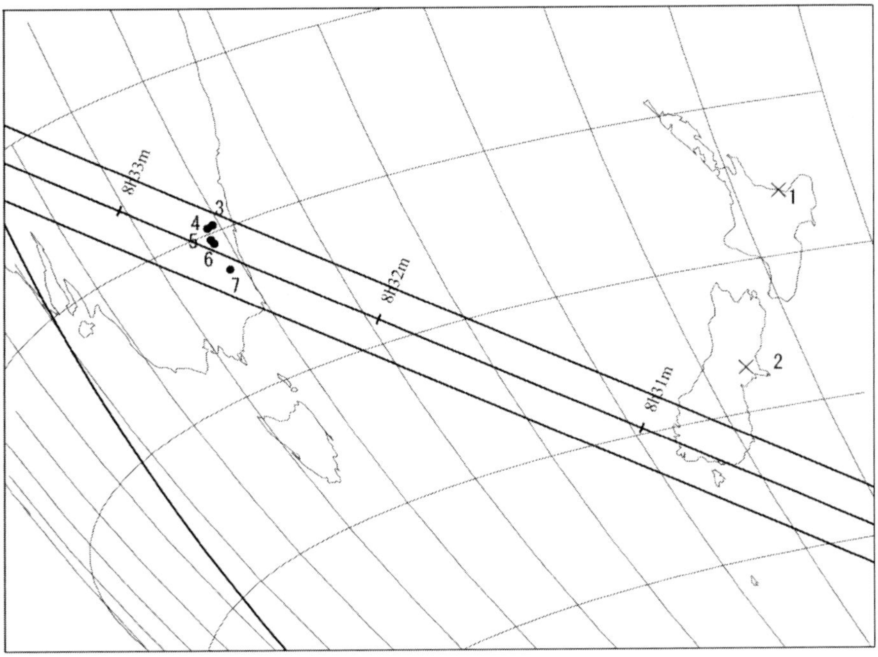

Figure 5. Occultation track and locations of the black circles observers of the occultation of HIP 059807 by (324) Bamberga on 2007 April 20. The black circles (#3 - 7) are sites where the occultation was successfully timed and the crosses (#1, 2) are sites where no occultation occurred.

Table 2. Observation data of the occultation of HIP 59807 by (324) Bamberga on 2007 April 20. Geodetic datum is WGS84 [20] .

No.	Observer	location	longitude	latitude	height	telescope	method	phen	UTC
1	D. Waton	Whakatane, NZ	E176°51'50.7"	S37°55'18.5"	3 m	20 cm			no occultation
2	B. Loader	Darfiled, NZ	E172°06'24.4"	S43°28'52.9"	210 m	25 cm			no occultation
3	D. Gault	Goulburn, NSW, Aus	E149°38'29.4"	S34°47'12.9"	702 m	20 cm		D	08 h 32 m 36.46 s
								R	08 h 32 m 47.66 s
4	A. Brakel	Gunning, NSW, Aus	E149°14'31.8"	S34°49'58.5"	620 m	20 cm		D	08 h 32 m 35.9 s ± 0.6 s
								R	08 h 32 m 49.1 s ± 0.4 s
5	M. Nelmes	Spence, ACT, Aus	E149°03'30.6"	S35°11'39.8"	702 m	10 cm		D	08 h 32 m 33.9 s ± 0.3 s
								R	08 h 32 m 48.2 s ± 0.3 s
6	P. Purcell	Red Hill, ACT, Aus	E149°06'51.8"	S35°19'30.6"	733 m	40 cm		D	08 h 32 m 33.1 s± 0.6 s
								R	08 h 32 m 47.73 s ± 0.4 s
7	David Herald	Cooma, NSW, Aus	E149°09'44.8"	S36°14'18.6"	835 m	20 cm		D	08 h 32 m 27.95 s
								R	08 h 32 m 41.48 s

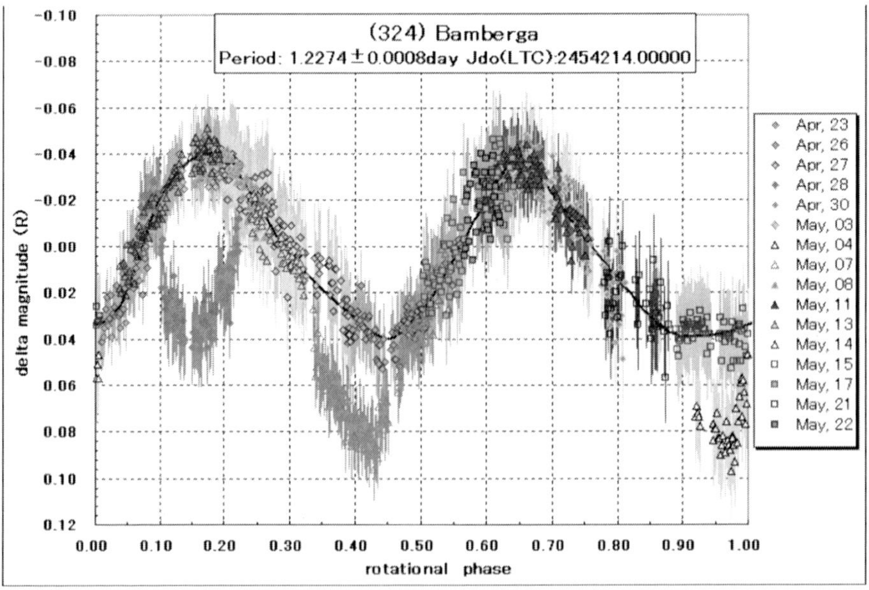

Figure 6. Composed light curve of (324) Bamberga from 2007 April 23 to May 22. The primary period is 1.2274 ± 0.0008 days. In addition to the primary period, secondary drops are shown on 2007 April 28, May 4 and 7.

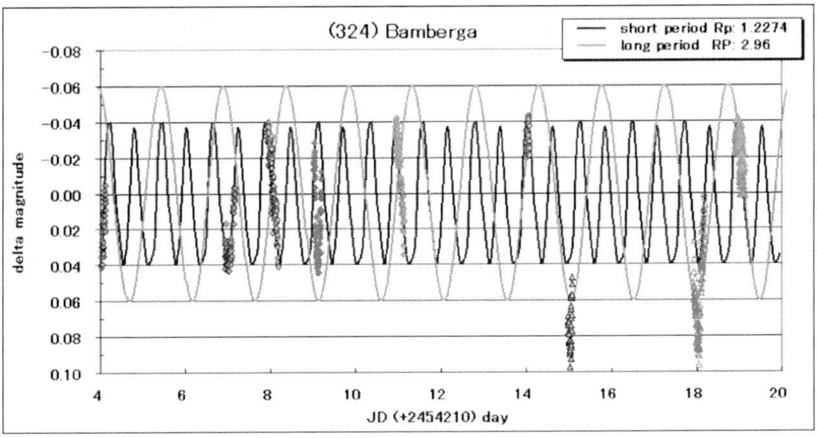

Figure 7. Expanded light curve of (324) Bamberga from 2007 April 23 to May 9. The short period sine curve (blue line) is 1.2274 days, and the long period sine curve (red line) is 2.96 days. The secondary drops on 2007 April 23, May 4 and 7 do not fall on the short period sine curve but are coinsident with the bottom of the long period sine curve.

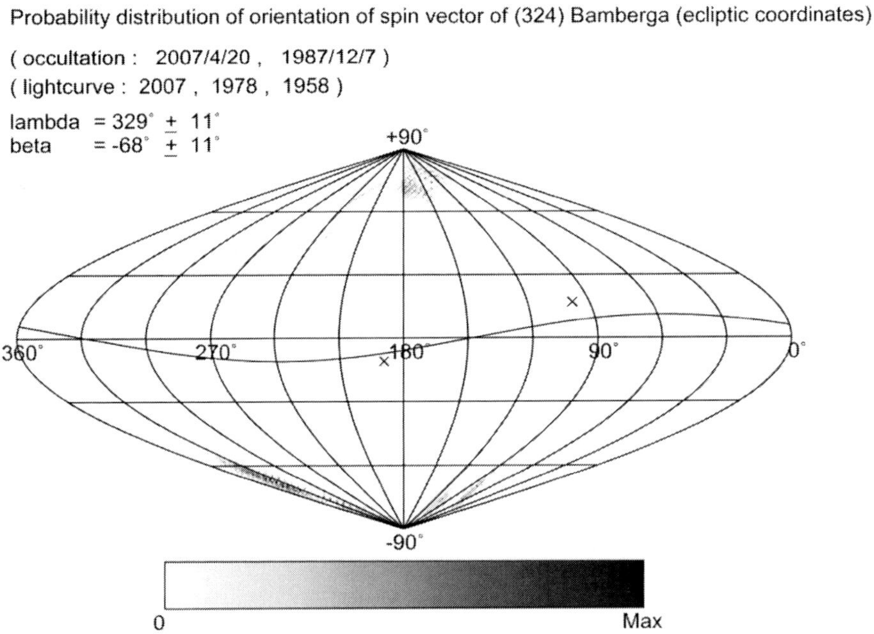

Figure 8. Probability distribution of the spin vector of (324) Bamberga. High probability regions are antipodal (λ, β) = (149° ± 11°, +68° ± 11°) (Ursa Major) or (λ, β) = (329° ± 11°, −68° ± 11°) (Eridanus). Two crosses indicate the positions of the occulted stars in 1987 and 2007 events, respectively.

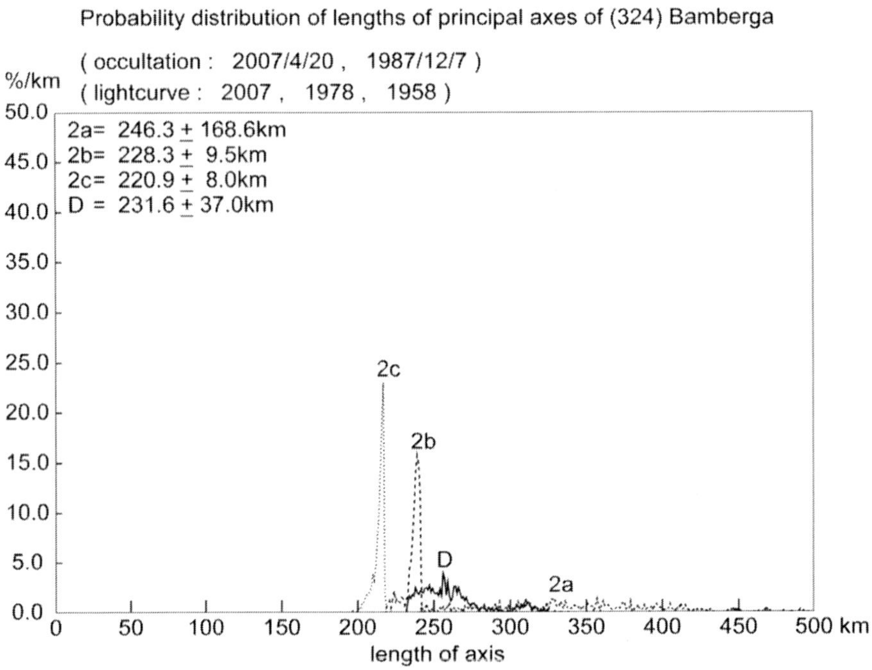

Figure 9. Probability distribution of the lengths of the three principal axes of (324) Bamberga. Since Bamberga is nearly a spheroid, the length of the longest axis a cannot be determined with sufficient accuracy. Assumed lengths are (2a, 2b, 2c) = (246.3 ± 168.6 km) × (228.3 ± 9.5 km) × (220.4 ± 8.0 km).

4. DUPLICITY OF (624) HEKTOR

The L4 Trojan (624) Hektor was discovered by A. Kopff in 1907. Hektor has a satellite S/2006 (624) 1 of diameter 15 km [8] . The stellar occultation by (624) Hektor of TYC057700887 (10.1 mag) was observed from two sites in Japan on 2008 January 24 (Figure 12). This is the first stellar occultation observation to be observed by Hektor. Tomioka at Hitachi, Ibaraki, detected an extinction for 4.4 seconds by video with GPS clock. Uehara at Tsukuba,

Ibaraki, observed two times extinction, 3.6 and 5.0 seconds, respectively, by visual (Table 3, Figure 13). His observation suggests the primary body is a close or contact binary. A possible unusual shape of the primary has been suggested by many authors [9] - [19] .

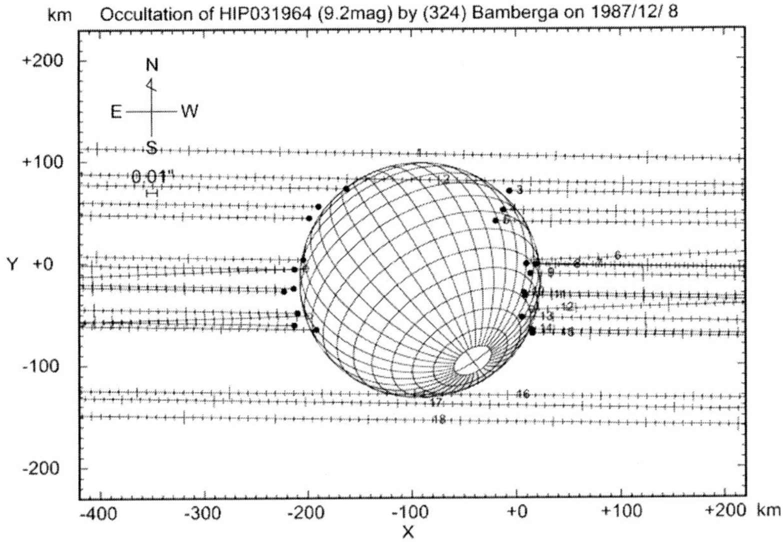

Figure 10. Reconstructed 3-D shape of (324) Bamberga during the occultation of HIP 031964 on 1987 December 8.

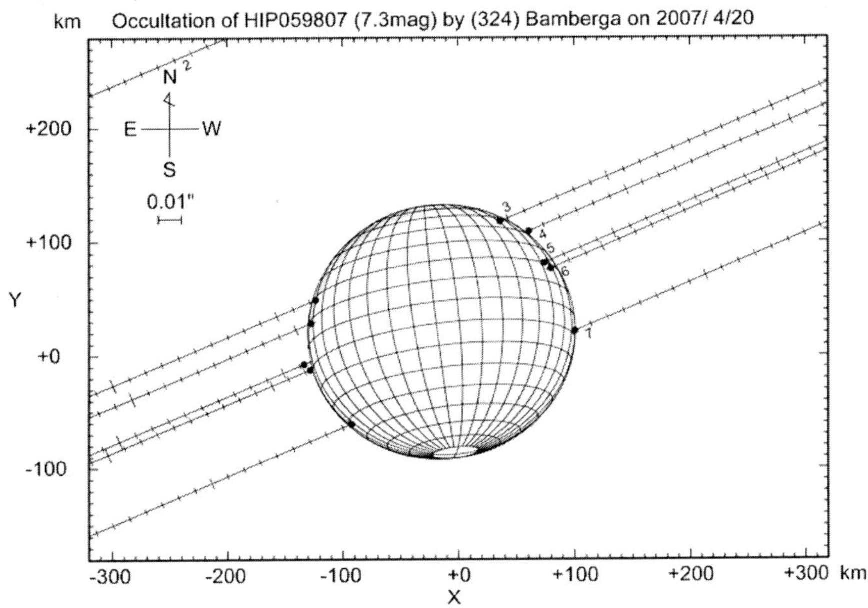

Figure 11. Reconstrunted 3-D shape of (324) Bamberga during the occultation of HIP 059807 on 2007 April 20.

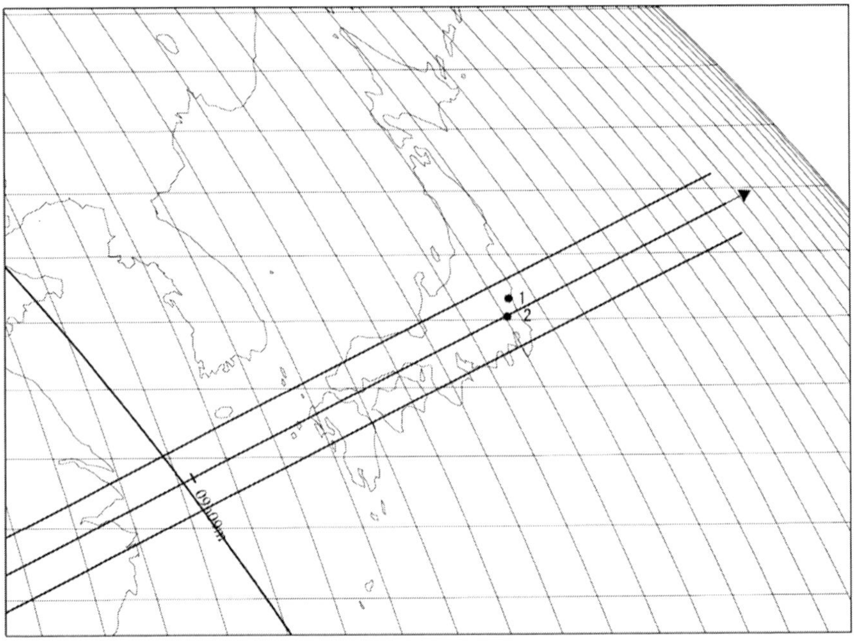

Figure 12. Occultation track and locations of observers of the occultation of TYC 057700817 by (624) Hektor on 2008 January 24. The black circles indicate sites where occultations were successfully timed.

Table 3. Observation data of the occultation of TYC 057700887 by (624) Hektor on 2008 January 24. Geodetic datum is WGS84.

No.	observer	location	longitude	latitude	thgieh	telescope	method	phen	UTC
1	Hiroyuki Tomioka	Hitachi, Ibaraki, Japan	E140°41'09"	N36°38'33"	33 m	30 cm C	GHS, TiVi, video	D	09 h 09 m 36.63 s ± 0.05 s
								R	09 h 09 m 41.03 s ± 0.05 s
2	Sadaharu Uehara	Tsukuba, Ibaraki, Japan	E140°07'04.2"	N36°05'11.2"	30 m	20 cm N, F = 4	visual	1D	09 h 09 m 28.8 s ± 0.5 s
								1R	09 h 09 m 32.4 s ± 0.3 s
								2D	09 h 09 m 35.2 s ± 0.6 s
								2R	09 h 09 m 40.2 s ± 0.4 s

5. POSSIBLE SATELLITE OF (657) GUNLÖD

(657) Gunlöd was discovered by A. Kopff in 1908. The planet is a main belt asteroid with a revolution period of 4.22 years. The stellar occultation of UCAC2 43078953 (12.0 mag) by (657) Gunlöd was observed from three sites in Japan (Table 4, Figure 14) on 2008 November 29. Hiroyuki Tomioka at Hitachi, Ibaraki, detected a 3.5 second extinction by video with GPS clock. Akira Yaeza at Hitachi, Ibaraki, also observed a 4.6 seconds visual extinction.

Two chords are coincident with a circle with a cross section of a diameter 45 ± 5 km compared with the assumed diameter of 42 km. These extinctions were caused by the primary. On the other hand, a short extinction of for 0.7 second was detected at Hamanowa Observatory by video with GPS clock located about 100 km north of the occultation track of the primary (Figure 15). Compared with the predicted extinction of 2.9 mag for 4.4 seconds, this extinction is possibly caused by an unknown satellite with a diameter of 7 km. Another possible interpretation is that the occulted star is double.

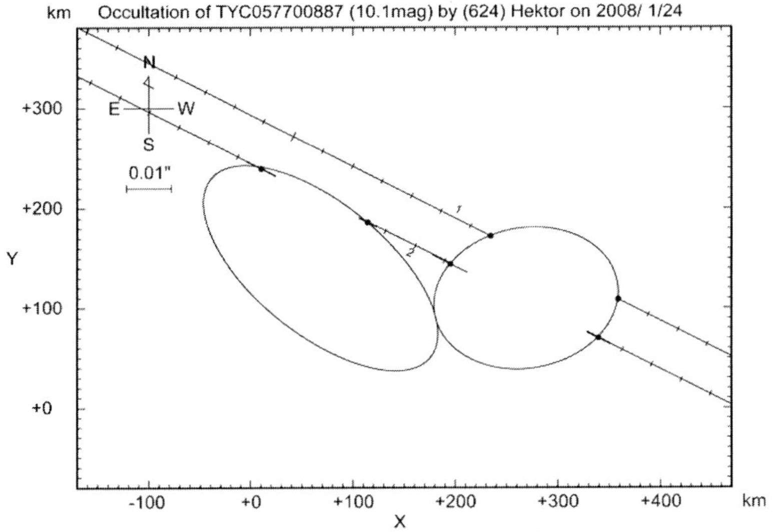

Figure 13. Result of the occultation of TYC 057700887 by (624) Hektor on 2008 January 24. Two extinctions were visually observed at #2 site. Since the main body is assumed to be a contact binary, two contact ellipses of (280 × 130 km) and (181 × 141 km) are fitted.

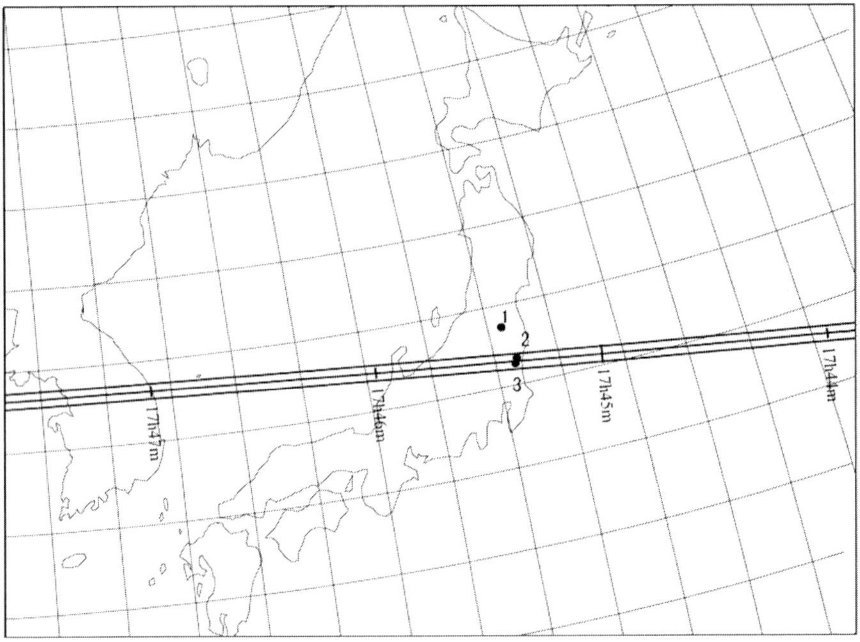

Figure 14. Occultation track and location of observers of the occultation of UCAC2 43078953 by (657) Gunlōd on 2008 November 29. The black circles indicate location where occultations were successfully timed. Sites #2 and 3 are where the star was occulted by the primary, and site #1 is where the occultation was by a possible satellite.

Table 4. Observation data of the occultation of UCAC2 43078953 by (657) Gunlöd on 2008 November 29. Geodetic datum is WGS84.

No.	Observer	location	longitude	latitude	height	telescope	method	phen	UTC
1	Hamanowa Observatory	Koriyama, Fukushima, Japan	E140°25'48.4"	N37°27'37.7"	260 m	40 cm N, F = 4.5	GHS, TiVi, video	D	17 h 45 m 21.34 s ± 0.06 s
									17 h 45 m 22.07 s ± 0.06 s
2	Hiroyuki Tomioka	Hitachi, Ibaraki, Japan	E140°41'09"	N36°38'33"	33 m	30 cm C	GHS, TiVi, video	D	17 45 m 21.0 s ± 0.5 s
								R	17 h 45 m 24.5 s ± 0.3 s
3	Akira Yaeza	Hitachi, Ibaraki, Japan	E140°36'11"	N36°31'28"	230 m	20 cm C, 133x	TEL visual	D	17 h 45 m 25.4 s ± 0.3 s
								R	17 h 45 m 30.0 s ± 0.3 s

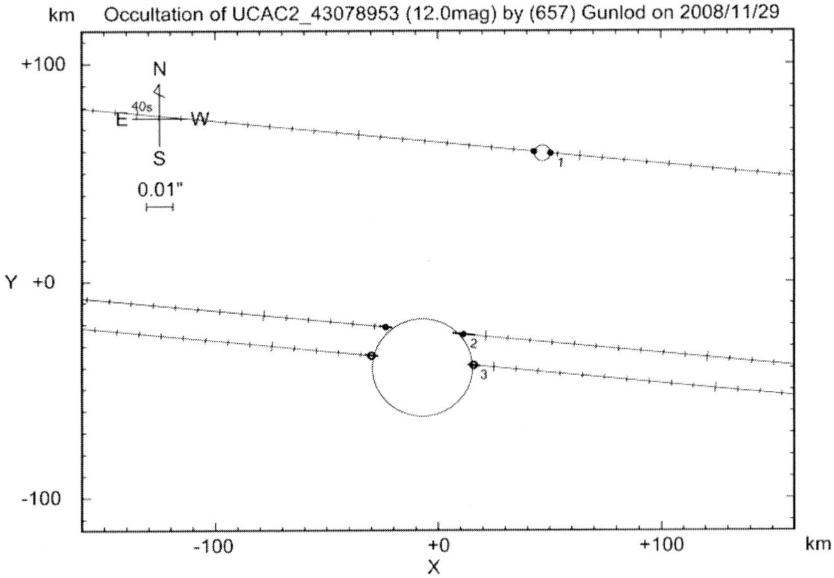

Figure 15. Result of the occultation of UCAC2 43078953 by (657) Gunlōd on 2008 November 29. Chord #1 and #2 are from timed video and #3 is from a visual timing.

6. POSSIBLE SATELLITE OF (739) MANDEVILLE

(739) Mandeville was discovered by J. H. Metcalf in 1913. The planet is a main belt asteroid with a revolution period of 4.53 years. The stellar occultation of PPM177093 = HIP052360 (9.4 mag) by (739) Mandeville was observed from two sites in Japan on 1980 December 10 (Table 5, Figure 16). The event was the first ever successfully observed asteroid occultation from Japan. Miyoshi Ida at Ohmi-hachiman, Shiga, observed an extinction for five or six second visually with 6.5 cm telescope, but he failed to record his voice, as a result the time and duration of the event are uncertain. The extinction corresponds to a chord from 69 to 83 km. On the other hands, Toshio Hirose at Yaita, Tochigi, visually observed an extinction of 2.1 seconds also with 6.5 cm telescope. The extinction corresponds to a chord of 29 km. Both Ida and Hirose are experienced lunar occultation

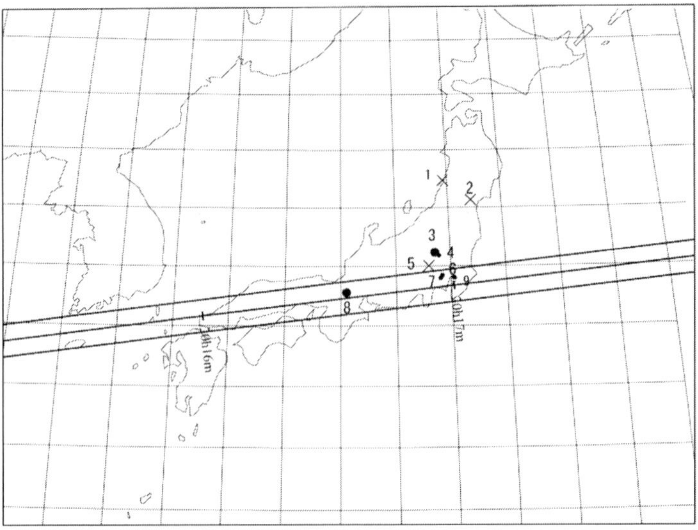

Figure 16. Occultation track and locations of observers of the occultation of HIP 052360 by (739) Mandeville on 1980 December 10. The black circle #3 is visual timing site, and #8 is an occultation from a site where no timing was observed. Crosses (#1, 2, 5) indicate sites where no occultation took place and points (#6, 7, 9) sites where clouds obscured any observation.

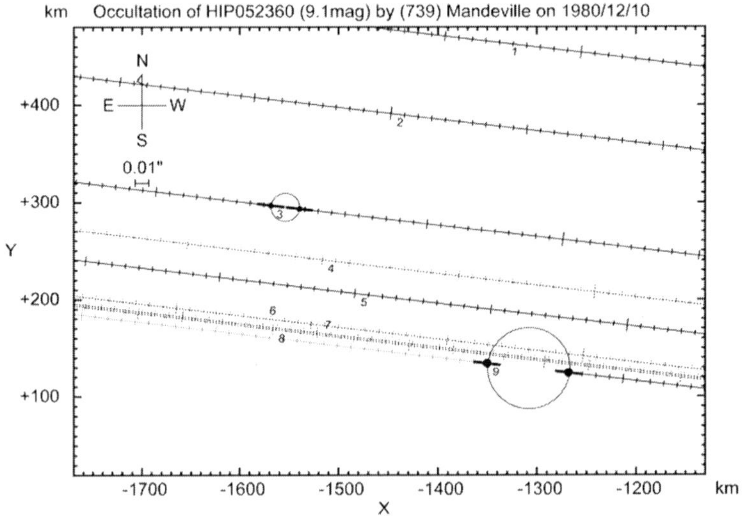

Figure 17. Result of the occultation of HIP 052360 by (739) Mandeville on 1980 December 10. Locations #3 and #8 are where visual occultation were timed but those times are inaccurate. But location #5 between them was no occultation.

Table 5. Observationdata od the occultation of PPM 157093 by (739) Mandeville on 1980 December 10. Geodetic datum is Tokyo Datum.

No.	Observer	Location	longitude	latitude	height	telescope	method	phen	UTC
1	Isao Satō	Sakata, Yamagata, Japan	E139°50'44.6"	N38°55'32.8"	4 m	13 cm N, F = 7.7	visual		no occultation
2	Sendai M. O.	Sendai, Miyagi, Japan	E140°51'56"	N38°15'22"	45 m	41 cm C, 15 cm R			no occultation
3	Toshio Hirose	Yaita, Tochigi, Japan	E139°25'48.4"	N36°27'37.7"	260 m	6.5 cm R	visual	D	20 h 16 m 38.5 s ± 1.0 s
								R	20 h 16 m 40.6 s ± 1.0 s
4	Kiyoshi Sakurai	Tanuma, Tochigi, Japan	E139°34'31.5"	N36°21'36.3"	63 m	12 cm N			cloudy
5	Dodaira A.O.	Tokikawa, Saitama, Japan	E139°11'42"	N36°00'22"	879 m	91 cm C	p.e.		no occultation
6	Yoshikazu Ogawa	Nakano, Tokyo, Japan	E139°40'31.1"	N35°41'01.6"	40 m	6.8 cm R			cloudy
7	Yoshihiro Musashi	Kawasaki, Kanagawa, Japan	E139°36'15.0"	N35°35'34.5"	31 m	10 cm N			cloudy
8	Shoji Kawana	Chiba, Chiba, Japan	E140°06'04.1"	N35°41'01.6"	16 m	8 cm R			cloudy
9	Miyoshi Ida	Ohmihachiman, Shiga, Japan	E136°11'53.3"	N35°06'38.5"	128 m	6.5 cm R	visual	D	20 h 16 m 30 s
								R	20 h 16 m 36 s

observers and they convinced the extinctions were real, but the timings were both visual in nature. However, Koichiro Tomita at Dodaira Station of the Tokyo Astronomical Observatory observed no occultation with a 91 cm reflector. The three observations are not consistent as is shown in Figure 17. A possible interpretation is duplicity of Mandeville, namely the extinction observed by Ida is by the primary body and the extinction observed by Hirose is by the secondary.

7. CONCLUSION

Possible satellites of (279) Thule, (324) Bamberga, (657) Gunlōd, (739) Mandeville, and duplicity of the primary body of the possible contact binary Trojan (624) Hektorare suggested from their stellar occultation and light curve observations. Confirmation is needed in order to verify of the nature of the observed extinctions.

CITE THIS PAPER

Isao Satō, Hiromi Hamanowa, Hiroyuki Tomioka, Sadaharu Uehara (2015) Possible Duplicities of Five Asteroids. *International Journal of Astronomy and Astrophysics*,**05**,193-207. doi: 10.4236/ijaa.2015.53024

REFERENCES

1. Neese, C., Ed. (2010) Asteroid Taxonomy V6.0. EAR-A-5-DDR-TAXONOMY-V6.0. NASA Planetary Data System.
2. Zappala, V., Di Martino, M., Cellino, A., Farinella, P., De Sanctis, G. and Ferreri, W. (1989) Rotation Properties of Outer Belt Asteroids. Icarus, 82, 354-368.
3. Millis, R.L., Wasserman, L.H., Franz, O.G., Bowell, E., Nye, R.A., Thompson, D.T., White, N.M., Hubbard, W.B., Eplee Jr., R.E., Lebofsky, L.A., Marcialis, R.L., Greenberg, R.J., Hunten, D.M., Reitsema, H.J., Qian, B., Dunham, D.W., Maley, P.D., Klemola, A.R. and Yeomans, D.K. (1989) Observation of the 8 December 1987 Occultation of AG +40° 0783 by 324 Bamberga. Astronomical Journal, 98, 1094-1099
4. Gehrels, T. and Owings, D. (1962) Photometric Studies of Asteroids IX. Additional Light Curves. Astrophysical Journal, 135, 906-924.
5. Scaltriti, F., Zappala, V., Stanzel, R., Blanco, C., Catalano, S. and Young, J.W. (1980) Lightcurves and phase relation of asteroid 324 Bamberga. Icarus, 43, 391-398.
6. Lagerkvist, C.I., Magnusson, P., Williams, I.P., Buontempo, M.E., Gibbs, P. and Morrison, L.V. (1989) Physical Studies of Asteroids. XIX—Phase Relations and Composite Lightcurves Obtained with the Carlsberg Meridian Circle. Astronomy & Astrophysics Supplement Series, 78, 519-532.
7. Satō I., Buie, M., Maley, P.D., Hamanowa, H., Tsuchikawa, A. and Dunham, D.W. (2014) A 3-D Shape of (704) Interamnia from Its Occultations and Lightcurves. International Journal of Astronomy and Astrophysics, 4, 91-118.
8. Marchis, F., Wong, M.H., Berthier, J., Descamps, P., Hestroffer, D., Vachier, F., Le Mignant, D.W.M. and de Pater, I. (2006) S/2006 (624) 1. IAU Circular No. 8732.
9. Dunlap, J.L. and Gehrels, T. (1969) Minor Planet III. Lightcurves of a Trojan Asteroid. Astronomical Journal, 74, 796-803.
10. . Hartmann, W.K. and Cruikshank, D.P. (1978) The Nature of Trojan Asteroid 624 Hektor. Icarus, 36, 353-366.
11. . Zappala, V. and Knezevic, Z. (1984) Rotation Axes of Asteroids: Results for 14 Objects. Icarus, 59, 436-455.
12. Magnusson, P. (1986) Distribution of Spin Axes and Senses of Rotation for 20 Large Asteroids. Icarus, 68, 1-39.
13. Michalowski, T. (1988) Photometric Astronomy Applied to Asteroids: 6, 15, 43, and 624. Acta Astronomica, 38, 455-468.
14. Dahlgren, M., Fitzsimmons, A., Lagerkvist, C.-I. and Williams, I.P. (1991) Differential CCD Photometry of Dubiago, Chron and Hektor. Monthly Notice of the Royal Astronomical Society, 250, 115-118.
15. De Angelis, G. (1992) New Asteroid Pole Determinations. In: Brahic, A., Gerard, J.-C. and Surdej, J., Eds., Observations and Physical Properties of Small Solar System Bodies, Universite de Liege, Institut d'Astrophysique, Liege, 195-201.
16. . Detal, A., Hainaut, O., Ibrahim-Denis, A., Pospieszalska-Surdej, A., Schils, P., Schober, H.J. and Surdej, J. (1994) Pole, Albedo and Shape of the Minor Planets 624 Hektor and 43 Ariadne: Two Tests for Comparing Four Different Pole Determination Methods. Astronomy & Astrophysics, 281, 269-280.

17. . De Angelis, G. (1995) Asteroid Spin, Pole and Shape Determinations. Planetary Space Science, 43, 649-682.
18. Hainaut-Rouelle, M.-C., Hainaut, O.R. and Detal, A. (1995) Light Curves of Selected Minor Planets. Astronomy & Astrophysics Supplement Series, 112, 125-142.
19. . Blanco, C. and Riccioli, D. (1998) Pole Coordinates and Shape of 30 Asteroids. Astronomy and Astrophysics Supplement Series, 131, 385-394.
20. Dunham, D.W., Herald, D., Frappa, E., Hayamizu, T., Talbot, J. and Timerson, B. (2014) Asteroid Occultations V12.0. EAR-A-3-RDR-OCCULTATIONS-V12.0. NASA Planetary Data System.

CHAPTER 7

A New Family of Extraterrestrial Amino Acids in The Murchison Meteorite

Toshiki Koga[1] & Hiroshi Naraoka[1,2]

[1] *Department of Earth and Planetray Sciences, Kyushu University, 744 Motooka, Nishi-ku, Fukuoka, 819-0395, Japan.*

[2] *Research Center for Planetary Trace Organic Compounds, Kyushu University, 744 Motooka, Nishi-ku, Fukuoka,*

ABSTRACT

The occurrence of extraterrestrial organic compounds is a key for understanding prebiotic organic synthesis in the universe. In particular, amino acids have been studied in carbonaceous meteorites for almost 50 years. Here we report ten new amino acids identified in the Murchison meteorite, including a new family of nine hydroxy amino acids. The discovery of mostly C_3 and C_4 structural isomers of hydroxy amino acids provides insight into the mechanisms of extraterrestrial synthesis of organic compounds. A complementary experiment suggests that these compounds could be produced from aldehydes and ammonia on the meteorite parent body. This study indicates that the meteoritic amino acids could be synthesized by mechanisms in addition to the Strecker reaction, which has been proposed to be the main synthetic pathway to produce amino acids.

INTRODUCTION

The extraterrestrial synthesis of amino acids is an intriguing discussion concerning the chemical evolution for the origins of life in the universe, because amino acids are fundamental building blocks of terrestrial life. The extraterrestrial amino acid distribution has been extensively examined using carbonaceous chondrites, which are the most chemically primitive meteorites containing volatile components such as water and organic matter, particularly the Murchison meteorite since it's fall in 1969. The Murchison meteorite is classified as a CM2 (Mighei-type) chondrite, moderately altered by aqueous activity on the parent body (e.g. ref. 1). Currently, a total of 86 amino acids have been identified in the Murchison meteorite as α, β, γ and δ amino structures with a carbon number between C_2 and C_9 including dicarboxyl and diamino functional groups[2,3,4]. Although the presence of C_{10} amino acids has been suggested[5], definitive C_{10} amino acid identification was not assigned to the molecular structure due to lack of the appropriate standards. The concentration and structural diversity of amino acids generally increase after hydrolysis of the water extract of the CM chondrites[6], even though some CR chondrites yielded more free amino acids than hydrolyzed amino acids[5]. The hydrolyzed amino acids are present as their precursors and the molecular occurrence of precursors relates to the sources and formation pathways of extraterrestrial amino acids[7,8]. It is generally considered that meteoritic amino acids could be formed in the meteorite parent bodies by the Strecker reaction, in which aldehyde or ketone reacts with cyanide and ammonia followed by hydrolysis to produce α-amino acid[9]. However, the Strecker reaction produces only α-amino acids (i.e. bearing amino and carboxyl group at the same carbon), and cannot explain the formation of other amino acids (β, γ and δ structures). It has been proposed that β-alanine could be formed by Michael addition of NH_3 to cyanoacetylene[10] and that γ- and δ-amino acids could be formed by hydrolysis of 5-membered and 6-membered lactams, respectively[9]. However, comprehensive formation mechanisms of extraterrestrial amino acids are not well understood.

Even though the Murchison meteorite has been studied for the occurrence of amino acids for almost 50 years, this study has revealed the presence of ten new amino acids, including a new family of nine hydroxy C_3 and C_4 amino acids. These new findings will expand our knowledge concerning the formation mechanism of meteoritic amino acids.

RESULTS

Newly discovered amino acids in Murchison

Thirty amino acids between C_2 and C_6 were identified in the hydrolyzed samples of the water extract and the extract residue of the Murchison meteorite (see the Supplementary Information, SI, for the detailed analytical methods and quantification) without consideration of their enantiomers (Table 1). Many other amino acids (especially for C_5-C_7amino acids) were reported from the Murchison meteorite[5,11], but could not be detected in this study due to the lack of appropriate standards. Glycine was the most abundant amino acid (~3 ppm in total), consistent with previous studies[6,11], even though α-aminoisobutyric acid (AIBA) or isovaline is sometimes the most abundant in the Murchison meteorite e.g. ref. 5. Such a distribution difference shows the heterogeneity of amino acids in the meteorite. The second most abundant amino acid was alanine (1.7 ppm) with a relative enrichment of L-alanine (Table 1). Because L-alanine is a common proteinogenic amino acid of life, and because small amounts of L-alanine, L-serine, L-threonine and L-leucine in addition to glycine were detected in the procedural blank (see SI), the L-predominance observed in this study may be due to contamination in the terrestrial environment since the fall of the Murchison meteorite in 1969. However, the L-preference of proteinogenic amino acids such as alanine and aspartic acid in the Murchison and Tagish Lake meteorites has been proposed to be indigenous[12,13]. Amino acids such as AIBA, β-alanine and isovaline were also present in relatively high concentrations (0.97, 1.7 and 1.7 ppm, respectively) in this study. These non-proteinogenic amino acids have been reported as indigenous amino acids in the Murchison meteorite as well as other CM chondrites by many previous studies[11,14,15,16].

Nine hydroxy amino acids (HAAs) including isoserine, homoserine, β-homoserine, α-methylserine, 4-amino-2-hydroxybutanoic acid (4-A-2-HBA), 4-amino-3-hydroxybutanoic acid (4-A-3-HBA), isothreonine, *allo*-isothreonine and 3-amino-2-(hydroxymethyl)-propanoic acid (3-A-2-HPA), as shown in Fig. 1 for their structures and the SI for their mass spectra, have been identified in the Murchison meteorite for the first time. Isoserine is a C_3 β-amino acid, the structural isomer of serine. Even though it was identified from Antarctic CR meteorites[16], this study revealed the first occurrence of isoserine in the Murchison meteorite. Eight additional HAAs are C_4 structural isomers including two α-, four β- and two γ-amino acids. Any organic compounds bearing –OH and –NH_2 at the same carbon are unstable and are unlikely to occur in the natural environment. Furthermore, four HAAs (one C_4 and three C_5) were found based on their mass spectra (see SI), but their structures could not be assigned due to the lack of appropriate standards. The newly discovered HAAs were present as near-racemic mixtures except for homoserine, if their DL forms can be separated by chromatography (see SI). In addition, a new C_5 dicarboxy amino acid, β-amino acid β-(aminomethyl)succinic acid, was identified in meteorites, although one C_6 β-amino dicarboxyl acid (3-aminoadipic acid) was reported in Murchison17.

Table 1: Amino acids identified in the Murchison meteorite and the experimental products.

Carbon number	Position of amino group	Amino acid	Murchison (ppb) (n = 2)			Experiments (ppm)		m/z**
			H_2O extract hydrolyzed	Residue hydrolyzed	Total (Gly = 100)	HCHO/CH_3CHO/NH_3 (Gly = 100) (n = 2)	HCHO/CH_3CHO/CH_2(OH)CHO/NH_3 (Gly = 100) (n = 3)	
2	α	Glycine	1373 ± 143	2192 ± 82	3566 ± 165 (100)	237 ± 26 (100)	288 ± 42 (100)	126
	α	D-Alanine	318 ± 59	252 ± 18	570 ± 62 (16)	14 ± 2 (5.9)	159 ± 26 (55)	140
	α	L-Alanine	398 ± 74	732 ± 8	1130 ± 75 (32)	14 ± 2 (5.9)	163 ± 27 (57)	140
	α	Sarcosine	201 ± 26	97 ± 5	298 ± 26 (8.4)	n.d.	n.d.	185
	α	D-Serine	267 ± 46	127 ± 17	394 ± 49 (11)	3.7 ± 1.0 (1.6)	23 ± 5 (8.0)	138 (280)
3	α	L-Serine	318 ± 72*	942 ± 89	1261 ± 114* (35)	5.4 ± 1.9 (2.3)	28 ± 7 (9.8)	138 (280)
	β	D-Isoserine[a]	tr.	57 ± 12	57 ± 12 (1.6)	14 ± 2 (2.3)	29 ± 7 (10)	138
	β	L-Isoserine[a]	tr.	57 ± 6	57 ± 6 (1.6)	15 ± 2 (6.3)	33 ± 7 (11)	138
	β	β-Alanine	1064 ± 151	658 ± 36	1721 ± 156 (48)	43 ± 6 (18)	198 ± 39 (69)	168
	α	α-AIBA	863 ± 154	152 ± 18	1015 ± 155 (28)	n.d.	n.d.	154
	α	D-α-ABA	532 ± 81	341 ± 12	873 ± 82 (24)	1.5 ± 0.7 (0.6)	63 ± 11 (22)	154 (107)
	α	L-α-ABA	541 ± 142	403 ± 70	944 ± 159 (26)	2.9 ± 0.9 (1.2)	62 ± 10 (22)	154 (107)
	α	D-Aspartic acid	108 ± 18	193 ± 33	301 ± 37 (8.4)	n.d.	1.3 ± 0.3 (0.45)	184
	α	L-Aspartic acid	267 ± 53	780 ± 117	1048 ± 129 (29)	tr.	2.5 ± 0.7 (0.87)	184
	α	D-Threonine	tr.	tr.	tr.	n.d.	n.d.	152
	α	L-Threonine	173 ± 23	519 ± 42	691 ± 48 (19)	tr.	6.9 ± 2.6 (2.4)	152
	α	D-allo-Threonine	n.d.	10 ± 1	10 ± 1 (0.28)	n.d.	n.d.	153
	α	L-allo-Threonine	n.d.	tr.	tr.	n.d.	n.d.	153
	α	DL-α-Methylserine[a]	79 ± 6*	65 ± 5*	144 ± 8* (4.0)	n.d.	74 ± 13 (26)	152 (184)
	α	D-Homoserine[a]	7.9 ± 1.0	14 ± 2*	22 ± 2* (0.62)	n.d.	9.0 ± 1.6 (3.1)	84 (152, 266)
	α	L-Homoserine[a]	13 ± 3	34 ± 6*	47 ± 6 (1.3)	n.d.	10 ± 2 (3.5)	84 (152, 266)
4	β	D-β-ABA	147 ± 17*	53 ± 3*	200 ± 17* (5.6)	19 ± 1 (8.0)	45 ± 9* (16)	140 (153, 182)
	β	L-β-ABA	146 ± 31*	48 ± 3*	193 ± 31* (5.4)	24 ± 2* (10)	41 ± 4* (14)	140 (153, 182)
	β	D-β-AIBA	150 ± 54	85 ± 4*	235 ± 54* (6.6)	n.d.	16 ± 2* (5.6)	182 (69)
	β	L-β-AIBA	131 ± 50*	74 ± 1*	204 ± 50* (5.7)	n.d.	15 ± 2*(5.2)	182 (69)
	β	DL-β-Homoserine[a]	12 ± 2*	8.3 ± 1.9*	20 ± 3* (0.56)	n.d.	20 ± 4 (6.9)	180 (294)
	β	DL-3-Amino-2-(hydroxy-methyl) propanoic acid[a]	33 ± 10	32 ± 8	65 ± 13 (1.8)	9.2 ± 1.8 (3.9)	114 ± 28 (40)	180 (197)
	β	DL-Isothreonine[a]	15 ± 1*	76 ± 2*	92 ± 3* (2.6)	16 ± 2 (6.8)	61 ± 15 (21)	152 (294)
	β	D-allo-Isothreonine[a]	tr.	59 ± 12*	59 ± 12* (1.7)	17 ± 4 (7.2)	42 ± 12 (15)	294 (266)
	β	L-allo-Isothreonine[a]	tr.	50 ± 5*	50 ± 5* (1.4)	21 ± 2 (8.9)	47 ± 14 (16)	294 (266)
	γ	DL-4-A-2-HBA[a]	28 ± 1	52 ± 12	80 ± 12 (2.2)	2.8 ± 0.7 (1.2)	21 ± 4 (7.3)	153
	γ	D-4-A-3-HBA[a]	21 ± 11*	11 ± 5*	32 ± 12* (0.90)	n.d.	n.d.	152
	γ	L-4-A-3-HBA[a]	12 ± 4	7.4 ± 3.1	19 ± 5 (0.53)	n.d.	n.d.	152
	γ	γ-ABA	1244 ± 242	638 ± 35	1882 ± 245 (53)	n.d.	tr.	182
	α	D-Valine	58 ± 1	30 ± 1	87 ± 1 (2.4)	n.d.	n.d.	55
	α	L-Valine	129 ± 15*	419 ± 62*	548 ± 64* (15)	n.d.	n.d.	55
	α	D-Norvaline	75 ± 19	55 ± 5	129 ± 20 (3.6)	n.d.	n.d.	168
	α	L-Norvaline	67 ± 12	52 ± 7	119 ± 14 (3.3)	n.d.	n.d.	168
5	α	DL-Isovaline	1418 ± 247	250 ± 20	1668 ± 248 (47)	n.d.	n.d.	168
	α	D-Glutamic acid	172 ± 27	142 ± 25	314 ± 37 (8.8)	n.d.	n.d.	180
	α	L-Glutamic acid	426 ± 85	788 ± 115	1214 ± 143 (34)	tr.	tr.	180
	β	D-β-(Aminomethyl)-succinic acid[a]	30 ± 5	16 ± 4	46 ± 7 (1.3)	n.d.	1.9 ± 0.6* (0.66)	226
	β	L-β-(Aminomethyl)-succinic acid[a]	27 ± 4*	17 ± 2*	44 ± 4* (1.2)	n.d.	2.4 ± 0.7* (0.83)	226
	α	D-α-Aminoadipic acid	115 ± 24	27 ± 4*	142 ± 25* (4.0)	n.d.	n.d.	194 (212)
	α	L-α-Aminoadipic acid	118 ± 35*	66 ± 9	184 ± 36* (5.2)	n.d.	n.d.	194 (212)
	α	D-Leucine	17 ± 1*	63 ± 1*	80 ± 1* (2.6)	n.d.	n.d.	140 (182)
	α	L-Leucine	170 ± 34	888 ± 13	1058 ± 37 (30)	n.d.	tr.	140 (182)
6	α	D-Isoleucine	101 ± 9	172 ± 1	273 ± 9 (7.7)	n.d.	n.d.	182
	α	L-Isoleucine	251 ± 60	696 ± 54	947 ± 81 (27)	n.d.	tr.	182
		Total	11636 ± 506	12500 ± 249	24136 ± 564	460 ± 28	1576 ± 82	
		HAA total[a]**	221 ± 17	523 ± 26	744 ± 31	95 ± 6	460 ± 41	

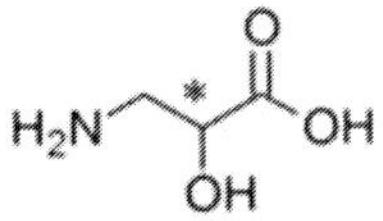

Isoserine

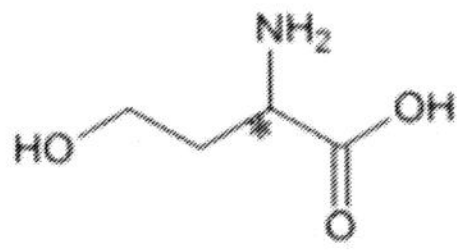

Homoserine

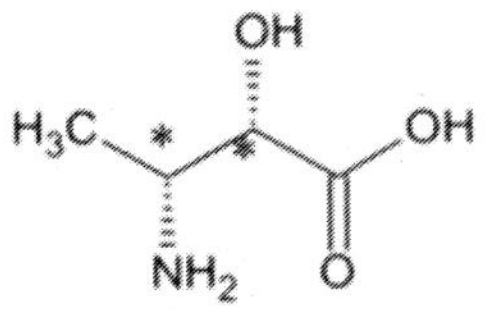

Isothreonine

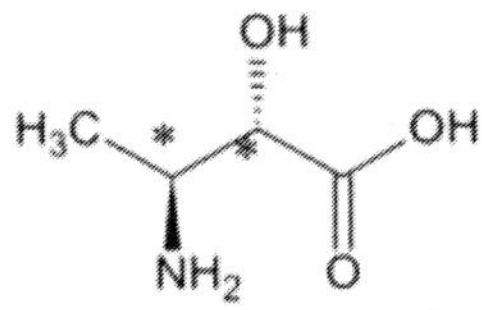

allo-Isothreonine

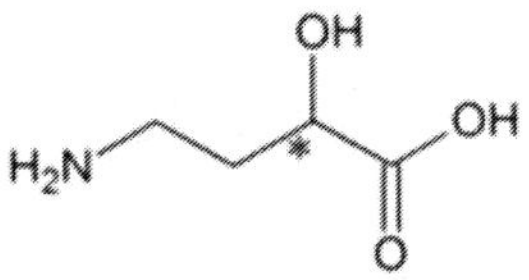

4-Amino-2-hydroxybutyric acid (4-A-2-HBA)

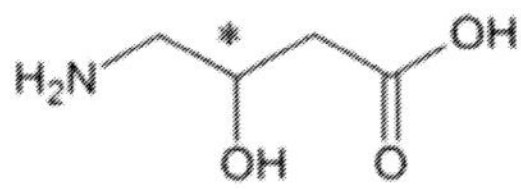

4-Amino-3-hydroxybutyric acid (4-A-3-HBA)

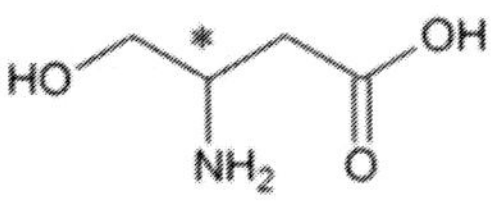

β-Homoserine

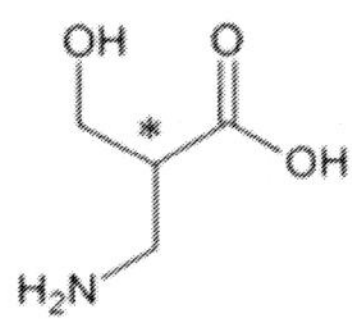

3-Amino-2-(hydroxymethyl)propanoic acid (3-A-2-HPA)

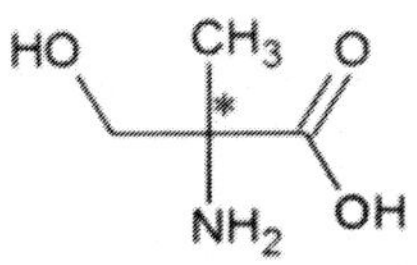

α-Methylserine

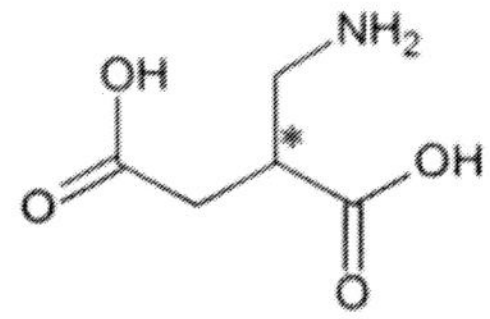

β-(Aminomethyl)succinic acid

Figure 1. Chemical structures of amino acids newly identified in the Murchison meteorite. *Asymmetric carbon.

The total concentration of the new HAAs ranged from ~20 to ~140 ppb (Table 1). Although the direct HCl hydrolysis of the extract residue is not commonly performed, the amino acid occurrence was different between the water extract and the extract residue. The HAAs were generally present in higher concentrations in the extract residue relative to the water extract. In particular, isoserine and *allo*-isothreonine were detected in trace amounts in the water extract (Table 1). In contrast, α-Methyl amino acids such as AIBA and isovaline were six times more abundant in the water extract relative to the extract residue.

Amino acids produced from aldehydes and ammonia in aqueous solution

As a complement to the analysis of amino acids in the Murchison meteorite, heating experiments were performed using a mixture of formaldehyde, acetaldehyde and ammonia in aqueous solution in the presence or absence of glycolaldehyde at 60 °C for 6 days (pH = 11.5 for the starting solution), to simulate possible chemical reaction during aqueous alteration of the meteorite parent body. Aldehydes such as formaldehyde and acetaldehyde are abundant in molecular clouds (e.g. ref. 18) and are probably the predominant carbon sources to produce meteoritic organic compounds such as sugars by the formose reaction[19, 20]. Glycolaldehyde was also used to evaluate the first reaction step, as glycolaldehyde can condense from two formaldehydes by the formose reaction. Ammonia has been detected in high concentrations in the Murchison meteorite[21] and is the principal nitrogen source for organic chemistry in meteorites.

Various amino acids including glycine and HAAs were synthesized from aldehydes and ammonia (Table 1). While glycine (C_2) was the most abundant amino acid, followed by alanine or β-alanine (C_3), α-methyl alkylated (no hydroxy) amino acids such as AIB and isovaline were not formed in the experiments. The number of amino acid species produced and their concentrations increased in the glycolaldehyde-present vs. glycolaldehyde-absent experiments. In the presence of glycolaldehyde, larger ($\geq C_4$) amino acids were found in more diverse structures. In particular, more C_4 HAAs and at higher concentrations, were produced when the starting material included glycolaldehyde (Table 1).

DISCUSSION

A new family of hydroxy amino acids in the Murchison meteorite

Previously only three HAAs (serine, threonine and *allo*-threonine) have been identified in the Murchison meteorite4. These are all α-amino acids, in which serine and threonine are proteinogenic amino acids and *allo*-threonine is a diastereoisomer of threonine, which have often been present as an L-rich signature. In particular, the presence of L-threonine detection (~700 ppb concentration) with only trace amounts of D-threonine (Table 1 and SI) is suggestive of terrestrial contamination. Non-proteinogenic HAAs such as isoserine and homoserine have previously been identified in a few Antarctic CR meteorites16, but to the best of our knowledge, they have not previously been reported from the Murchison meteorite. In this study, nine new $C_3$and C_4 HAAs were discovered in the Murchison meteorite, comprising a new family of meteoritic amino acids. Three C_4 additional HAAs (α-methylisoserine, N-methylserine and N-methylisoserine) were not positively identified currently due to the lack of the appropriate standard or the low intensity of mass peaks (see SI). The finding of most structural isomers of C_3 and C_4 HAAs expands the structural diversity of soluble meteoritic organic compounds, particularly for small amino acids ($\leq C_4$). The three C_5 HAAs were identified by mass spectra that are characteristic of their structures (see SI), although these identifications are somewhat tentative, as the complete structures were not determined due to the necessary standards being unavailable.

It is notable that these HAAs have never been found before in the Murchison meteorite, despite numerous rigorous surveys of amino acids over almost 50 years. Although the C_4 HAAs were present in concentrations less than the major amino acids such as alanine and aminobutyric acids, the concentrations were comparable to those of the C_5 amino acids (Table 1). We suggest that these HAAs have not been identified before as they are non-proteinogenic amino acids and it is relatively difficult to prepare their standards. Moreover, the hydroxyl amino acids were generally present in higher concentrations in the extract residue vs. the water extract. For example, isoserine and *allo*-isothreonine were only detected in trace amounts in the water extract, but were present in significant concentrations in the extract residue (Table 1), possibly due to the hydroxyl group being intimately adsorbed or chemically bound with clay minerals, and hence being poorly extracted by water. The direct HCl treatment of the extract residue may have yielded the HAAs due to the dissolution of the clay minerals and/or due to protonation of the clay mineral surfaces. The detection of isoserine and homoserine in the water extract of CR meteorites14, which are less aqueously altered than CM chondrites1, may be due to the presence of less abundant clay minerals in the CR chondrites.

The chirality of HAAs was very different between proteinogenic (serine and threonine) and the newly identified non-proteinogenic HAAs in this study. Serine and threonine showed strong L-enrichment, while non-proteinogenic HAAs were present as nearly racemic mixtures. One exception (homoserine) apparently showed the L-enrichment when considering the fragment ion with *m/z* 84. However, when considering the *m/z* 152 and *m/z* 267 fragment ions for DL resolution, the chromatograms showed D-enrichment (see SI). Because the amino acid fraction was a complex mixture, the L-preference of homoserine in this study is not conclusive. Possible racemization during the analytical procedures was minimized, demonstrated by D-threonine being detected only in trace amounts. The small amount of D-allo-threonine could be formed from L-threonine, since L-threonine is known to be epimerized to the D-allo-threonine e.g. ref. 22. The racemization experiment of L-HAA standards also showed that the racemization was minimal (up to 1.6%) during the analytical procedure (see SI). Therefore, the occurrence of racemic HAAs implies non-asymmetric synthesis in the formation pathway(s).

Comparison of amino acid occurrence between the Murchison meteorite and experiments

The relative abundance of amino acids was normalized to glycine abundance, the simplest and most abundant amino acid in the Murchison meteorite as well as the experiments (Table 1). Although the C_3 α-amino acids (e.g. alanine) were abundant in both the Murchison meteorite and the experiments, the experimental products were relatively depleted in larger ($\geq C_4$) alkylated α-amino acids vs. the Murchison extracts. In particular, α-methyl alkylated (OH-free) amino acids such as AIBA and isovaline were present only in the Murchison meteorite, and were not detected in the experimental products. The larger alkylated and α-methyl amino acids were likely synthesized using the corresponding aldehydes or ketones by the Strecker reaction. However, the experimental result clearly indicates that the α-amino acids can be synthesized from aldehydes and ammonia, and that this reaction does not need cyanide, which is required for the Strecker reaction to proceed. Therefore, a certain fraction of the meteoritic α-amino acids appear to be produced from aldehydes under an NH_3-rich condition.

The HAAs newly identified in the Murchison meteorite were effectively synthesized by the experiments, except for 4-A-3-HBA. In Fig. 2, the relative abundance of HAAs was compared between the Murchison meteorite and the experimental product in the presence of glycolaldehyde (α-methylserine = 1), which shows a similarity between the serine-derivatives (isoserine, homoserine and β-homoserine). However, other β-HAAs have larger abundances in the experiment, while γ-HAAs have larger abundances in the Murchison meteorite. In particular, no 4-A-3-HBA was produced in the experiment, which probably reflects the extent of carbon elongation depending upon the starting materials and/or regulation by minerals in the meteorite.

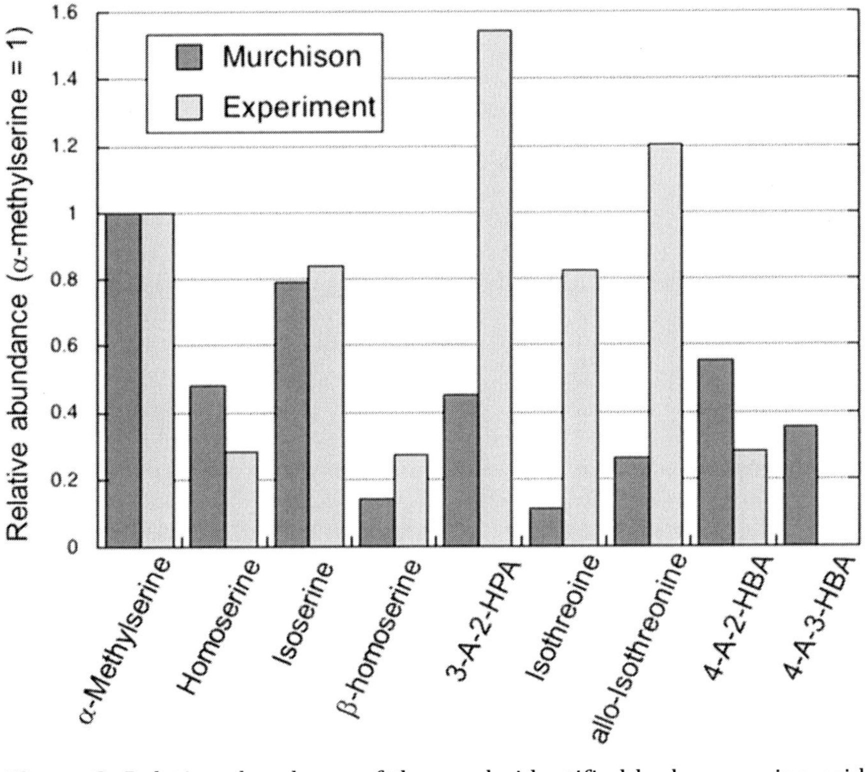

Figure 2. Relative abundance of the newly identified hydroxy amino acids between the Murchison meteorite and the experiment.

In contrast, threonine and its epimer allo-threonine were not detected in the experimental products except for small amount of L-threonine which is likely a contaminant because of its small occurrence in the procedural blank. D-threonine and L-allo-threonine were detected only in trace amounts in the Murchison meteorite, and a small amount of D-allo-threonine could be the product of the epimerization of the large amount of L-threonine present. Even though the L-threonine enrichment has been proposed to be indigenous to the meteorites13, the occurrence of threonine and allo-threonine was consistent between the meteorite and the experiment in this study if L-threonine is considered to be a contaminant. In addition to HAAs, of the β-amino acids identified, β-alanine and β-(2-)amino butanoic acid were relatively abundant in both the Murchison meteorite and the experiments, suggesting a common formation pathway other than the Strecker reaction. The newly identified dicarboxy-β-amino acid, i.e. 2-(aminomethyl)succinic acid, also suggests a similar mechanism for the production of β-amino acids from aldehydes and ammonia (see below).

Formation pathways of meteoritic amino acids

The Strecker reaction has long been considered as the principal mechanism to produce meteoritic α-amino acids4, 11. Hydroxy α-amino acids such as serine, threonine, homoserine and α-methylserine can be produced by electric discharge using CH_4, NH_3, H_2 and H_2O23. These are all hydroxy α-amino acids, suggesting the Strecker reaction as a possible formation mechanism. The occurrence of β and γ amino acids suggests the possibility of other mechanisms to produce meteoritic amino acids. In particular, our experimental results suggest that the meteoritic HAAs could be produced from aldehydes and ammonia through a formose reaction and aldol condensation. The total concentration of newly identified HAAs was 744 ppb relative to the bulk meteorite and approximately 40 ppm relative to the total C, since the bulk C concentration of the Murchison meteorite ranged from 1.6 to 2.5 wt% (e.g. ref. 24). In contrast, the concentration of the HAAs in the experiment was 460 ppm relative to the total C of the starting material. Therefore, the concentration of HAAs in the Murchison meteorite is comparable (about 10%) relative to the concentration of the HAAs in the experiment. Hence, it is quantitatively assumed that the meteoritic HAAs could be produced through a synthesis using aldehydes and ammonia.

A possible formation pathway for HAAs is proposed in Fig. 3. Initially formaldehyde disproportionates under alkaline conditions, known as the Cannizzaro reaction, to produce formic acid (a). The formic acid produces formamide by reaction with ammonia, followed by production of the carbomyl anion (b). The carbomyl anion then reacts with formaldehyde and acetaldehyde, producing glycine and alanine respectively after hydrolysis (c). Similarly, serine and isoserine can be produced by the reaction of the carbomyl anion with glycolaldehyde (d). Addition of the carbomyl anion to C_3 keto-enol components, produced by aldol condensation, then yields various C_4 HAAs (e). These processes may have multiple routes through rearrangement and/or cyclisation. Furthermore, 2-(aminomethyl)succinic acid can be produced from succinimide, which was identified in the Murchison meteorite7, resulting from the formose reaction with ammonia (see SI). It is known that the formation of glycolaldehyde accelerates further condensation of aldehydes to lengthen the carbon chains25. Therefore, the first step of formose reaction using two formaldehydes is the rate-determining step for production of the larger amino acids, as shown in the glycolaldehyde-present experiment. Hence, formation mechanism(s) other than the Strecker reaction are implied for β, γ and δ meteoritic amino acids, and are also possible for some α-amino acids.

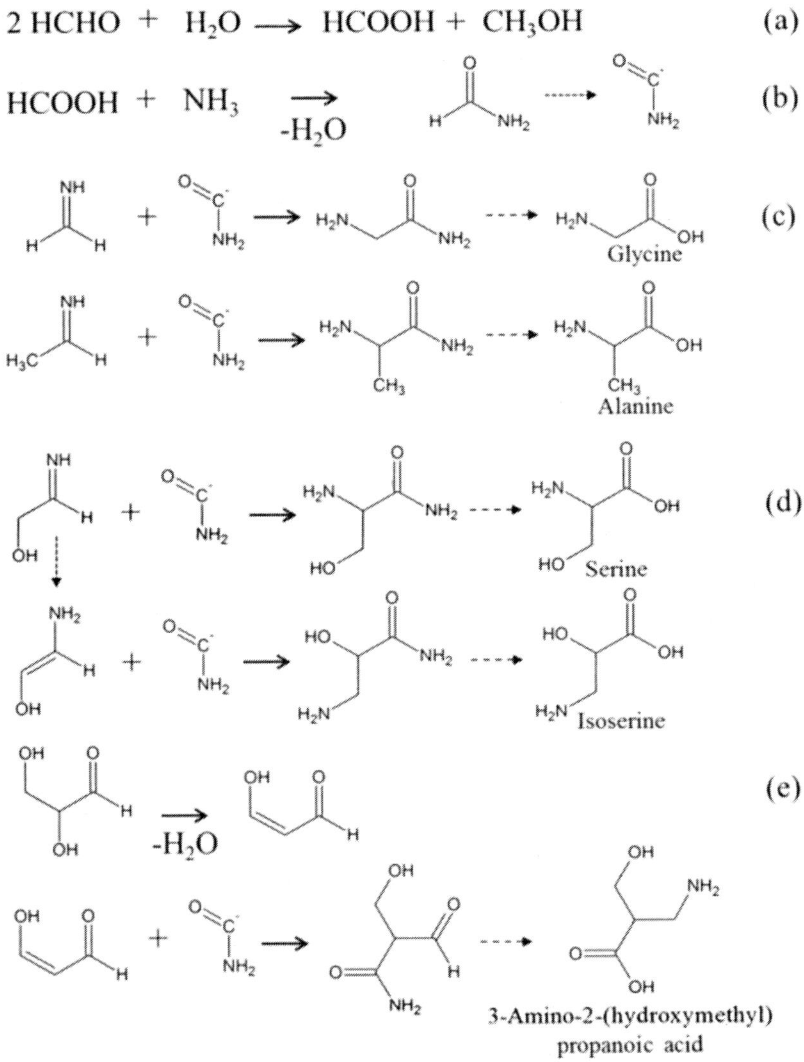

Figure 3. Reaction scheme producing amino acids (glycine, alanine and hydroxy amino acids) from aldehydes and ammonia.

The proposed formation pathways are also consistent with the occurrence of other soluble organic compounds, including the C_3-C_6sugar-related compounds19 and alkylated (up to C_{26}) pyridines26 in the Murchison meteorite. The sugar-related compounds can be produced from formaldehyde through the formose reaction under alkaline conditions. Moreover, the alkylated pyridines can be synthesized by the aldol condensation and imine formation from aldehydes under alkaline conditions with ammonia. Furthermore, it is suggested that chondritic insoluble organic matter was produced through polymerization of formaldehyde followed by condensation upon heating probably on the parent body27,28. The parent body of many carbonaceous chondrites, including the Murchison meteorite, suffered aqueous alteration at 20–100 °C1,29. This aqueous reaction can produce clay minerals, such as serpentine, by alteration of anhydrous silicates (e.g. olivine and pyroxene), which consumes protons and results in alkaline conditions. Therefore, the minerals present in meteorites may also have important roles as catalysts to control organic reactions.

In prebiotic chemistry in the primitive asteroid, simple aldehydes such as formaldehyde and acetaldehyde could be the principal carbon sources to synthesize various organic compounds through their polymerization in the presence of NH_3 under alkaline conditions. Hence, further investigation is needed to reveal possible organic-mineral interactions for the better understanding of chemical evolution in the solar system.

METHODS

Amino acid analysis of Murchison

The detailed analytical procedure is described in the Supplementary Information (SI). Briefly, powdered Murchison meteorite (425 mg) was extracted with water at 100 °C for 20 h. The supernatant (water extract) and the extract residue were subjected to acid hydrolysis with 3 M and 6 M HCl, respectively, at 105 °C for 20 h. After removing ether-soluble organic components, the resulting solution was desalted using an ion exchange column. The purified amino acids were converted to trifluoroacetyl-amino acid-isopropyl ester derivatives to be analyzed by gas chromatography/mass spectrometry with a Chirasil-L-Val capillary column. The amino acids were identified based on their retention times and the mass spectra of standard amino acids, and quantified by comparison of the peak area using characteristic fragment ions of each amino acid. The amino acid concentration was calculated vs. the bulk meteorite (ppb). A pre-heated sea sand was used for the procedural blank, and less than 1.3% of L-serine, glycine, L-leucine, L-alanine, L-threonine, L-glutamic acid and L-aspartic acid were detected relative to the corresponding amino acids in the Murchison meteorite (see SI in detail). The possible racemization was evaluated using the L-hydroxy amino acid standards, suggesting a small racemization (up to 1.6%) during the analytical procedure (see SI in more detail).

Experimental amino acid synthesis

An aqueous solution (300 μL) containing H_2O/ammonia/ formaldehyde /acetaldehyde (1000/100/10/1 molar ratio) or H_2O/ammonia /formaldehyde/acetaldehyde/glycolaldehyde (1000/100/10/1/1 molar ratio) was heated at 60 °C for 6 days in a N_2-purged glass ampoule. The molar ratio of H_2O/ammonia/formaldehyde was adapted from the observed interstellar and cometary ice compositions18. The reaction products were hydrolyzed followed by derivatization, and analyzed using the same procedure as described above. The amino acid concentration was calculated relative to the total initial carbon concentration of aldehydes (ppm).

ACKNOWLEDGEMENTS

We thank S. Poulson, S. Pizzarello and two anonymous reviewers for providing invaluable comments to improve an earlier version of the manuscript. This work was supported financially by JSPS KAKENHI (Grant Number 15H05749) and MEXT KAKENHI (Grant Number 25108006).

INFORMATION

Publisher's note: Springer Nature remains neutral with regard to jurisdictional claims in published maps and institutional affiliations.

REFERENCES

1. Brearley, A. J. The action of water. In Meteorites and the Early Solar System II, Arizona University Press. p.p. 587–624 (2006).
2. Pizzarello, S., Cooper, G. & Flynn, G. The nature and distribution of the organic material in carbonaceous chondrites and interplanetary dust particles. In Meteorites and the Early Solar System II, p.p. 625–651 Univ. Arizona Press (2006).

3. Meierhenrich, U. J. et al. Identification of diamino acids in the Murchison meteorite. Proc. Natl Acad. Sci. USA 101, 9182–9186 (2004).
4. Burton, A. S., Stern, J. C., Elsila, J. E., Glavin, D. P. & Dworkin, J. P. Understanding prebiotic chemistry through the analysis of extraterrestrial amino acids and nucleobases in meteorites. Chem. Soc. Rev. 41, 5459–5472 (2012).
5. Glavin, D. P., Callahan, M. P., Dworkin, J. P. & Elsila, J. E. The effect of parent body processes on amino acids in carbonaceous chondrites. Meteorit. Planet. Sci. 45, 1948–1972 (2011).
6. Cronin, J. R. & Moore, C. B. Amino acid analyses of the Murchison, Murray, and Allende carbonaceous chondirites. Science 172, 1327–1329 (1971).
7. Cronin, J. R. Acid-labile amino acid precursors in the Murchison meteorite II: A search for peptides and amino acyl amides. Origins Life 7, 343–348 (1976).
8. Cooper, G. & Cronin, J. R. Linear and cyclic aliphatic carboxamides of the Murchison meteorite: Hydrolyzable derivatives of amino acids and other carboxylic acids. Geochim. Cosmochim. Acta 59, 1003–1015 (1995).
9. Cronin, J. R., Cooper, G. W. & Pizzarello, S. Characteristics and formation of amino acids and hydroxy acids of the Murchison meteorite. Adv. Space Res. 15, 91–97 (1995).
10. Miller, S. L. The mechanism of synthesis of amino acids by electric discharges. Biochim. Biophys. Acta 23, 480–489 (1957).
11. Cronin, J. R., Pizzarello, S. & Cruikshank, D. P. Organic matter in carbonaceous chondrites: In Meteorites and the Early Solar System p.p. 819–857 Univ. Arizona Press (1988).
12. Engel, M. H. & Macko, S. A. Isotopic evidence for extraterrestrial non-racemic amino acids in the Murchison meteorite. Nature389, 265–268 (1997).
13. Glavin, D. P. et al. Unusual nonterrestrial L-proteinogenic amino acid excesses in the Tagish Lake meteorite. Meteorit. Planet. Sci.47, 1347–1364 (2012).
14. Cronin, J. R. & Chang, S. Organic matter in meteorites: Molecular and isotopic analyses of the Murchison meteorite. In The Chemistry of Life's Origin p.p. 209–258 (1993).
15. Glavin, D. P. & Dworkin, J. P. Enrichment of the amino acid L-isovaline by aqueous alteration on CI and CM meteorite parent bodies. Proc. Natl Acad. Sci. USA 106, 5487–5492 (2009).
16. Pizzarello, S. et al. Large enantiomeric excesses in primitive meteorites and the diverse effects of water in cosmochemical evolution. Proc. Natl Acad. Sci. USA 109, 11949–11954 (2012).
17. Pizzarello, S., Huang, Y. & Fuller, M. The carbon isotopic distribution of Murchison amino acids. Geochim. Cosmochim. Acta68, 4963–4969 (2004).
18. Tielens, A. G. G. M. The physics and chemistry of the interstellar medium. p.p. 495. Cambridge Univ. Press (2005).
19. Cooper, G. et al. Carbonaceous meteorites as a source of sugar-related organic compounds for the early Earth. Nature 414, 879–883 (2001).
20. Cooper, G. & Rios, A. C. Enantiomer excesses of rare and common sugar derivatives in carbonaceous meteorites. Proc. Natl Acad. Sci. USA 113, E3322–E3331 (2016).
21. Pizzarello, S. & Williams, L. B. Ammonia in the early solar system: An account from carbonaceous meteorites. Astrophys. J. 749, #161 (2012).
22. Schroeder, R. A. & Bada, J. L. Kinetics and mechanism of the epimerization and decomposition of threonine in fossil foraminifera. Geochim. Cosmochim. Acta 41, 1087–1095 (1977).
23. Bada, J. L. New insights into prebiotic chemistry from Stanley Miller's spark discharge experiments. Chem. Soc. Rev. 42, 2186–2196 (2013).
24. Kerridge, J. F. Carbon, hydrogen and nitrogen in carbonaceous chondrites: Abundance and isotopic compositions in bulk samples. Geochim. Cosmochim. Acta 49, 1707–1714 (1985).
25. Breslow, R. On the mechanism of the formose reaction. Tetrahedron Lett. 21, 22–26 (1959).

26. Yamashita, Y. & Naraoka, H. Two homologous series of alkylpyridines in the Murchison meteorite. Geochem. J. 48, 519–525 (2014).
27. Cody, G. D. et al. Establishing a molecular relationship between chondritic and cometary organic solids. Proc. Natl. Acad. Sci. 108, 19171–19176 (2011).
28. Kebukawa, Y., Kilcoyne, A. L. D. & Cody, G. D. Exploring the potential formation of organic solids in chondrites and comets through polymerization of interstellar formaldehyde. Astrophys. J.771, 19,
29. Guo, W. & Eiler, J. M. Temperatures of aqueous alteration and evidence for methane generation on the parent bodies of the CM chondrites. Geochim. Cosmochim. Acta 71, 5565–5575 (2007).

CHAPTER 9

A Presumptive Fossilized Bacterial Biofilm Occurring in a Commercially Sourced Mars Meteorite

Milton Wainwright[1]*, Christopher E Rose[1], Tareq Omairi[1], Alexander J Baker[2], Chandra Wickramasinghe[3] and Fawaz Alshammari[4]

[1]*Department of Molecular Biology and Biotechnology, University of Sheffield, UK*

[2]*Leonardo Centre for Tribology, University of Sheffield, UK*

[3]*Centre for Astrobiology, University of Buckingham, UK*

[4]*College of Applied Medical Sciences, University of Hail, Saudi Arabia*

ABSTRACT

A commercially sourced Mars meteorite sample was found to contain a presumptive fossilized bacterial biofilm composed of a wide variety of bacteria-like morphologies. We conclude this presumptive biofilm, and other biomorphs, provide preliminary evidence for the occurrence of fossilised bacteria in a Mars meteorite and therefore that life once existed on Mars.

Keywords: Martian meteorite; Microfossils; Panspermia

INTRODUCTION

A number of papers have reported the presence of bacteria-like features in Martian and other meteorites [1-4]; the most famous of these being the claimed nanobacteria-like, fossilized structures found in the Allen Hills (ALH 84001) meteorite [5]. To date none of these claims has been fully substantiated, largely because meteorites contain mineral features which closely mimic bacteria and other microorganisms, making it difficult to distinguish between potential fossilized bacteria and non-biological artefacts [6]. The aim of the study reported here was to use scanning electron microscopy (SEM) to examine a sample of a commercial sourced Martian meteorite of known provenance to look for bacteria-like structures and where found, to provide evidence to show that such bacteriamorphs are not mineral artefacts.

MATERIALS AND METHODS

Details of the meteorite examined

The single sample used here was cut from a larger meteorite sample, listed as Northwest Africa 4925 NWA 4925; it was obtained from Meteorites for Sale.com. The authenticity of the sample is guaranteed by fellow members of the International Meteorite Collectors Association. It was found in Erfoud, Morocco, in 2007 and is classified as an anachondrite, i.e. a Martian, olivine-phyric shergottite). One fragment partly covered by fusion crust weighing 282.3 g was found. The meteorite displays a porphyritic texture with large chemically zoned olivine megacrysts set into a fine-grained groundmass composed of pyroxene and maskelynite. Minor phases include chromite, sulphides, phosphates, and small Fe-rich olivines. The olivine megacrysts often contain melt inclusions and small chromites. Its mineral composition (EMPA) is: Olivine, Fa27.6–46.8; pyroxene, Fs20.0–37.7Wo3–14.8; maskelynite, An67–69. It is classified as an achondrite (Martian, olivine-phyric shergottite); severely shocked with some melt pockets; moderately weathered.

Sterilization and lysis of potential modern biofilms

A single meteorite sample was used as sent by the supplier and was not subjected to further cutting. It was immersed in 80 per cent ethanol for 1h, washed twice with sterile deionized water and transferred to a sterile plastic petri dish before being examined under the SEM. All water used was sterilized by autoclaving at 120°C for 20 min. and then filtered through a 0.1 micron micropore filter (Nalgene).

Scanning electron microscopy

The meteorite sample was placed in a staging chamber with the face to be analysed flush to the base. Konductomet phenolic mounting compound (20-3375-016) was used to stage the sample. Usually surface grinding and then polishing of the sample surface is undertaken at this point however in this case only an instantaneous grinding process was done. This was done to remove any build up that might be present atop the surface to be studied and ensure that only fresh sample material be exposed. The coarseness of pile used was 120 microns using a Bueler Automet 250 for 5 seconds with a touch force of 20N, a head speed of 50 RPM and a Platen speen of 140 RPM A second sample was prepared presenting the outside surface of the 'meteorite'. This was staged atop a conductive carbon tab. Due to the relatively low conductive nature of the samples, in order to minimise charging effects and optimise image acquisition the sample was coated using an Emscope gold sputter coater. The sample was coated for a deposition-duration of 1 minute at 15 milliamps. Ahead of introduction into the SEM the samples were placed in a vacuum chamber overnight to remove any remaining moisture from the porous sample. No chemicals or concentrated alcohols were introduced at any stage as a cleaning step. The samples were all delicately irrigated using de-ionised water.

RESULTS AND DISCUSSION

Figure 1 shows the meteorite sample under the scanning electron microscope (SEM); the sample has clearly been cut from a larger piece, so that the exterior surfaces would originally have been inside the meteorite. The surface of the sample was carefully scanned and imaged. Figure 2 shows one of two areas of the surface showing the presence of a presumptive bacterial biofilm consisting of a number of forms which look remarkably like bacteria and which had they been observed on the surface of a terrestrial soil or rock sample would be regarded by most microbiologist as comprising a distinct bacterial biofilm. Details of the individual bacterial types are shown in Figure 3 which apparent rods, cocci, spiral shapes bacteria-like chains, and individual rods occurring in star-shaped and more complex groupings. The chemical composition of both the presumptive biofilm area and an adjacent non-biofilm region, as determined by EDX, is shown to be essentially identical and being made up principally of silicon iron, magnesium, calcium and oxygen, i.e. a composition typical of a meteorite.

Figure 1: The Meteorite sample. The images shown below were taken from the main face which, before cutting, would have been inside the meteorite sample.

We suggest that (Figures 2 and 3) shows a fossilized bacterial biofilm made up of typical bacterial forms, which would have originally been located inside the original meteorite from which this sample was cut. It could be argued that the image is of a) an Earth-derived biofilm which fossilized after the meteorite had landed or b) a modern biofilm made up of living bacteria. The fact that the bacteria show no signs of lysis, after having been exposed to 80% ethanol suggests that this is not the case. The possibility that a modern terrestrial biofilm might have undergone mineralization during the period when the meteorite resided on Earth, seems highly unlikely. Similarly it is highly unlikely that a modern terrestrial biofilm could have formed from an air-derived bacterial inoculum, while, the meteorite was kept in storage.

EDX analysis (Figure 4) shows that the presumptive biofilm region is mineralized and has essentially the same chemical composition of an adjacent area of the meteorite which is free of the presumptive biofilm. These findings, we maintain, confirm, that the presumptive biofilm is part of the Mars meteorite and is not a modern biofilm made up of contaminating terrestrial bacteria.

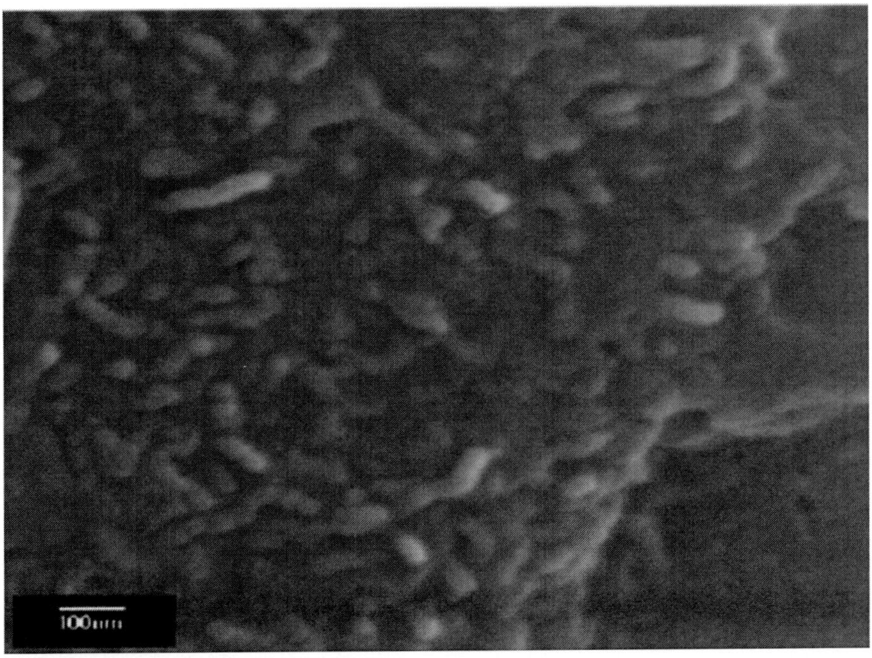

Figure 2: The presumptive Martian bacterial biofilm showing cocci, rods and other bacteria-like morphologies (bar represents 100 nm).

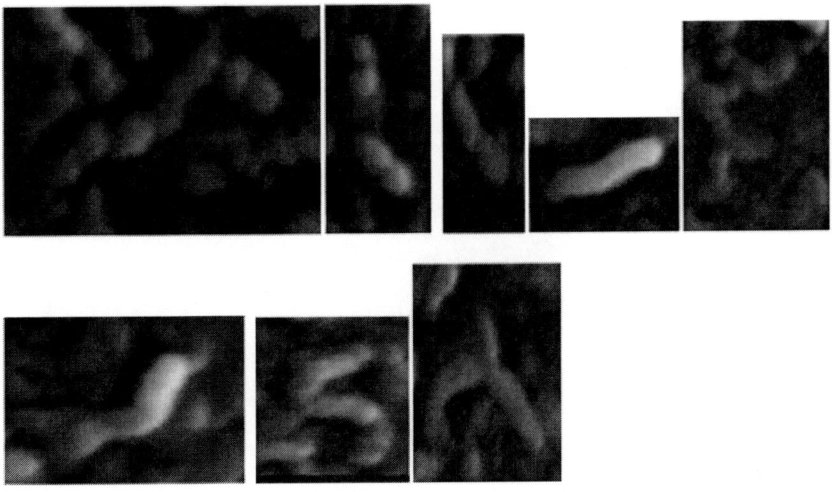

Figure 3: Details of the presumptive bacteria-like found in the Mars meteorite biofilm showing various bacteria-like forms and groupings (Figure 2 for dimensions).

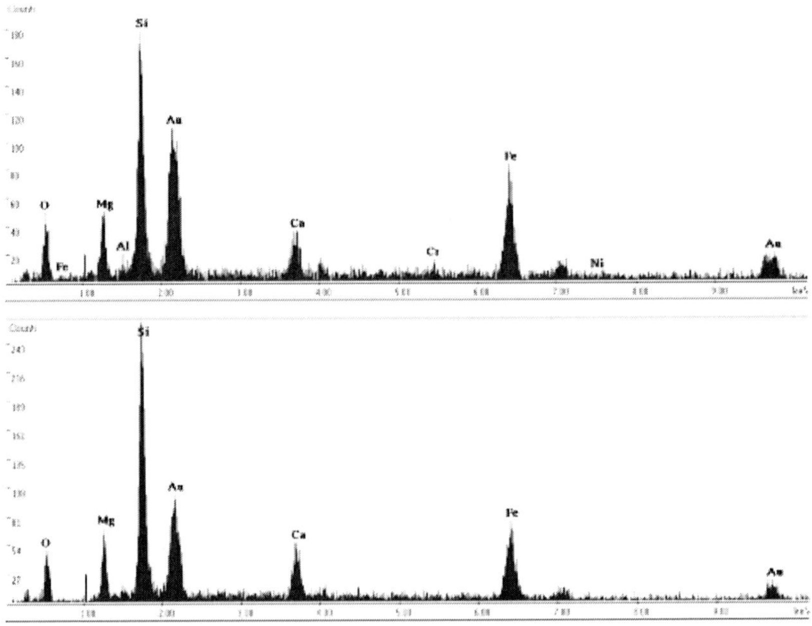

Figure 4: EDX of A, the "biofilm" region and B an area of the meteorite away from the "biofilm".

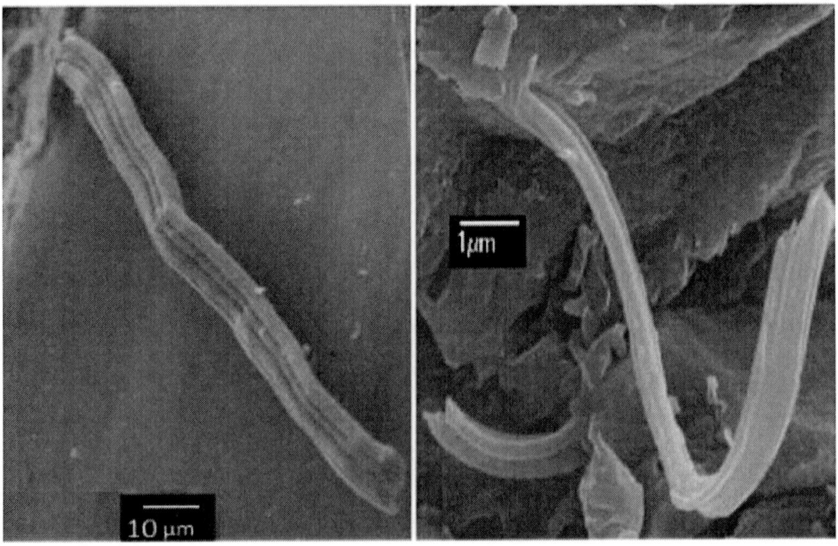

Figure 5: Filaments on surface of the Mars meteorite, A, a filament bundle showing folding at the top (A); B, Attachment of filament bundle to the body of the meteorite (b).

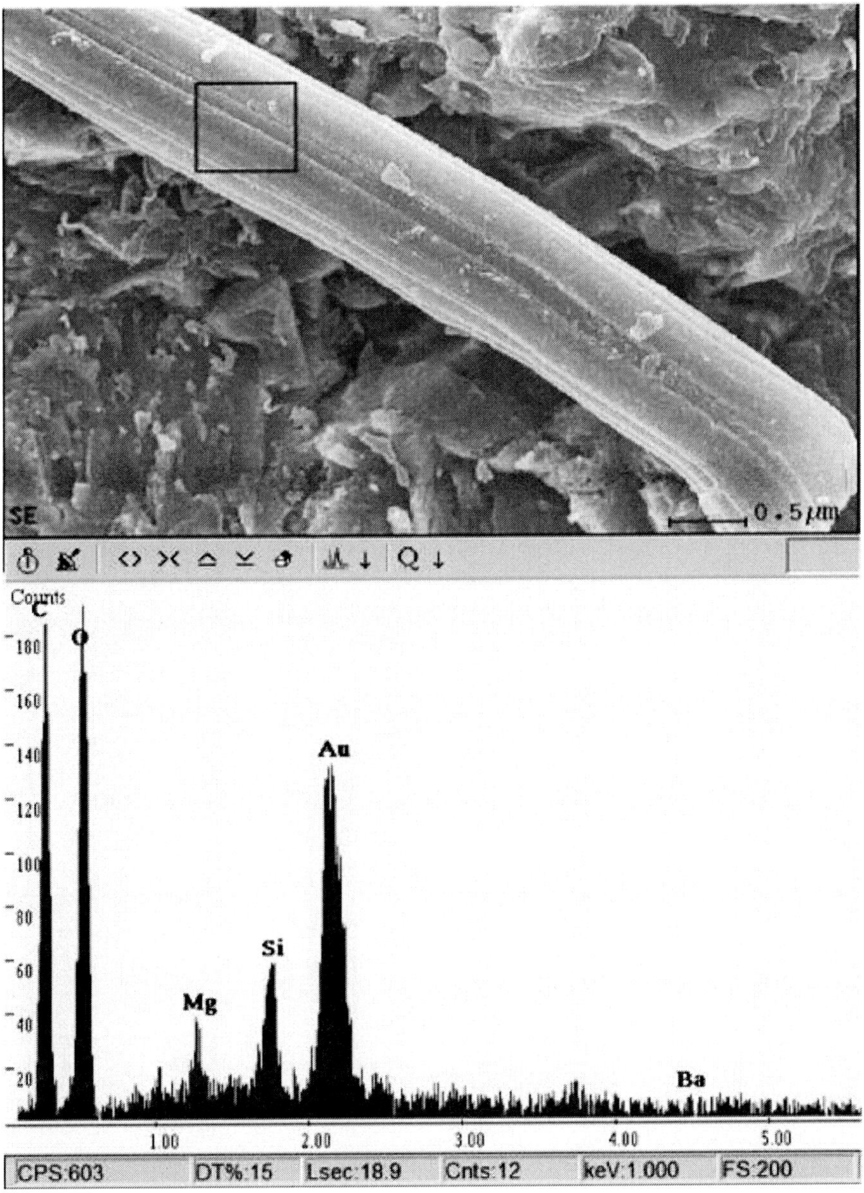

Figure 6: EDX analysis of a fibre bundle (square in upper image, indicates area of EDX analysis), showing preponderance of carbon and oxygen.

It could also be argued that the bacteria-like structures seen here are mineral artefacts which simulate bacteria. Our response would be that, if this were the case, it would be a remarkable coincidence that such a wide variety of apparent bacterial forms would appear so closely packed into an apparent biofilm. Unlike the claimed bacterial form seen in the Allen Hills meteorite, the bacteria-like forms seen here are not nanosized but, at around 0.2 micron, are similar in size to terrestrial bacteria seen growing in naturally occurring, nutrient-limited environments found on Earth [5]. Again, unlike the Allen Hills form, the putative bacteria found here are large enough to avoid arguments over whether or not they could contain a complete bacterial genome.

A final argument against the bacteria-like forms, seen here being, Mars–derived bacteria is that they appear too much like terrestrial bacteria. However, this argument is only valid if it is assumed that bacteria from a non-terrestrial source would, of necessity, differ in morphology from those found on Earth. If, as we think likely, both Earth and Mars derived their bacteria from a common source, this criticism will not hold [7].

Figure 5 shows another presumptive biomorph present in the meteorite under study, occurring in the shape of bundles of long fibres which are folded back on themselves and are attached to the body of the meteorite. Analysis of a fibre bundle by EDX (Figure 6) shows a preponderance of carbon and oxygen, with smaller amounts of silicon and magnesium, a composition which is suggestive of a mineralized biological entity. The fact that the bundles are attached to the body of the meteorite shows that the bundles are not contaminating fibres of natural origin (e.g. cotton), nor man-made fibres. In terms of morphology the bundles appear to be made up of filamentous microbes, possibly fungi or filamentous algae or cyanobacteria. The filaments are analogous to similar filamentous structures claimed by Hoover [3] to occur in the Orgueil carbonaceous meteorite, which the author claims are fossilized cyanobacterial filaments of Mars origin.

The nanobacteria-like features found in the Allen Hills meteorite have not yet been finally confirmed to be fossilized Martian bacteria [6], this despite the fact that this meteorite is possibly the most intensively studied of all geological samples. Final confirmation that the biofilm and filaments described here are biological and originate from Mars will presumably be likewise very difficult to achieve [8]. For the moment then, we can only present what we claim is provisional evidence for the presence of a fossilized bacterial biofilm on the inside cut surface of a Mars meteorite sample. Since biofilms form in environments rich in water, we suggest that the presumptive bacterial biofilm, provided here (and the associated filament bundles), formed in a watery environment on Mars. This conclusion appears consistent with recent findings from the NASA Mars

Curiosity Rover of fine-grained sedimentary rocks which were interpreted to represent an ancient lake preserving evidence of an environment that would have been suited to support life [9]. It is also relevant to note that a recent re-evaluation of the results from the Labelled Release Experiment on the Mars-Viking probe of 1976 has concluded that extant Martian surface microbiology was most likely to have been discovered at this time [10].

REFERENCES

1. Claus G, Nagy B (1961) A microbiological examination of some carbonaceous chondrites. Nature 192: 594-596.
2. Pflug HD (1984) Ultrafine Structure in the Organic Matter in Meteorites, in Fundamental Studies and the Future of Science.
3. Hoover RB (2010) Fossils of Cyanobacteria in C11 carbonaceous meteorites: implications to life on comets, Europa and Enceladus. J. Cosmol. 11: 3500-3549.
4. Sears DWG, KralTA (1998) Martian microfossils in lunar meteorites. Meteorics. Plan. Sci. 33: 791-794.
5. McKayDS, GibsonEK, Thomas-Keptra, Hojatollah V, Christopher SR, et al. (1996) Possible relic biogenic activity in Martian meteorite ALH 84001. Science 273: 924-930.
6. McSweenHY, HarveyRP (2010) An evaporation model for formation of carbonates in AHL 84001 Martian meteorite. Intern. Geol.Rev. 40: 774-783.
7. Hoyle F, Wickramasinghe NC (1981) Evolution from Space.
8. SteeleA, GoddardDT, StapletonD, Topoorsk JKW,Peters V, et al. (2000) Investigations into an unknown organism in the Allen Hills 8400001. Meteorics Plan. Sci. 35: 237-241
9. Grotzinger JP, Summer DY, Kah LC, Gupta S, Lewis K, et al. (2014) A Habitable Fluvio-Lacustrine Environment at Yellowknife Bay, Gale Crater, Mars.Science343
10. Bianciardi G, Miller JD, Straat PA, Levin GV (2012) Complexity Analysis of the Viking Labeled Release Experiment. Int. J. Aeronautical and Space Science 13: 14-26.

CHAPTER 10

Stable Chromium Isotopic Composition of Meteorites and Metal–Silicate Experiments: Implications for Fractionation During Core Formation

P. Bonnand [a], H.M. Williams [b], I.J. Parkinson [c,d], B.J. Wood [a], A.N. Halliday [a]

[a] *Department of Earth Sciences, University of Oxford, South Parks Road, Oxford, OX1 3AN, United Kingdom*

[b] *Department of Earth Sciences, Durham University, Sciences Labs, Durham, DH1 3LE, United Kingdom*

[c] *School of Earth Sciences, University of Bristol, Bristol, BS8 1RJ, United Kingdom*

[d] *Department of Environment, Earth and Ecosystems, CEPSAR, The Open University, Walton Hall, Milton Keynes, MK7 6AA, United Kingdom*

ABSTRACT

We present new mass independent and mass dependent Cr isotope compositions for meteorites measured by double spike thermal ionisation mass spectrometry. Small differences in both mass independent ^{53}Cr and ^{54}Cr relative to the Bulk Silicate Earth are reported and are very similar to previously published values. Carbonaceous chondrites are characterised by an excess in ^{54}Cr compared to ordinary and enstatite chondrites which make mass independent Cr isotopes a useful tool for distinguishing between meteoritic groups. Mass dependent stable Cr isotope compositions for the same samples are also reported. Carbonaceous and ordinary chondrites are identical within uncertainty with average $\delta^{53}Cr$ values of -0.118 ± 0.040‰ and -0.143 ± 0.074‰ respectively. The heaviest isotope compositions are recorded by an enstatite chondrite and a CO carbonaceous chondrite, both of which have relatively reduced chemical compositions implying some stable Cr isotope fractionation related to redox processes in the circumstellar disk. The average $\delta^{53}Cr$ values for chondrites are within error of the estimate for the Bulk Silicate Earth (BSE) also determined by double spiking. The lack of isotopic difference between chondritic material and the BSE provides evidence that Cr isotopes were not fractionated during core formation on Earth. A series of high-pressure experiments was also carried out to investigate stable Cr isotope fractionation between metal and silicate and no demonstrable fractionation was observed, consistent with our meteorites data. Mass dependent Cr isotope data for achondrites suggest that Cr isotopes are fractionated during magmatic differentiation and therefore further work is required to constrain the Cr isotopic compositions of the mantles of Vesta and Mars.

Keywords: Cr isotopes; core formation; metal–silicate equilibrium; meteorites; differentiation

1. INTRODUCTION

Variations in chemical composition in the Solar System reflect early nebular processes as well as planetary accretion and differentiation. Metal–silicate differentiation (or core formation) is a particularly dramatic process that transforms the budgets of slightly- to strongly-siderophile elements in the residual bulk silicate Earth (BSE). These chemical changes in the Earth have been replicated experimentally and modelled over many years, as reviewed elsewhere (e.g. Newsom, 1990 ; Wood *et al.*, 2008). During Earth accretion and core formation, the elements were distributed between the mantle (silicate phases) and the core (Fe-rich phases) according to their bulk distribution coefficients ($D = X_{metal}/X_{silicate}$). Variably siderophile ($D > 0$) as opposed to strictly lithophile ($D = 0$) elements are respectively depleted and enriched (by removal of metal) in the mantle. Non-chondritic elemental ratios in the residual terrestrial mantle, such as Mg/Si, have been used to argue that light elements like Si became slightly siderophile ($D > 0$) and were incorporated into the Earth's core (Allègre *et al.*, 1995 ; Drake and Righter, 2002). A new tool to assess the removal of elements into the core is to utilise the fractionation of stable isotopes between metal and silicate, which also provide insights into the reactions occurring during core formation and constraints on key physical parameters such as temperature, pressure and oxygen fugacity. Such mass dependent isotopic variations have already been utilised in this way for Si (Georg *et al.*, 2007; Armytage *et al.*, 2011 ; Zambardi *et al.*, 2013), Fe (Williams et al., 2012) and Cr (Moynier et al., 2011).

Stable isotopes can be fractionated between phases by both kinetic processes and equilibrium fractionation processes. In the latter case, the fractionation is dependent on speciation, coordination and bonding environment, where the isotopically heavier species are generally associated with a lower coordination and higher oxidation states of the element of interest (Schauble, 2004). It has been proposed that metal–silicate equilibrium could lead to an isotopic shift due to the difference in bonding between metal and silicate liquids (Georg et al., 2007). If the metal fraction is enriched or depleted relative to the silicate fraction then the bulk silicate Earth (BSE) should be isotopically fractionated relative to the starting material (bulk undifferentiated chondrites). Therefore stable isotopes of variably siderophile elements have great potential to better constrain core formation processes (Georg et al., 2007). Laboratory experiments have been performed to replicate and understand mass dependent fractionation between metal and silicate melts (Hin *et al.*, 2013 ; Shahar *et al.*, 2011). However, the isotopic fractionations at high temperature are small and high precision measurements are needed to resolve these variations.

Chromium is an element of great interest to core formation because it becomes more siderophile under more reducing conditions (iron-wüstite (IW) −1.5 to −5 log units) and at high temperatures (1480 to 2000 °C) with a D value of 2.5–3.5 (Wood et al., 2008), but is less sensitive to pressure effects. It has been argued that ~60% of the Earth's total Cr is in the core such that the mantle is depleted in Cr relative to chondrites (Allègre *et al.*, 1995 ; McDonough, 2003). Therefore it is possible that during core formation stable Cr isotopes are fractionated. Indeed, it has previously been proposed that the Cr isotopic composition of the BSE is fractionated by metal–silicate fractionation relative to chondrites with the stable Cr isotopic composition of meteorites reported to be lighter than the BSE value (Moynier et al., 2011). The same study presented ab-initio calculations that demonstrated that the metal fraction should be isotopically enriched in ^{52}Cr (that is lighter) compared to the silicate phase, in agreement with their interpretation of Cr isotope fractionation during core formation (Moynier et al., 2011).

Earlier studies of Cr isotopes in meteorites have focussed on radiogenic ($\varepsilon^{53}Cr$) and nucleosynthetic ($\varepsilon^{54}Cr$) variations (e.g. Birck and Allègre,

1988; Shukolyukov and Lugmair, 2006; Trinquier *et al.*, 2007; Trinquier *et al.*, 2008a; Moynier *et al.*, 2007 ; Qin *et al.*, 2010). Variations in $\varepsilon^{53}Cr$ are due to the decay of ^{53}Mn (half-life 3.7 Ma) and are related to the Mn/Cr ratio and time, whereas variations in $\varepsilon^{54}Cr$ are due to the production of this neutron rich isotope in a type 1a supernova (e.g. Hartmann et al., 1985). It is also possible that Cr isotopes are influenced by spallation reactions during cosmic ray exposure (Leya et al., 2003). However, spallation effects are believed to be negligible for samples with low Fe/Cr, including the samples we discuss in this paper.

While variations in radiogenic and nucleosynthetic Cr isotopes are not directly relevant to core formation these studies have utilised high precision thermal ionisation mass spectrometry (TIMS) techniques that allow assessment of variations in $\varepsilon^{53}Cr$ and $\varepsilon^{54}Cr$ at the 10 ppm (0.1ε unit) level (e.g. Trinquier et al., 2008b). By contrast, mass dependent Cr isotopes have the potential to characterise Cr behaviour during core formation on Earth and other planetary bodies, but studies have used less precise multi-collector inductively-coupled-plasma mass spectrometry (MC-ICP-MS) techniques with precisions of ~50 ppm (Schoenberg *et al.*, 2008; Moynier *et al.*, 2011 ; Schiller *et al.*, 2014). The composition of the BSE has been determined by several studies using the double-spike technique yielding a $\delta^{53}Cr - 0.12 \pm 0.10$‰ (Fig. 1, Schoenberg *et al.*, 2008 ; Farkaš *et al.*, 2013). This relatively uniform value indicates that magmatic processes do not significantly fractionate stable Cr isotopes, although data for lunar samples indicates extensive magmatic differentiation can produce $\delta^{53}Cr$ compositions ~0.1‰ lighter than the lunar and terrestrial mantle (Fig. 1, Bonnand et al., in press). By contrast, Moynier et al. (2011) and Schiller et al. (2014) use a standard-sample bracketing technique and report significant isotopic difference in $\delta^{53}Cr$ between chondritic meteorites and the BSE, with the meteorites being 0.3‰ lighter. However, a recent study by Qin et al. (2015) using the double-spike technique reveals no such difference. Clearly a resolution of these differences is required.

Here we utilise high-precision TIMS measurements and the double-spike technique to extend the high precision Cr isotope meteorite dataset. These data provide assessment of both the mass independent and mass dependent variations in meteorites. We also investigate the Cr isotope fractionation between metal and silicate phases in high pressure experiments. The data are compared to published datasets and discrepancies between the different studies are discussed. We then attempt to understand the origins of Cr isotope variations in the meteorite dataset and discuss the use of Cr isotopes to study core formation processes.

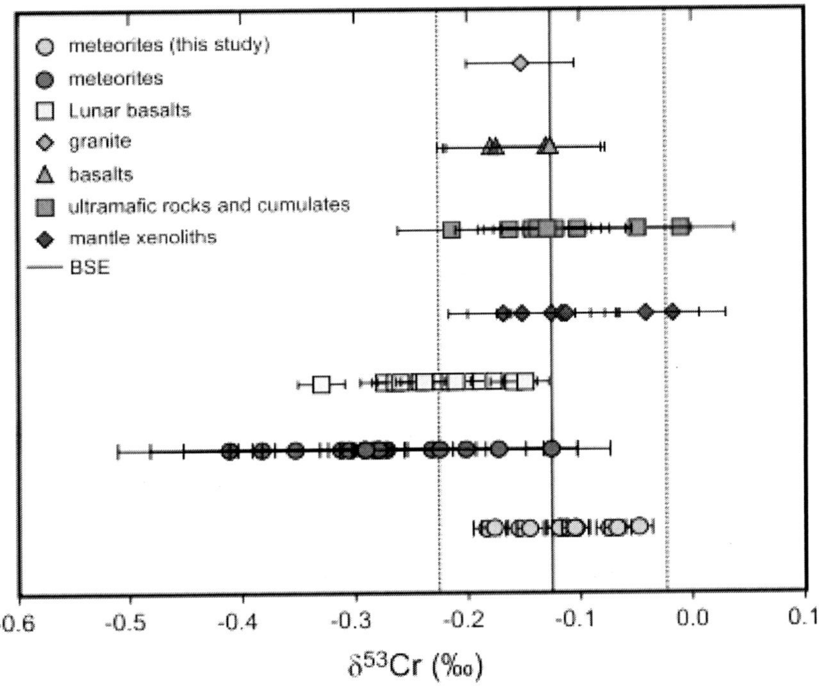

Figure 1. Compilation of Cr data in high temperature rocks. The granite, basalts and ultramafic rocks and cumulates data are from Schoenberg et al. (2008), the meteorite data is from Moynier et al. (2011), Schiller et al. (2014) and this study. The lunar samples data is from Bonnand et al. (in press). The thick vertical line represents the mean and the thin vertical lines ±2 s.d. of the terrestrial data from Schoenberg et al. (2008).

2. SAMPLES

We have determined the Cr-isotope compositions of 19 meteorites. Meteorite samples include four carbonaceous chondrites (Allende (CV3), Orgueil (CI), Murchison (CM2) and Ornans (CO3)), six ordinary chondrites (Bremervörde (H/L3), Kernouvé (H6), Barratta (L4), Bruderheim (L6), Parnallee (LL3) and Saint-Séverin (LL6)), two enstatite chondrites (Indarch (EH4) and Khaipur (EL6)), one martian meteorites (Nakhla) and four HED meteorites (Kapoeta (howardite), Pasamonte (eucrite), Juvinas (eucrite) and Johnstown (diogenite)). We also present data for a terrestrial peridotite (JP-1). Five experimental isotope equilibrium runs have been performed and for each run, metal and silicate have been analysed (Table 1).

Table 1. Chromium concentration and Cr isotopic composition for metal and silicate fraction for the five experiments performed in this study. The Cr concentrations were analysed by ICP-MS at the University of Oxford. The 2 s.d. for $\delta^{53}Cr$ is the external reproducibility on the JP-1 measurement (see text for details, Bonnand et al., in press).

Sample name	Capsule	Run time (min)	Fraction	Cr ($\mu g\,g^{-1}$)	$D_{met\text{-}silicate}$	$\delta^{53}Cr$	2 s.d.	$\Delta^{53}Cr$	2 s.d.
A171	Si	70	metal	8723	1.02	−0.166	0.022	0.010	0.031
			silicate	8584		−0.175	0.022		
A176	Mg	10	metal	11 880	1.18	−0.132	0.022	−0.008	0.031
			silicate	10 082		−0.124	0.022		
B100	Mg	40	metal	13 658	2.00	−0.039	0.022	0.045	0.031
			metal (rep)			−0.024	0.022	0.060	0.031
			silicate	6836		−0.084	0.022		
B97	Si	43	metal	5872	0.87	−0.195	0.022	−0.024	0.031
			silicate	6788		−0.170	0.022		
B98	Si	10	metal	8165	0.97	−0.245	0.022	−0.008	0.031
			silicate	8398		−0.237	0.022		

3. ANALYTICAL METHODS

3.1. Metal–silicate experiments

The experiments designed to study Cr isotopic fractionation between metal and silicate were similar to those used for Zn (Bridgestock et al., 2014). Briefly, the starting materials were prepared from high purity oxide (70%) and metal (30%) powders. The silicate starting material approximates the 1.5 GPa eutectic composition in the system anorthite–diopside–forsterite ($An_{50}Di_{28}Fo_{22}$, Presnall et al., 1978) and was doped with 2 wt.% Cr (as Cr_2O_3). Experiments were carried out in either polycrystalline MgO or silica glass capsules with graphite furnaces. The experiments were performed at 1.5 GPa in a piston cylinder apparatus at the Department of Earth Sciences, University of Oxford. In order to ensure complete melting of both metal and silicate, the run temperature was 1650 °C and was monitored using $W_{95}Re_5$–$W_{74}Re_{26}$ thermocouples. Experimental run times varied from 10 to 70 min (Table 1). The experiments were quenched by turning off the power, after which the pressure was slowly released. After quenching, the charges were lightly hand-crushed and the two phases were handpicked under a binocular microscope.

3.2. Sample preparation for meteorites

In order to obtain accurate mass dependent, each sample was analysed for its radiogenic and nucleosynthetic Cr isotope compositions and an aliquot of the same solution was spiked and analysed for its mass dependent Cr isotopic composition. Spiking occurred before any separation chemistry and the unspiked and spiked samples were processed in the same way. Protocols for the separation of Cr from geological and environmental samples have been previously described (e.g. Ball and Izbicki, 2004; Trinquier *et al.*, 2008b; Bonnand *et al.*, 2011; Bonnand *et al.*, 2013; Bonnand *et al.*, in press ; Schiller *et al.*, 2014). A summary of the method used in this study is given below.

For silicate samples, approximately 20 mg of whole rock powder were digested using concentrated HNO_3–HF–HCl acid mixtures. The fully dissolved sample was evaporated to dryness and then dissolved in 6 M HCl. Two aliquots of ~2 μg of chromium was dried and dissolved in 6 M HCl and one of them was spiked with the requisite amount of double spike (^{50}Cr–^{54}Cr). Chromium separation from silicate matrices requires two columns. The first separates Cr from the main matrix cations (e.g. Ca, Na, Mg) and a full description is given in Bonnand et al. (2011). For the purposes of this study, a disposable Bio-Rad Poly-Prep 10 ml column was filled with 1 ml of AG50 X8 (200–400 mesh chloride form resin) and the sample was loaded and eluted in 0.5 M HCl. The Cr fraction and small quantities of other minor elements such as Ti pass straight through this column and were collected during this first elution. In order to clean the Cr fraction from isobaric interferences (Ti, V and Fe) a second column chemistry was performed. The Cr fraction was dried and dissolved in 0.5 M HNO_3. A disposable Bio-Rad Poly-Prep column was filled with 1 mL of AG50 X8 200–400 mesh resin. The resin was first cleaned with 20 ml 6 M HCl and 20 mL MQ H_2O and then preconditioned with 10 mL of 0.5 M HNO_3. The sample was then loaded in 4 mL 0.5 M HNO_3. In a weak HNO_3 medium, Cr(III) will stick to the cation resin. The column is then washed with 4 mL 0.5 M HNO_3, then 6 mL 0.5 M HNO_3 + 0.25 M HF, then 7 mL 0.5 M HCl. Finally, the cleaned Cr fraction is eluted with 6 mL 3 M HCl. The second step of the separation protocol has a yield better than 96%, although the total procedure yield for the two separations is about 80%. The total blank of this chemical process is ~0.2 ng, which is negligible compared to the 2000 ng processed through the columns.

The protocol used for the separation of the Cr fraction from Fe rich samples is different, involving three columns. The first column is designed to remove Fe from the sample and has been described in Bonnand et al. (2013). To this end, a disposable Bio-rad Poly-Prep 10 ml column was filled with 1 ml of AG1 X8 and the sample loaded in 7 M HCl. The Fe sticks to the column whereas Cr is eluted. After this first column the samples are treated in the same way as silicate samples.

3.3. Mass spectrometry

The chromium isotope analyses of the meteorite samples were performed with a ThermoFisher Triton Thermal Ionisation Mass Spectrometer (TIMS) in static multicollection mode at the University of Oxford (UK). Chromium samples were loaded onto an outgassed zone refined Re filament in 1 μL 6 M HCl, along with 3 μL of a mixture of silicic and boric acids. The filament was slowly heated at 0.65 A to dry down the sample. The current was then increased until a dull red glow was observed. Amplifier gains were performed daily and the baseline was measured after each block of 10 cycles, with amplifier rotation after each block. Samples and standards were run in static mode with a typical beam current of $6-8\times10^{-11}$ A of ^{52}Cr for 54 blocks of 10 cycles (with an integration time per cycle of 8.4 s). Interferences were monitored on ^{56}Fe, ^{49}Ti and ^{51}V, such that ^{54}Cr (^{56}Fe) and ^{50}Cr (^{50}Ti and ^{50}V) signals were corrected for isobaric interferences. These corrections were always negligible. For un-spiked runs, Cr isotope ratios were corrected for mass fractionation using an exponential law and $^{50}Cr/^{52}Cr$ = 0.051859 (Shields et al., 1966) (see Section 3.4).

The external reproducibility and accuracy of $\varepsilon^{53}Cr$ and $\varepsilon^{54}Cr$ were assessed by 2 standard deviation of NIST SRM 979 processed through chemical purification and is 0.00 ± 0.10 and 0.00 ± 0.19 respectively (2 s.d., $n=15$, Table 2). The accuracy of the mass dependent variation ($\delta^{53}Cr$) was assessed by repeat measurements of a NIST SRM 979 Cr standard and is -0.005 ± 0.009‰ (2 s.d., $n=15$, Table 2). The external reproducibility was assessed with repeated measurements of Orgueil sample and is -0.145 ± 0.012‰ (2 s.d., $n=6$, Table 2). This is at least a four times improvement compared to previous studies (Moynier et al., 2011). The uncertainties associated with the individual $\delta^{53}Cr$ data reported in Table 2 incorporate errors propagated from the unspiked and spiked analyses.

Chromium stable isotope analyses of the experimental samples were performed with a ThermoFisher Neptune Multi collector ICP-MS at the Open University. The analytical protocols have been previously described (Bonnand *et al.*, 2011; Bonnand *et al.*, in press). The external reproducibility obtained on this instrument was assessed with replicate measurements of JP-1 standard and yield a value of -0.128 ± 0.022‰ (2 s.d., $n=14$, Bonnand et al., in press).

3.4. Deconvolution procedure

We need to correct for mass fractionation that occurs during chemical separation and mass spectrometry analyses. To do this we employ two methodologies to deconvolve the isotope data in this paper, and the rational for this is as follows. For terrestrial samples, there is no variation in mass independent ^{53}Cr and ^{54}Cr (e.g., Trinquier *et al.*, 2008a; Qin *et al.*, 2010) and therefore all samples (including NIST SRM 979) lie on a single mass fractionation trend in four isotope space, so that only a double-spiked analyses is required to derive the stable isotope composition of Cr isotopes (e.g. Schoenberg *et al.*, 2008; Bonnand *et al.*, 2011). The metal–silicate experiments samples were deconvolved in this way using the NIST SRM 979 standard value as the natural composition as in previous studies for terrestrial samples (e.g. Schoenberg et al., 2008).

Table 2. Mass independent and mass dependent Cr isotopic compositions for meteorites and terrestrial samples analysed in this study. The Cr concentrations are from Qin et al. (2010), Trinquier et al. (2008a) and Shukolyukov and Lugmair (2006).

Sample name	Group	Cr ($\mu g\, g^{-1}$)	$\varepsilon^{53}Cr$	$2\sigma^{a}$	$\varepsilon^{54}Cr$	$2\sigma^{a}$	n^{b}	$\delta^{53}Cr$ (‰)	$2\sigma^{c}$	n^{d}
NIST SRM979		n/a	0.00	0.10	0.00	0.19	15	0.000	0.009*	15
JP-1		2780	0.11	0.10	0.28	0.19	3	−0.102	0.012	5
Carb. Chondrite										
Allende	CV3	3600	0.14	0.10	1.04	0.19	1	−0.113	0.012	4
Orgueil	CI	2650	0.53	0.10	1.45	0.19	6	−0.145	0.012	6
Murchison	CM2	3100	0.36	0.10	0.98	0.19	1	−0.119	0.012	1
Ornans	CO3	3300	0.19	0.10	1.03	0.19	1	−0.068	0.012	1
Average CC								−0.111	0.065	
Ord. Chondrite										
Bremervörde	H/L3	3600	0.36	0.10	0.36	0.19	1	−0.146	0.012	1
Kernouve	H6	3700	0.27	0.10	0.27	0.19	1	−0.111	0.012	1
Barratta	L4	3740	0.29	0.10	−0.10	0.19	1	−0.118	0.012	1
Bruderheim	L6	3400	0.30	0.10	−0.26	0.19	1	−0.112	0.012	1
Parnallee	LL3	3200	0.38	0.10	−0.08	0.19	1	−0.117	0.012	1
Saint-Severin	LL6	3970	0.38	0.10	0.38	0.19	1	−0.104	0.012	1
Average OC								−0.118	0.029	
Enst. Chondrite										
Indarch	EH4	3400	0.37	0.10	0.15	0.19	1	−0.103	0.012	1
Khairpur	EL6	4060	0.27	0.10	0.14	0.19	1	−0.048	0.012	1
Average EC								−0.076	0.078	
HED										
Kapoeta	How.	3600	0.32	0.10	−0.71	0.19	1	−0.156	0.012	1
Pasamonte	Eucrite	1750	1.03	0.10	−0.45	0.19	1	−0.176	0.012	1
Juvinas	Eucrite	2100	0.93	0.10	−0.55	0.19	1	−0.184	0.012	1
Johnstown	Diog.	5700	0.32	0.10	−0.56	0.19	1	−0.074	0.012	1
Mars										
Nakhla	SNC	1770	0.27	0.10	−0.35	0.19	1	−0.156	0.012	1

Note: Carb.: carbonaceous, Ord.: ordinary, Enst.: enstatite, How.: howardite, Diog.: diogenite.
[a] 2 standard deviations of repeated measurements of the NIST 979 standard.
[b] Number of unspiked runs.
[c] 2 standard deviations of repeated measurements of Orgueil sample.
[d] Number of spiked runs.
[*] 2 standard deviations of the NIST 979 (see text for details).

For extra-terrestrial samples this assumption is not true as there are resolvable variations in mass independent Cr isotopes, such that samples do not lie on a single mass fractionation trend. Therefore for meteoritic samples the unspiked runs were first internally normalised using an exponential law to the canonical $^{50}Cr/^{52}Cr$ ratio of 0.051859 (Shields et al., 1966) to reveal any mass independent $\varepsilon^{53}Cr$ and $\varepsilon^{54}Cr$ variation. By definition this removes any mass dependent information but provides data comparable to other studies (Birck and Allègre, 1988; Shukolyukov and Lugmair, 2006; Moynier *et al.*, 2007; Trinquier *et al.*, 2007; Trinquier *et al.*, 2008a; Yin *et al.*, 2007; Qin *et al.*, 2010). Finally the unspiked run is also combined with the spiked run for each sample in the double spike calculation and allows the determination of the mass dependent Cr stable isotope composition (and the true isotope composition in four isotope space).

4. RESULTS

Seventeen meteorite samples were analysed. All data are reported in Table 2. Radiogenic and nucleosynthetic variations are reported using the equation:

$$\varepsilon^{X}Cr = \left({}^{X}Cr/{}^{52}Cr_{sample}/{}^{X}Cr/{}^{52}Cr_{SRM979} - 1\right) \times 10000$$

where ${}^{X}Cr/{}^{52}Cr_{sample}$ /${}^{52}Cr_{sample}$ is the mass fractionation corrected isotope ratios (${}^{53}Cr/{}^{52}Cr$ and ${}^{54}Cr/{}^{52}Cr$). The mass dependent variations are reported using the equation below:

$$\delta^{53}Cr = \left({}^{53}Cr/{}^{52}Cr_{sample}/{}^{53}Cr/{}^{52}Cr_{SRM979} - 1\right) \times 1000$$

where the ratio ${}^{53}Cr/{}^{52}Cr_{sample}$ is the ratio obtained after the double spike deconvolution procedure and is corrected for mass dependent fractionation occurring during sample preparation and mass spectrometry. The ${}^{53}Cr/{}^{52}Cr$ and ${}^{54}Cr/{}^{52}Cr$ isotopic ratios for the SRM 979 standard used in this study for normalisation are 0.1134563 and 0.0282115 respectively.

Internally normalised isotope ratios (${}^{53}Cr/{}^{52}Cr$ and ${}^{54}Cr/{}^{52}Cr$) for the NIST SRM 979 standards are not correlated and this indicates that a double normalisation using ${}^{54}Cr/{}^{52}Cr$ ratio is not necessary to correct for mass isotope fractionation (i.e. there is no residual mass fractionation variation in the data at the level of analytical uncertainty). Radiogenic $\varepsilon^{53}Cr$ values for meteorites analysed in this study range from 0.14 to 1.03 and nucleosynthetic $\varepsilon^{54}Cr$ values between −0.71 and 1.45. Carbonaceous chondrites show excess ${}^{54}Cr$ compared to ordinary chondrites (Fig. 2). There is no clear relationship between $\varepsilon^{53}Cr$ and $\varepsilon^{54}Cr$ (Fig. 2), however. Overall, the data obtained in this study are in agreement with previously published data for non-mass dependent Cr isotope variations in meteorites (Trinquier *et al.*, 2007; Trinquier *et al.*, 2008a ; Qin *et al.*, 2010) when the offset due to normalisation is taken into account.

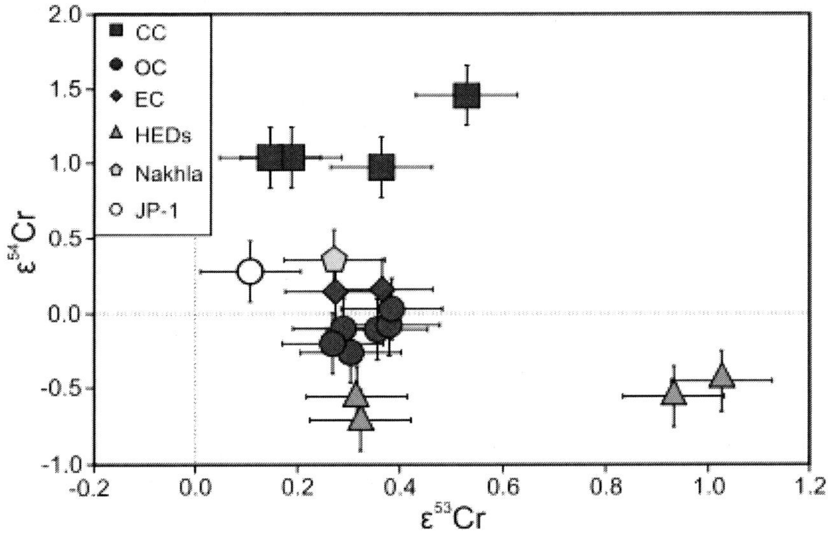

Figure 2. Mass independent chromium isotopic compositions for bulk meteorites analysed in this study. CC: carbonaceous chondrites, OC: ordinary chondrites, EC: enstatite chondrites, HED: howardite–eucrite–diogenite.

For mass dependent fractionation the samples analysed in this study are characterised by $\delta^{53}Cr$ values ranging from −0.184 to −0.048‰. The data are presented in Table 2 and Fig. 3. The carbonaceous chondrites have an average $\delta^{53}Cr$ value of −0.111±0.065‰ (n=4) and display some variations with the CO sample having the heaviest $\delta^{53}Cr$ values. The ordinary chondrites are characterised by very little spread in $\delta^{53}Cr$ values with a mean $\delta^{53}Cr$ value of −0.118±0.029‰ (n=6). Enstatite chondrites display greater spread in $\delta^{53}Cr$ but only two samples have been measured with Khaipur being the heaviest sample in this study ($\delta^{53}Cr$=−0.048‰). Finally, HED and martian meteorites have $\delta^{53}Cr$ values ranging from −0.166 to −0.070‰.

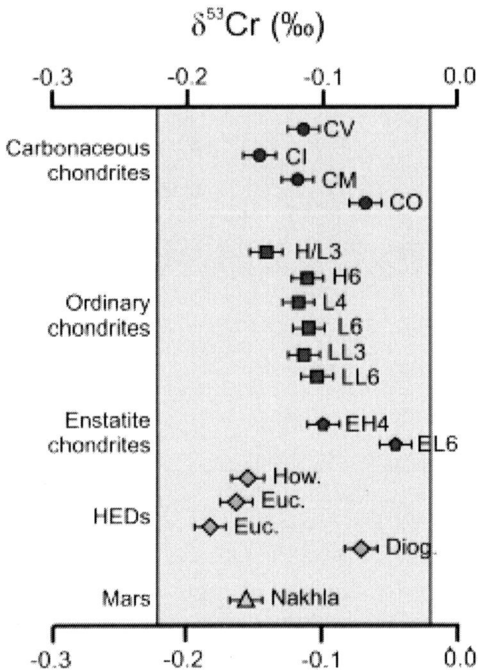

Figure 3. Stable Cr isotopic compositions in meteorites analysed in this study. The grey vertical bar is the current estimate (mean ± 2 s.d.) for the Bulk Silicate Earth (BSE) from Schoenberg et al. (2008).

The results for five metal–silicate experiments are presented in Table 1. Overall, metal and silicate pairs from the five experiments have slightly different Cr stable isotope compositions. However, the Cr isotopic composition of the metal and the silicate fractions are in all cases identical or barely distinguishable; B100 has $\Delta^{53}Cr$ ($\Delta^{53}Cr = \delta^{53}Cr_{metal} - \delta^{53}Cr_{silicate}$) of 0.045 ± 0.031 (Table 1). There is a good correlation between the Cr stable isotope compositions of the metal and silicate pairs and experimental run duration for both Mg and Si capsule experiments (Fig. 4).

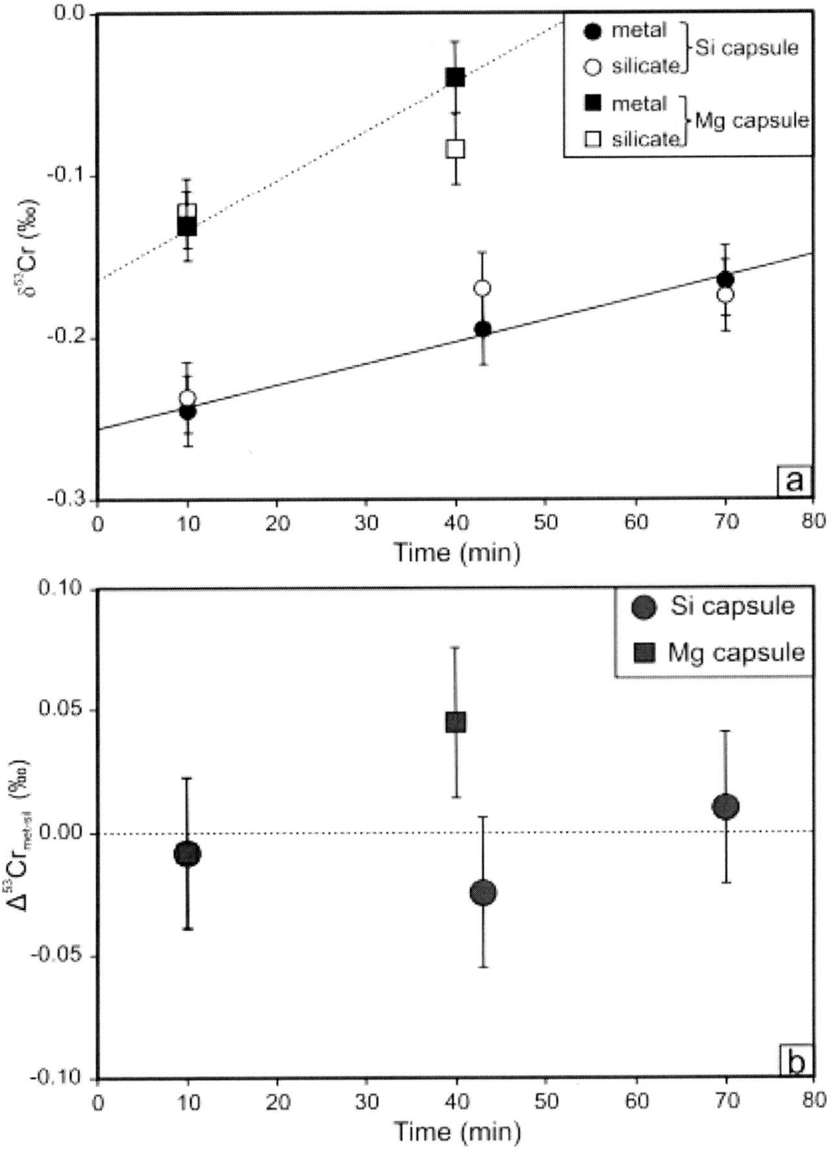

Figure 4. (a) Chromium isotopic compositions of metal–silicate experiments versus time. The solid and dashed lines track the Cr isotopic evolution of the phases through time in the silica (Si) and magnesium (Mg) capsules respectively. (b) Fractionation factors ($\Delta^{53}Cr_{met\text{-}sil} = \delta^{53}Cr_{metal} - \delta^{53}Cr_{silicate}$) versus time for the five experiments analysed in this study. The dashed line is $\Delta^{53}Cr_{met\text{-}sil} = 0$.

5. DISCUSSION

In this section we first discuss mass independent variations in Cr isotopes. We then estimate the mass dependent fractionation during metal segregation and discuss the implication for the utility of stable Cr isotopes to study core formation on Earth. Finally, we discuss the potential explanations for mass dependent variations in the meteorite dataset relative to other published meteorite datasets.

5.1. Variation in $\varepsilon^{53}Cr$ and $\varepsilon^{54}Cr$

The meteorites analysed in this study display significant variations in both $\varepsilon^{53}Cr$ and $\varepsilon^{54}Cr$. All chondrites show either a weak positive or no anomaly in $\varepsilon^{53}Cr$ compare to terrestrial samples and this is due to variations in the Mn/Cr ratio is primordial materials. In the literature, Orgueil and Allende samples show small excesses in $\varepsilon^{53}Cr$ and have been used to estimate timescales for the formation of the Earth (Shukolyukov and Lugmair, 2006; Trinquier *et al.*, 2008a; Moynier *et al.*, 2007 ; Qin *et al.*, 2010). Given the relatively small isotopic variations reported in this study and the reproducibility of our measurements, we have not attempted to use our carbonaceous chondrites data to reconstruct a unique initial $^{53}Mn/^{55}Mn$ ratio of the solar system. Ordinary chondrites display a smaller spread in their $\varepsilon^{53}Cr$ values.

The HED achondrites are characterised by large $\varepsilon^{53}Cr$ variations and this suggests that differentiation of Vesta was extremely rapid. With a half-life of 3.7 Ma for ^{53}Mn and the differences between the HED meteorites can only reasonably be explained by early fractionation of the Mn/Cr ratios during Vesta's differentiation. To be recorded by the $^{53}Mn/^{53}Cr$ system, the differentiation must have happened in the first 20 Ma of the asteroid history. This is in agreement with previously published data on eucrite parent body (Trinquier et al., 2008a).

Among all chondrite groups, carbonaceous chondrites show the largest ^{54}Cr anomalies with values ranging from 0.98 to 1.45. The two enstatite chondrites analysed show similar ^{54}Cr excesses to those of carbonaceous chondrites. Finally, ordinary chondrites show a small spread in $\varepsilon^{54}Cr$ values but are depleted in ^{54}Cr relative to other chondrites and terrestrial samples. It was previously proposed (e.g. Qin et al., 2010) that variations in $\varepsilon^{54}Cr$ could be used to identify meteorite groups or parent bodies and the data presented in this study agree with this interpretation.

5.2. Isotope fractionation during metal–silicate partitioning

Two experiments of 10 min and 40 min duration were performed in MgO capsules and three further experiments of 10 min, 45 min and 70 min duration were carried out in silica capsules (Table 1). The experimental products were analysed for their Cr concentration and isotopic compositions in order to calculate Cr metal–silicate distribution coefficients ($D_{met\text{-}sil}=[Cr]_{met}/[Cr]_{silicate}$) and stable isotope fractionation factors $\Delta^{53}Cr_{met\text{-}sil}=\delta^{53}Cr_{met}-\delta^{53}Cr_{sil}$ (Table 1).

The $D_{met\text{-}silicate}$ values for the silica capsules experiments are very consistent at 0.87 to 1.02 (Table 1), in agreement with results from previous high-pressure experimental studies (Tuff et al., 2011). The two experiments in MgO capsules have slightly different $D_{met\text{-}silicate}$, of 1.18 (A176) and 2.00 (B100) which might reflect small differences in oxygen fugacity between them. The silicate fraction of experiment B100 was, however, extremely difficult to separate from the capsule and it is likely that the silicate fraction of this sample has been contaminated to some extent by capsule material. This is the likeliest explanation for the lower Cr concentration obtained in the silicate fraction and the higher apparent $D_{met\text{-}silicate}$ of this experiment. Therefore, experiment B100 should not be used to assess whether isotope fractionation occurred during metal–silicate partitioning. The consistency of the $D_{met\text{-}silicate}$ values obtained for experiment durations ranging from 10 min to 70 min provides strong evidence for the attainment of Cr equilibrium between metal and silicate during the experiments. This is also consistent with earlier observations on equilibrium times for metal–silicate partitioning (Tuff *et al.*, 2011 ; Wood and Halliday, 2010).

In terms of Cr isotope partitioning, the metal and silicate phases in all experimental products have very similar Cr stable isotope compositions. The Cr isotopes compositions of both metal and silicate phases in all experiments show a correlation with experiment duration (Fig. 4), both metal and silicate phases becoming increasingly heavy with time. This can be interpreted in terms of loss of isotopically light Cr to the capsule. The metal and silicate phases from MgO capsule experiments have heavier $\delta^{53}Cr$ values than those from silica capsules (Fig. 4). This is probably due the greater tendency of crystalline MgO to dissolve Cr, facilitating loss of Cr to the capsule walls. These observations are in agreement with previous experimental studies (Tuff et al., 2011), which demonstrated that Cr is indeed lost to the capsule but that this loss is relatively small. The co-evolution of both silicate and metal Cr isotopic compositions with time provides further evidence for attainment of Cr equilibrium between metal and silicate during the experiments.

The Cr metal–silicate isotope fractionation $\Delta^{53}Cr_{met-sil}$ in the silica capsule experiments ranges from $-0.024\pm0.031‰$ (43 min) to $+0.010\pm0.031‰$ (70 min) and in the MgO capsule experiment $\Delta^{53}Cr_{met-sil}$ is $-0.008\pm0.031‰$ (10 min). The time series experiments conducted in silica capsules show no observable correlation between metal–silicate fractionation and experiments duration (Fig. 4). Therefore, although some Cr is lost to the capsule, the metal and silicate phases appear to have reached equilibrium before this point, and they remain at equilibrium during the experiment, even as minor Cr is lost to the capsule. Importantly, in all of the experiments $\Delta^{53}Cr_{met-sil}$ is within uncertainty of zero indicating negligible Cr isotope fractionation between molten metal and silicate.

The experiments carried out for this study were performed at 1.5 GPa and 1650 °C which correspond to a depth of about 45 km on Earth. Metal segregation and core formation on Earth undoubtedly happened at considerably higher pressure and temperature (Wade and Wood, 2005). Although the effect of pressure is unknown, the increase of temperature will have an impact on the isotope fractionation (as $1/T^2$), such that the magnitude of $\Delta^{53}Cr_{met-sil}$ should decrease. Given our results therefore, it is unlikely that metal segregation at higher pressure and temperature is associated with any isotope fractionation. We conclude that there is no evidence of significant Cr isotope fractionation during metal–silicate partitioning on Earth but that further experiments under conditions more closely approximating those of core formation would enable more stringent testing of this conclusion.

5.3. Mass dependent chromium isotopic fractionation in meteorites

The chondrites analysed in this study are characterised by double-spike derived $\delta^{53}Cr$ values ranging from $-0.146\pm0.012‰$ to $-0.048\pm0.012‰$ (Fig. 5). Although the total range of fractionation is small, it is three times larger than previously reported for radiogenic variations indicating that mass dependent processes dominate the variation. The data reported for chondrites are not in agreement with previously published meteorite data (Moynier *et al.*, 2011 ; Schiller *et al.*, 2014). For example, Orgueil has been analysed several times in the course of this study and is characterised by a Cr isotopic composition of $-0.145\pm0.012‰$ (2 s.d., $n=6$), which is different from the previously published value ($-0.41\pm0.18‰$, 2 s.d., $n=10$; Moynier et al., 2011). There are several possible reasons for this discrepancy and critically there is the difference in method used to determine the Cr isotopic composition. We use a double spike TIMS method whereas Moynier et al.

(2011) and Schiller et al. (2014) used a standard sample bracketing MC-ICPMS technique. The double spike method has two clear advantages over standard sample bracketing. Firstly, it corrects for isotopic fractionation that occurs during sample preparation and analytical measurements. The chemical separation technique we and other groups use (e.g. Trinquier *et al.*, 2008b ; Bonnand *et al.*, 2011) demonstrably does not have a 100% column yield. This is corrected for successfully using a double-spike (e.g. Bonnand et al., 2013), but sample-standard bracketing requires 100% yield, although Moynier et al. (2011) and Schiller et al. (2014) suggest they meet this requirement. Secondly, the double-spike technique corrects for instrumental mass fractionation for each individual sample, without recourse to bracketing standards and so deals with any small amounts of residual matrix elements in the final analytical solution that would induce mass dependent fractionation in standard bracketing measurements. The issue of residual matrix elements is much harder to assess for standard-sample bracketing methods unless standards doped with the same matrix as the samples are run through the chemistry and then analysed (see Bonnand *et al.*, 2011 ; Bonnand *et al.*, 2013). This is the most likely discrepancy between the data sets and we note that organic material can be particularly problematical for standard-sample bracketing studies. It is interesting to note that the organic-rich CI1 chondrite Orgueil, has the most displaced $\delta^{53}Cr$ value (Moynier et al., 2011) from our new data. Additionally a possible origin of the discrepancy is incomplete sample dissolution. In this case, spinel is the likely mineral left after dissolution and is one of the principal Cr-bearing minerals in these samples. The technique used in this study is robust and the data can be directly compared to the BSE values obtained also by double spike techniques (Schoenberg et al., 2008). The terrestrial sample (JP-1, $\delta^{53}Cr = -0.102 \pm 0.012$‰) measured in this study is similar to previously analysed mantle xenoliths and within error of the BSE value. The data reported for the meteorite samples are also in agreement with a recent contribution by Qin et al. (2015).

Given the external reproducibility of our technique it is possible to test the hypothesis of Cr isotopic fractionation during core formation. Carbonaceous chondrites are characterised by $\delta^{53}Cr$ values ranging from −0.145 to -0.068‰ which is within error of the currently used estimate for the BSE (-0.12 ± 0.10‰, Schoenberg et al., 2008, Fig. 3). Our results yield no resolvable difference between the Cr isotopic composition of chondrites and the BSE implying that Cr isotopes were not fractionated during core formation processes on Earth. Our experiments on metal–silicate partitioning of Cr are in agreement with this conclusion.

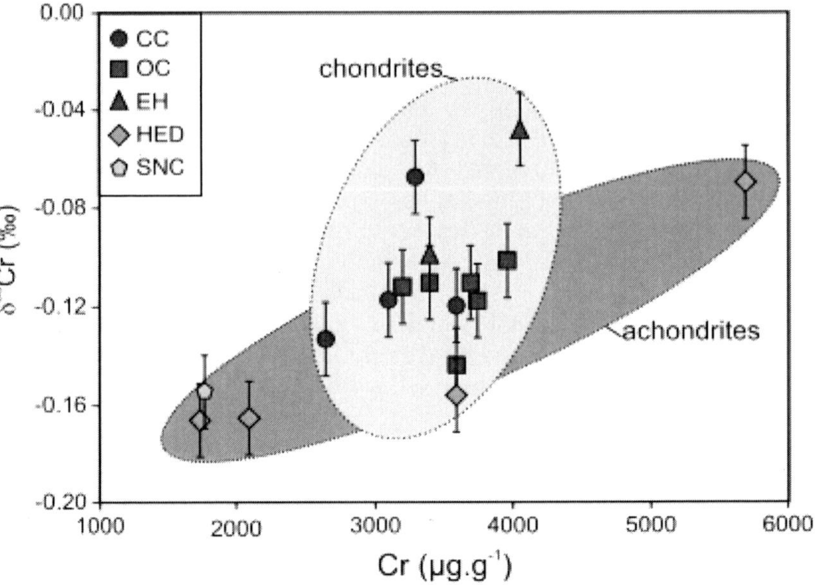

Figure 5. Mass dependent chromium isotopic compositions versus chromium concentrations in bulk meteorites analysed in this study. SNC: shergottites–nakhlites–chassignites.

5.4. Mass dependent Cr stable isotope variations in chondrites

The results presented in this study suggest that Cr isotopes cannot be used to constrain core formation processes on Earth. However, small variations within the chondritic data set can help us understand the fundamental differences between the groups of meteorites. First of all, the ordinary chondrites are characterised by small variations compared to both carbonaceous chondrites and enstatite chondrites. Within the carbonaceous chondrites, CO is the heaviest which could relate to the redox state of CO chondrites as they are believed to be more reduced than other carbonaceous chondrites (Rubin, 1997). Similarly, one of the enstatite chondrites (the most reduced chondrite group) has the heaviest $\delta^{53}Cr$ of all the samples we have measured. However, there is no difference in the $\delta^{53}Cr$ among LL, L and H group ordinary chondrites, which follow progressive more reduced conditions, suggesting that any redox effects on Cr isotope fractionation are subtle. However, it is noteworthy that the variability we see in only two enstatite chondrites is reflected in other stable isotope systems, notably Si, which have different compositions from other chondrite groups (Armytage *et al.*, 2012 ; Savage and Moynier, 2013).

Perhaps the most striking feature of the chondritic data is the overall lack of variability compared with the BSE. Unlike the study of Moynier et al. (2011), we do not see convincing correlations with other compositional data such as $\Delta^{17}O$ and Mg/Cr ratios (not plotted), which indicates there is rather limited variability in the $\delta^{53}Cr$ composition of material that was mixed in the early solar nebular. It also suggests that magmatic processes in the Earth may have produced the greater, albeit small, range of $\delta^{53}Cr$ found in high temperature terrestrial rocks. More data are required to assess whether small variations in stable Cr isotopes in meteorites can tease out more information about the formation of primitive material in the Solar System.

5.5. Mass dependent variations in achondrites

HEDs and Martian meteorites have also been analysed in this study. They are characterised by a greater range in stable Cr isotope composition relative to those displayed by other meteorite groups. These samples are believed to be basalts from Vesta (Pasamonte, Juvinas), associated cumulates from Vesta (Johnstown) and cumulate from Mars (Nakhla). The aim was to determine the Cr stable isotope composition of Vesta and Mars mantles. However, it was recently demonstrated that Cr isotopes are fractionated by magmatic processes (Bonnand et al., in press) and therefore a direct comparison of the data and chondrite values is not possible. The correlation between Cr concentration and Cr stable isotope composition for the achondrites suggests that the samples depleted in Cr are isotopically lighter than the Cr enriched samples (Fig. 5). This is in agreement with mare basalts fractionation (Bonnand et al., in press). A suite of chemically related samples should be analysed before the Cr isotopic compositions of both Vesta and Mars' mantles can be assessed.

6. CONCLUSIONS

Seventeen meteorites samples were analysed for their radiogenic and nucleosynthetic as well as mass dependent Cr isotopic compositions. Chondrites are characterised by excesses in radiogenic ^{53}Cr as previously demonstrated. The larger variations in $\varepsilon^{53}Cr$ are within the HEDs achondrites. Large variations in $\varepsilon^{54}Cr$ are also reported with carbonaceous chondrites showing ^{54}Cr excesses compared to terrestrial samples while ordinary chondrites are ^{54}Cr depleted. Finally, enstatite chondrites have $\varepsilon^{53}Cr$ and $\varepsilon^{54}Cr$ values similar to that of the Earth. These results are similar to previously published dataset for meteoritic samples.

Five metal–silicate experiments have been performed to study the equilibrium Cr isotope fractionation between metal and silicate. There is no isotopic difference between metal and silicate, which provides evidence that Cr isotopes are unlikely to be fractionated during metal segregation. This is in agreement with Cr stable isotope compositions of chondrites which range from −0.146 to $-0.048‰$, essentially identical to the current estimate of the Bulk Silicate Earth ($-0.12\pm0.10‰$, Schoenberg et al., 2008).

Small mass dependent Cr isotope variations are reported in bulk rock meteorites both in chondrites and achondrites. The variations in chondrites could be related to volatility and/or redox conditions but more data are required to test those hypotheses, although the total variation in chondritic meteorites suggests the $\delta^{53}Cr$ of early solar nebular material was small ($<0.1‰$). Mass dependent Cr isotope composition is potentially a very powerful tool to constrain the chemical composition of the building blocks of the Earth. Achondrites seem to be affected by magmatic differentiation and therefore the Cr isotopic composition of Mars and Vesta cannot be fully assessed until a more extensive suite of data is acquired for these bodies, although their composition does overlap that of the bulk silicate Earth.

ACKNOWLEDGMENTS

This research was supported by STFC grants (ST/M001318/ 1 and ST /G00272X/1) to AH and by Open University funding to IJP. HW acknowledges funding from the ERC Starting Grant ("Habitable Planet" 306655) and NERC Advanced Fellowship NE/F014295 /1. We would like to thank two anonymous reviewers for their useful comments that helped to improve the manuscript. The research materials supporting this publication can be accessed by contacting PB.

REFERENCES

1. Allègre, C.J., Poirier, J.P., Humler, E., Hofmann, A.W., 1995. The chemical composition of the Earth. Earth Planet. Sci. Lett. 134, 515–526.
2. Armytage, R.M.G., Georg, R.B., Savage, P.S., Williams, H.M., Halliday, A.N., 2011. Silicon isotopes in meteorites and planetary core formation. Geochim. Cosmochim. Acta 75, 3662–3676.

3. Armytage, R.M.G., Georg, R.B., Williams, H.M., Halliday, A.N., 2012. Silicon isotopes in lunar rocks: implications for the Moon's formation and the early history of the Earth. Geochim. Cosmochim. Acta 77, 504–514.
4. Ball, J.W., Izbicki, J.A., 2004. Occurrence of hexavalent chromium in ground water in the western Mojave Desert, California. Appl. Geochem. 19, 1123–1135.
5. Birck, J.L., Allègre, C.J., 1988. Manganese–chromium isotope systematics and the development of the early Solar System. Nature 331, 579–584.
6. Bonnand, P., James, R.H., Parkinson, I.J., Connelly, D.P., Fairchild, I.J., 2013. The chromium isotopic composition of seawater and marine carbonates. Earth Planet. Sci. Lett. 382, 10–20.
7. Bonnand, P., Parkinson, I.J., James, R.H., Karjalainen, A.-M., Fehr, M.A., 2011. Accurate and precise determination of stable Cr isotope compositions in carbonates by double spike MC-ICP-MS. J. Anal. At. Spectrom. 26, 528–535.
8. Bonnand, P., Parkinson, I.J., Anand, M., in press. Mass dependent fractionation of stable chromium isotopes in mare basalts: implications for the formation and the differentiation of the Moon. Geochim. Cosmochim. Acta.
9. Bridgestock, L.J., Williams, H., Rehkaemper, M., Larner, F., Giscard, M.D., Hammond, S., Coles, B., Andreasen, R., Wood, B.J., Theis, K.J., Smith, C.L., Benedix, G.K., Schoenbaechler, M., 2014. Unlocking the zinc isotope systematics of iron meteorites. Earth Planet. Sci. Lett. 400, 153–164.
10. Drake, M.J., Righter, K., 2002. Determining the composition of the Earth. Nature 416, 39–44.
11. Farkaš, J., Chrastný, V., Novák, M., Cadkova, ˇ E., Pašava, J., Chakrabarti, R., Jacobsen, S.B., Ackerman, L., Bullen, T.D., 2013. Chromium isotope variations (δ53/52Cr) in mantle-derived sources and their weathering products: implications for environmental studies and the evolution of δ53/52Cr in the Earth's mantle over geologic time. Geochim. Cosmochim. Acta 123, 74–92.
12. Georg, R.B., Halliday, A.N., Schauble, E.A., Reynolds, B.C., 2007. Silicon in the Earth's core. Nature 447, 1102–1106.
13. Hartmann, M., Woosler, S.E., El Eid, M.F., 1985. Nucleosynthesis in neutron-rich supernova ejecta. Astrophys. J. 297, 837–845.
14. Hin, R.C., Burkhardt, C., Schmidt, M.W., Bourdon, B., Kleine, T., 2013. Experimental evidence for Mo isotope fractionation between metal and silicate liquids. Earth Planet. Sci. Lett. 379, 38–48.
15. Leya, I., Wieler, R., Halliday, A.N., 2003. The influence of cosmic-ray production on extinct nuclide systems. Geochim. Cosmochim. Acta 67, 529–541.
16. McDonough, W.F., 2003. Compositional model for the Earth's core. In: Holland, H.D., Turekian, K.K. (Eds.), Treatise on Geochemistry, vol. 2. Elsevier–Pergamon, Oxford, pp. 547–568.
17. Moynier, F., Yin, Q.-Z., Jacobsen, B., 2007. Dating the first stage of planet formation. Astrophys. J. 671, 181–183.
18. Moynier, F., Yin, Q.-Z., Schauble, E., 2011. Isotopic evidence of Cr partitioning into Earth's core. Science 331, 1417–1420. Newsom, H.E., 1990. Accretion and core formation in the Earth: evidence from siderophile elements. In: Newsom, H.E., Jones, J.H. (Eds.), Origin of the Earth. Oxford University Press, New York, NY, pp. 273–288.
19. Presnall, D.C., Dixon, S.A., Dixon, J.R., Odonnell, T.H., Brenner, N.L., Schrock, R.L., Dycus, D.W., 1978. Liquidus phase relations on join diopside–forsterite–anorthite from 1 atm to 20 kbar – their bearing on generation and crystallization of basaltic magma. Contrib. Mineral. Petr. 66, 203–220. Qin, L., Alexander, C.M.O.D., Carlson, R.W., Horan, M.F., Yokoyama, T., 2010. Contributors to chromium isotope variation of meteorites. Geochim. Cosmochim. Acta 74, 1122–1145. Qin, L., Xia, J., Carlson, R.W., Zhang, Q., 2015. Chromium stable isotope composition of meteorites. In: 46th Lunar Planet. Sci. Conf. Rubin, A.E., 1997. Mineralogy of meteorite groups. Meteorit. Planet. Sci. 32, 231–247. Savage, P.S., Moynier, F., 2013. Silicon isotopic variation in enstatite meteorites: clues to their origin and Earth-forming material. Earth Planet. Sci. Lett. 361, 487–496. Schauble, E.A., 2004. Applying stable isotope fractionation

theory to new systems. Rev. Mineral. Geochem. 55, 65–111. Schiller, M., Van Kooten, E., Holst, J.C., Olsen, M.B., Bizzarro, M., 2014. Precise measurement of chromium isotopes by MC-ICPMS. J. Anal. At. Spectrom. 29, 1406–1416. Schoenberg, R., Zink, S., Staubwasser, M., von Blanckenburg, F., 2008. The stable Cr isotope inventory of solid Earth reservoirs determined by double spike MC-ICPMS. Chem. Geol. 249, 294–306. Shahar, A., Hillgren, V.J., Young, E.D., Fei, Y., Macris, C.A., Deng, L., 2011. Hightemperature Si isotope fractionation between iron metal and silicate. Geochim. Cosmochim. Acta 75, 7688–7697. Shields, W.R., Murphy, T.J., Catanzar, E.J., Garner, E.L., 1966. Absolute isotopic abundance ratios and atomic weight of a reference sample of chromium. J. Res. Nat. Bur. Stand. Section a-Physics and Chemistry A 70, 193–197. Shukolyukov, A., Lugmair, G.W., 2006. Manganese–chromium isotope systematics of carbonaceous chondrites. Earth Planet. Sci. Lett. 250, 200–213. Trinquier, A., Birck, J.-L., Allègre, C.J., 2007. Widespread Cr-54 heterogeneity in the inner Solar System. Astrophys. J. 655, 1179–1185. Trinquier, A., Birck, J.L., Allègre, C.J., Gopel, C., Ulfbeck, D., 2008a. 53Mn–53Cr systematics of the early Solar System revisited. Geochim. Cosmochim. Acta 72, 5146–5163. Trinquier, A., Birck, J.-L., Allègre, C.J., 2008b. High-precision analysis of chromium isotopes in terrestrial and meteorite samples by thermal ionization mass spectrometry. J. Anal. At. Spectrom. 23, 1565–1574. Tuff, J., Wood, B.J., Wade, J., 2011. The effect of Si on metal–silicate partitioning of siderophile elements and implications for the conditions of core formation. Geochim. Cosmochim. Acta 75, 673–690. Wade, J., Wood, B.J., 2005. Core formation and the oxidation state of the Earth. Earth Planet. Sci. Lett. 236, 78–95. Williams, H.M., Wood, B.J., Wade, J., Frost, D.J., Tuff, J., 2012. Isotopic evidence for internal oxidation of the Earth's mantle during accretion. Earth Planet. Sci. Lett. 321, 54–63. Wood, B.J., Wade, J., Kilburn, M.R., 2008. Core formation and the oxidation state of the Earth: additional constraints from Nb, V and Cr partitioning. Geochim. Cosmochim. Acta 72, 1415–1426. Wood, B.J., Halliday, A.N., 2010. The lead isotopic age of the Earth can be explained by core formation alone. Nature 465, 767–770. Yin, Q.-Z., Jacobsen, B., Moynier, F., Hutcheon, I.D., 2007. Toward consistent chronology in the early Solar System: high-resolution Mn53–Cr53 chronometry for chondrules. Astrophys. J. 662, 43–46. Zambardi, T., Poitrasson, F., Corgne, A., Meheut, M., Quitte, G., Anand, M., 2013. Silicon isotope variations in the inner Solar System: implications for planetary formation, differentiation and composition. Geochim. Cosmochim. Acta 121, 67–83.

CHAPTER 11

Concentration of Platinum Group Elements during the Early Earth Evolution: A Review*

Arkady Pilchin[1], Lev Eppelbaum[2]

[1]*Universal Geoscience and Environment Consulting Company, Willowdale, Ontario, Canada*

[2]*Department of Geosciences, Faculty of Exact Sciences, Tel Aviv University, Tel Aviv, Israel*

ABSTRACT

Numerous unique geological processes [1] took place during the early Earth evolution; several of them, especially those occurring in the Hadean—Early Archean and later, are reflected in the modern geological (geophysical, geochemical, etc.) pattern. One such significant enigmatic feature is the preservation of extremely dense and heavy platinum group elements (PGEs): Pt, Pd, Rh, Ru, Ir, Os. Concentration of PGEs during this period could have taken place in two ways: 1) presence of particular matter capable of preserving PGEs near the earth's surface, 2) transportation of PGEs by magma flows from deep lithospheric (asthenospheric) layers (slabs) to the subsurface. Clearly, much of the dense and heavy PGEs did not sink through to the Earth's mantle (core) at the time of the magma-ocean, and occur near Earth's surface in abundances for formation of ore deposits with PGE concentrations found to be 2 - 3 orders of magnitude greater than those in their host media. Their enrichments are associated in numerous cases with such enigmatic phenomena as formation of anorthosites and anorthosite-bearing layered magmatic intrusions. PGE deposits and mineralization zones are also found in associations with chromitites, dunites and serpentinites. In this review, problems related to the initial concentration and preservation of PGEs, their association with anorthosites, and formation of layered intrusions are discussed in detail. The main aim of this article is analysis of the requirements—initial concentration and preservation of PGE and PGM (Platinum Group Minerals) during the early Earth evolution, as well as examination of the distribution behavior of some PGEs in different ore deposits and meteorites. It is supposed that meteoritic bombardment of Earth has played a significant role in formation of PGEs deposits. Some conclusions made in this article may be useful for developing and enhancing strategies of prospecting for PGEs deposits.

Keywords: Platinum Group Elements, Early Earth Evolution, Magma-Ocean, Earth's Bombardment, PGE Deposit Formation

1. INTRODUCTION

A significant enigmatic feature of early Earth evolution is the preservation of extremely dense and heavy platinum group elements (PGE): Pt, Pd, Rh, Ru, Ir, Os, which somehow managed not to sink to the Earth's core throughout the duration of the magma-ocean, and instead were accumulated at relatively small depths in quantities yielding formation of economic ore deposits. In numerous cases, their enrichments are associated with such enigmatic rock formations as anorthosites, and such inscrutable magmatic structures as layered magmatic intrusions (layered mafic-ultramafic intrusions). Despite that PGEs mining has a very long history, specific aspects of PGE ore formation still remain the subject of great controversy (e.g., [2] [3]).

Any genesis model of ore deposits usually includes such components as source, transport and trap (e.g., [4]). However, certain authors (e.g., [5]) indicate that formation of any ore deposit includes four basic geological requirements: 1) source of metals and ligands (e.g., sulfur and chlorine), 2) mechanism which either transports these source components or removes non-ore components to allow residual concentration of metals, 3) trap (depositional mechanism for formation of the ore body as ore minerals and associated gangue), and 4) conditions necessary for preservation of the ore deposit. Most such models are usually oversimplified, since different ore deposits have different specifications for their formation, various classifications, different types of mineralization, etc. (e.g., [5]). However, for formation of economic PGE ore deposits, another requirement must be emphasized. A mechanism is necessary for the enrichment of ore metal(s) concentration in ore deposits of up to 3 - 4 orders of magnitude greater than their typical crustal abundances (e.g., [3]). The concentration of platinum in areas exposed to earth's surface is about 0.4 ppb [3] . This means that such requirement as process (processes) of extraction of PGE metals from the source rocks and minerals must also be taken into consideration.

Experimental data show that at equilibrium conditions, the concentration of all PGEs is at least 10,000 times higher in sulfide melt than in the coexisting silicate melt, making sulfide an extremely potent agent for the collection and segregation of PGEs from magmas [3] . This means that known associations of PGEs with sulfide deposits [6] are not accidental and should rather be expected. For example, in anorthosite-bearing layered mafic and ultramafic intrusions (ALMI) of Australia stratabound layers of disseminated Ni-Cu sulfides enriched in PGEs, stratabound chromitite layers (the Panton Sill in the Halls Creek orogen, with PGE-rich sulfides) and basal segregations of Ni-Cu-Co-PGE sulfides have been reported in anorthosite-bearing layered mafic and ultramafic intrusions (ALMI) of Australia [7] . Moreover, it has been proposed that chromitites owe their high Pt (and Pd) content to segregation of the magmatic sulfides during chromite precipitation [8] . However, on other hand, the most important deposits occur as relatively narrow stratiform reefs, which tend to be sulfide-poor (e.g., [2]); and the rocks contain no visible sulfides in many chromitites, magnetitites and silicate-hosted ores [2] . Chromitites from dunites in the Uktus Ural- Alaskan type complex (the Urals, Russia) are characterized by high Pt contents, even though there is no evidence of attaining sulfur saturation during their formation [9] . Maier [2] also points to low sulfide contents of many PGE-rich intrusions. PGEs are known to be present in rocks as inclusions of native metals and alloys, as well. A sulfide-free upper zone is reported in the South Kawishiwi Intrusion of the Duluth Complex (Minnesota, USA) [10] . PGE mineralization occurs within a sulfide-free magmatic environment in the Mid-Arc- hean Bangur Gabbro (Orissa, India) [11] . Sulfur-poor intrusions are known in the Ural Pt-belt (Russia) [12] . PGE deposits of Sukhoy Log, Bodaybinski area (Irkutsk region, Russia) were discovered in the black schists with PGEs concentrations of up to 1.7 g/t [13] [14] .

Chernyshov [15] reported a number of structural complexes of different age (from Early Archean to Early Proterozoic) containing Au-PGE mineralization in association with the high carbon formations (e.g., graphitized gneisses, carbon-rich schists) in the Voronezh Crystalline

Massif (Russia). Here the ore deposits include Pd-containing (Pd up to 1.2 g/t and Pt up to 0.3 g/t) ferruginous quartzites of Kursk series.

Chernyshov and Chernyshova [16] described gold-PGEs mineralization (Oboyansky type) in the Early Archean graphitized gneisses of the Voronezh Crystalline Massif with concentration of PGEs of ~5 g/t and Au of 1.3 - 1.5 g/t.

Houlé and Lesher [17] show that from four major volcanic episodes involving komatiites in the Abitibi Greenstone Belt (Superior Province, Canada) (2760 - 2735, 2723 - 2720, 2720 - 2710, and 2710 - 2704 Ma), the last two alone host nearly the entire Ni-Cu-(PGE) inventory of the belt.

This is a common problem with development of any model of the formation of PGE ore deposits. The Bushveld Complex in South Africa has highest content of PGEs and platinum group minerals (PGM) in the world (e.g., [3] [18]), known to contain overwhelming deposits of PGEs and chromium (over 80% of the world's deposits of each) [18] [19] . USGS [20] states that ~88.7% of the global PGE reserves are concentrated in the South Africa and most of them belong to the Bushveld Igneous Complex. These facts contradict any standard model of ore deposit formation, because it is very difficult to explain why these vast amounts of PGEs and Cr should be collected in a single area of the Earth.

Another problem of PGE ore deposit formation relates to the fact that most of siderophile (Fe-lowing) elements (e.g., Sn, Mo, Au, and the PGEs) may be locked away in the core, causing their crustal abundances to be much less than those of the global bulk (e.g., [21]). This problem is especially significant for formation of PGE ore deposits, because the densities of PGEs are extremely high (Table 1). This creates another requirement for PGE ore deposit formation: the initial concentration of the PGEs and PGMs during the early Earth evolution and their preservation as sources for the formation of PGE ore deposits. It is self-evident that the initial concentration of PGEs and PGMs is one of the most important processes for formation of PGE deposits, which allowed for later formation of PGE ore deposits.

In many cases, extremely high concentrations of PGEs found in different re gions of the Earth have no relations to any ore deposits. For example, PGE resources in China are relatively poor, and are mainly concentrated in the Jinchuan Cu-Ni deposit where in disseminated ores ΣPGE ranges within 98 - 7159 ppb but does not reach a level of economic deposit [24] . Such a situation prompts a question: how is it possible that PGE concentrations within the same deposit varies from poor (98 ppb) to very rich (7159 ppb)?

Table 1. Densities and melting points at 0.1 MPa (after [22]) and condensation temperatures at 10^{-5} MPa (after [23]) of PGE.

Element	Pd	Pt	Rh	Ru	Ir	Os
Density, kg/m³	12,000	21,500	12,400	12,100	22,500	22,590
Melting point, K	1827.95	2031.35	2236.15	2606.15	2719.15	3306.15
Condensation temperature, K	1324	1408	1392	1551	1603	1812

All this is highly controversial and problematic for any modeling of formation of economic PGE deposits, and renders the problem of generation of PGE ore deposit as enigmatic.

The main aim of this investigation is not related to the creation of any new model of PGE ore deposit formation, but rather to the analysis of the requirements—initial concentration and preservation of PGEs and PGMs during the early Earth evolution, and analysis of certain patterns in the distribution of PGEs in different ore deposits and meteorites.

It is clear that all matter on Earth came from within the solar nebula by the in-falling of nebular material onto the evolving planet during its accretion, at the end of which Earth was covered with a magma-ocean about 1000 km deep [1]. This means that in order to accumulate significant amounts of PGEs and PGMs near Earth's surface, there must have been certain mechanisms of their concentration in near-surface layers, their trapping, and preservation for the future formation of PGE ore deposits.

2. ON THE FORMATION OF PGES AND PGMS AND THEIR CONTENT IN ROCKS AND METEORITES

It is widely accepted that PGEs, like any other elements with an atomic number greater than that of iron (A > 26), were formed by the supernova explosion (e.g., [25]). This means that the real source of PGEs was a supernova explosion. Other than Palladium, all PGEs including PPGEs (Palladium (Pd), Platinum (Pt), Rhodium (Rh)) and IPGEs (Ruthenium, Iridium and Osmium) are refractory elements, most of which were condensed in refractory mineral alloys [23] at temperatures (see Table 1) within the range of the formation of calcium-aluminum inclusions (CAIs) (1473 - 2273 K; [26]), and those above the formation of chondrules (over 1473 K; [26]).

The extraordinarily high melting points of PGEs (see Table 1) make it impossible for most of them to be present in molten state in any kind of magmatic rocks. Only Pd can be melted by some Mg-rich komatiites, and both Pd and Pt may be present in the molten form in some chromites and molten forsterite. This means that the most PGMs would accrete on the Earth or any other terrestrial object in solid state. However, the initial formation of PGMs (mostly as alloys and solid solutions) requires their components to be molten and then cooled, and solidifying very quickly because of the huge difference in melting points. This could have taken place only in two ways: 1) during the cooling of matter from the supernova explosion, when the temperatures dropped to those of the melting points or condensation temperatures of PGEs and certain other metals; or 2) during the beginning of solar accretion, when previously cooled matter within the solar nebula was once again heated to the melting points or condensation temperatures of PGEs and certain other metals, with a subsequent ejection of the formed alloys into the range of the forming terrestrial planets and asteroid belt, similar to the process of the formation and transportation of CAIs described in [27] .

For metals having such a vast difference in melting points, the unusual polygonal intergrowth of osmium, iridium and ruthenium with Pt-Fe alloy observed within grains in the Aikora River area of Papua New Guinea [28] indicates their formation took place with rapid cooling of the molten metals. Inclusions of PGEs and their alloys within the Pt-Fe alloys were reported from Florence Creek, Yukon (Ir-dominant alloys; [29]) and Novoseltsi placers of eastern Bulgaria (Os and Os-dominant alloys; [30]). A comparison of the melting points of PGEs reveals (see Table 1) that these inclusions of Ir- and Os-dominant alloys were added to the Pt-Fe alloy melt in solid state.

PGE contents in different kinds of terrestrial rocks indicate (Table 2) that their content in the main magmatic rock composing the Earth crust and upper mantle is extremely low.

This renders it very difficult to assess the source of the PGEs for known ore mineralization and deposits with concentrations as high as: average grades of 3 - 20 ppm for PGEs + Au in the PGE reefs [38] ; up to 18 ppm in the Monchegorsk area of the Kola Peninsula (Russia) [39] ; up to 61 ppm in the Kytlym area of the Northern Urals (Russia) [40] ; up to over 60 ppm of Pt + Pd in Cliff from Shet land ophiolite complex (Scotland) [41] ; up to 25 ppm in Veria of the northern Greece [35] ; average of 10.03 ppm in the Noril'sk region (Russia) [42] ; average of 5.418 ppm in the Great Dyke of the South Africa [42] ; average of 5.670 ppm for the Bushveld complex of the South

Table 2. PGE contents in different kinds of terrestrial rocks.

Rocks	Concentration, in ppb						References
	Pt	Pd	Rh	Ru	Ir	Os	
Primary mantle	7.1	3.9	0.9	5	3.2	3.4	[31]
Bulk crust	1.5	1.5	1.1	0.6	0.037	0.041	[32]
Upper mantle peridotite	6.5	5.7	0.6	5.8	3.5	3.5	[33]
Basaltic andesite	18	11	<0.5	2	0.32	0.5	[34]
Tholeiitic basalt	14	11	-	1.5	0.18	<0.5	[34]
Komatiitic lavas	5	18	-	7	0.7	4	[35]
MORB	0.89	2.05	-	0.11	0.03	-	[36]
Alkali basalt	0.23	0.37	-	0.11	0.04	-	[36]
Peridotite xenoliths, S. Africa*	4.3	1.84	0.93	6.54	3.67	-	[37]
Xenoliths, Karelian Craton*	4.84	2.28	0.79	6.58	3.75	-	[37]

*Xenoliths from kimberlites.

Africa [42] ; average of 20.699 ppm for the Stillwater area (Montana, USA) [42] , etc. What makes the problem even more inexplicable—it is the fact that in numerous cases rocks which usually hosting PGE ore deposits (chromites/chromitites, volcanogenic massive sulfide deposits (VMS)) have very low contents or even practical absence of PGE. For example, PGE in the Uzelginsk VMS deposit (southern Urals, Russia) are found mostly below the detection limits [43] ; low values of PGE were discovered in chromites from the Mt. Albert (Quebec, Canada) [44] , very low PGE content was identified in some layers of the Bushveld Complex (South Africa) [45] ; extremely low PGE content was revealed in the upper portions of the Bushveld Complex [46] , etc. All these facts raisesthe questions: "What was the source of PGEs?" and "What was the mechanism of PGEs enrichment in thousands and tens of thousand times?" in the PGE ore deposits.

Table 3 shows comparison of PGE contents in the meteorites of different origin. It is clear from Table 3 that enstatite achondrites, eucrites, Martian meteorites and Lunar meteorites have PGE concentrations pretty close to those in typical Earth's rocks (see Table 2). However, PGE concentrations in CAI inclusions in meteorites, carbonaceous chondrites (CI, CM, CV, CK, CR, CH), chondrules in CR2 chondrites, enstatite chondrites, LL6 ordinary chondrites are 2 to 4 orders of magnitude higher than those in typical Earth's rocks from Table 2. For more information about mean PGE concentrations in different chondrite groups see Table 1 in [47] .

Table 3. PGE contents in meteorites of different origin and in different kinds of meteorites and meteorite inclusions.

Meteorites	Concentration, in ppb						Refe-rences
	Pt	Pd	Rh	Ru	Ir	Os	
enstatite achondrites	8.8 - 13.5	9.4 - 12.2	1.4 - 1.9	3.3 - 4.1	2.6 - 3.2	-	[48]
eucrites	-	0.4 - 8.5	-	-	0.035 - 0.36	0.003- 0.565	[49]
Martian meteorites	5.40 - 8.65	4.67 - 6.05	0.88 - 1.30	1.60 - 4.37	0.59 - 3.65	-	[50]
Martian meteorites	0.16 - 0.5	0.15 - 30	-	-	0.057 - 3.9	0.007- 4.4	[51]
Lunar meteorite	9.96	2.43	0.55	1.35	0.93	-	[52]
Lunar meteorite	12.4	3.9	0.55	1.2	0.93	-	[53]
CAI Inclusions, Allende	≤24,000	≤2290	≤2840	≤20,000	≤19,800	≤19,000	[54]
coarse-grained inclusions	-	-	-	≤16,900	≤12,980	≤12,100	[55]
mean CI Chondrites	947	558	139	686	469	493	[23]
mean CM chondrites	112	649	159	909	605	675	[47]
mean CV chondrites	1443	707	209	1150	758	807	[47]
mean CK chondrites	1560	749	213	1121	772	825	[47]
mean CR chondrites	1208	734	-	926	622	655	[47]
CR2, bulk for chondrules	≤1570	-	-	-	≤930	≤950	[56]
enstatite chondrites	598 - 1080	389 - 872	92 - 179	457- 897	293 - 526	-	[48]
LL6 ordinary chondrite	478	361	81	351	234	-	[57]

Interestingly, PGEs content of terrestrial rocks and meteorites is usually compared to that of CI carbonaceous chondrites, because their chemical composition has some similarities with that of the solar photosphere and this is the reason why they viewed in many cases as standards representing composition of solar nebula during its earliest evolution [23] [58] [59] . CI (Ivuna type) group represents the smallest group of carbonaceous chondrites, which in turn making a small fraction (~5%) of all chondrites [60] . CI chondrites are very poor in content of chondrules (<1 vol%), CAIs (<1 vol%), and contain no metal [61] . They contain serpentine-like minerals, which were produced from magnesium-rich olivines and pyroxenes by aqueous alteration [62] . CI carbonaceous chondrites are chemically the most primitive meteorites whose primary mineralogy and petrography were erased by extensive aqueous alteration at ~323 - 423 K on their parent body, and which subsequently experienced thermal metamorphism [63] .

However, comparison of chemical composition of the Earth's rocks with the different kinds of meteorites led Drake and Righter [64] to conclusion that no primitive material similar to the Earth's mantle is currently represented in any of meteorite collections, and chemical composition of the Earth's mantle is distinct from any extant primitive meteorite type. On the top of that it is well-known fact that the Earth, in contrast to chondrites, passed both melting stage (formation of magma-ocean) and complete differentiation [1] [65] [66] . This means that the chemical composition of CI carbonaceous chondrites should be accepted as composition close to initial composition of solar nebula in regions of formation of terrestrial planets and asteroid belt. It is in complete agreement with analysis of PGEs content in the different kinds of meteorites. Really, high content of PGEs in carbonaceous chondrites could be explained by presence of significant amounts of chondrules and CAIs both of which are rich in PGE (see Table 3). Since CI carbonaceous chondrites are the most primitive and least altered among meteorites, PGEs containing in their chondrules and CAIs escaped re- distribution during their evolution that actually is the reason of their high PGEs content. In contrast, achondrites, representing 8% of meteorites overall [67] , are known to pass both stages of melting and differentiation [68] . Their chemical composition and principal minerals (e.g., [69] [70]) are similar to those of magmatic rocks of Earth's crust and upper mantle, and contain PGE amounts (see Table 2 and Table 3) about two orders of magnitude lower than CI carbonaceous chondrites (e.g., [71]) which have about the same level of the PGE content that the Earth's upper mantle rocks. Interestingly, that achondrites do not have chondrules, which most likely were re-worked during melting, differentiation and alteration processes passed by achondrites. Such facts as extremely high Ir content in IIAB, IIIAB, IVA and IVB irons (from single units up to tens of ppm) [72] , very low content of Ir (up to 42 and 69 ppb), and extremely high content of Pd (up to 5.82 and 7.23 ppm) in IIIB irons and pallasites, respectively [73] , as well as high content of Ir (up to 1 - 2 ppm) in some achondrites—Acapulcoites, Lodranites and Winonaites [74] , points at importance of influence of re-work- ing, differentiation and processes of alteration to the final PGE content. These facts clearly prove that chemical composition of CI carbonaceous chondrites may represent only initial composition of part of solar nebula at about 4.5 - 4.6 Ga, but it cannot characterize present chemical composition of any matter of CI carbonaceous chondrites, because present chemical composition of PGEs in rocks of Earth's crust and upper mantle depends on the last 4.5 - 4.6 Ga of the Earth evolution.

The formation of PGE compounds is under control of their high density and extremely high melting points (see Table 1). This makes formation of PGE compounds depending on conditions preferable by density (e.g., Pt-Ir and Pt-Ir- Os alloys), by melting point (e.g., Pt-Pd, Pd-Rh, Pt-Rh and Ir-Os-Ru alloys, and Pt-Pd sulfides and arsenides) and by both density and melting (e.g., Os-Ir alloys). Some researches show that formation of alloys and sulfidation in chondrites are of primary and secondary origin, respectively (e.g., [75] [76]). The same effects can also be used for analysis of PGE formation of sulfides in the Earth. As it was shown in [1] [66] [78] , at the time of magma-ocean and its initial solidification almost all sulfur was within the early Earth atmosphere forming sulfur layer much denser than water layer. Thus, until re-distribution of sulfur layer at the end of Early Archean—beginning of Middle Archean, sulfur and its oxides have been interacted only with the Earth surface.

An important point in formation of the PGE compounds in solar nebula is the fact that except Pd all of these compounds are refractory elements, and some of them (Ru, Ir and Os compounds) condense within the temperature range (see Table 1) of formation of CAIs (1473 - 2273 K; [26]), and above temperatures of chondrules formation (>1473 K; [26]). Lodders et al. [23] noted that most of PGEs condense as refractory mineral alloys. Such important minerals of CAIs as hibonite and melilite at a total nebular pressure of 10^{-4} MPa first condense at temperatures of 1743 K and 1628 K, respectively [27] and anorthite condensed at temperature of 1387 K under pressure of 10^{-5} MPa [58] . This means that PGEs alloys were formed very early during the accretion process along with the formation of CAIs. It should be stated that Pd is the only element from PGEs, which condenses at temperature slightly less than the condensation temperatures of Fe (1334 K) and Ni (1353 K), and in solar nebula it is concentrated within the Fe-alloys [23] .

3. THE "LATE VENEER" HYPOTHESIS OF PGES ENRICHMENT

One of main problems related by numerous researchers to obvious preservation of PGEs in subsurface is the fact that siderophile ("iron-loving") elements heavier than Fe and Ni, and especially such heavy elements as PGEs (see Table 1) and gold, during the accretion and evolution of magma-ocean must have sank to the Earth core (e.g., [79]) and should not be present in the uppermost layers (slabs) of the Earth. At the same time, it is known that the Earth's upper mantle has an overabundance of highly siderophile elements (e.g., PGEs, gold and rhenium) (e.g., [79] [80] [81]) by ~three orders of magnitude higher than expected from experiments [82] . This fact makes obvious presence of PGEs ore deposits on and near the Earth's surface at present even more controversial and enigmatic. To avoid this controversy a new hypothesis was introduced. According to this hypothesis, PGEs, as well as gold and other siderophile elements, numerous ore deposits of which are well known around the world, were added to the Earth not at the time of accretion, but later, after the Earth's core separation (e.g., [79] [83]) known as the "late veneer" (e.g., [83] [84] [85]) and caused by meteorite bombardment during 4.5 - 3.8 Ga [85] or late heavy bombardment at 3.9 - 3.8 Ga [82] , including accretion of asteroids and comets [84] . The authors [84] estimated amount of extraterrestrial bodies struck the Earth after core formation at (0.7 - 2.7) × 10^{22} kg. Taking into account not only masses, but also velocity of cosmic bodies bombarding the Earth, they must produce significant signatures at the Earth's surface and in the subsurface. Geologically and geophysically such signatures (indicators) may be considered (in combination with other factors) as existence of ring structures (RS) of different diameters. Such RS may be identified visually and by the use of special geophysical methodologies oriented to delineation of RS anomalies against the significant noise background (e.g., [86] [87] [88] [89]).

However, there are some problems of the late veneer coordination with geologic data. The heavy bombardment by debris during planetary formation, which lasted from ~4.5 to 3.8 Ga, and especially during the most intense "Late Heavy Bombardment" at about 3.9 - 3.85 Ga [90] , could have led to additional heating of Earth's crust and mantle, as well as the formation of local magma-oceans. Such bombardment could have caused heating of upper layers of Earth to the depth of up to 1200 km to the melting point during the last stage of the Earth's formation [91] . During the Hadean and beginning of the Archean, the Earth's surface temperature was very high and the surface was represented by either a planet-wide ocean or local oceans of magma for a significant time period. Vacquier [92] suggests that events such as runaway accretion, the formation of the Moon, and the impact that tilted Earth's axis of rotation could possibly have created conditions such that the Earth could have melted and solidified three times. All this means that at the end of the "Late Heavy Bombardment" at about 3.85 - 3.9 Ga, the whole Earth could have been covered with either entire or number of local magma oceans up to 1200 km deep. It is obvious that greater bodies bombarding Earth would release greater amounts of energy and cause more voluminous amounts of melt and deeper local magma-oceans, but those magma-oceans would be formed in places of the most heavy bombardment and late veneer matter would be within those magma-oceans. This makes no difference from initial sinking of extremely dense PGEs within the initial magma-ocean formed as result of the accretion, because delivery of PGEs from the mantle depths of about 1200 km is not much easier than delivery of PGEs from the core. This means that according to this point of view, PGEs must be presented in the upper mantle and crust only in insignificant amounts.

Another problem is the timing of core separation, which is highly speculative. It is obvious that it would not take place before the end of the Earth's accretion. However, Earth's accretion has lasted up to 100 million years [78] and energy released during the accretion was enough to melt the entire Earth. This means that core separation could have taken place only after the accretion finishing. At the same time, at the end of accretion, all the Earth was entirely covered by magma-ocean which could have been existed up to 100 - 200 million years [93] [94] or even longer [95] . This makes any estimation of the core separation timing very speculative one. Moreover, even after the core separation, sinking PGEs within the magma-ocean to depths of about 1000 - 1200 km would also move them out of the Earth's subsurface. On other hand, it is obvious that initial deep magma-ocean could have been existed until at least 4.3 Ga and it means that all PGEs from the cosmic objects involved in the heavy bombardment between ~4.5 Ga and 4.3 Ga would have sunk to the depths of up to 1000 km or deeper. But there is a huge possibility that the heavy bombardment would not let magma-ocean to solidify and, as it was shown above, may have lasted until 3.85 Ga. In such a case the hypothesis of late veneer addition of siderophile elements to the Earth is lack of evidence.

According to late veneer hypothesis, absolute majority of extremely heavy PGEs must have sunk to the core during the accretion. In such a case, concentrations of PGEs in metal core must be high and maximal. In contrary, in some iron meteorites, which are viewed as analogy of metal core matter, concentrations of PGEs are quite low. For example, content of Ir in some iron meteorites could be as low as: 10 ppb in IIb irons [61] ; 10 ppb for IIAB irons [96] ; generally between 10 ppb and 100 ppb in IIIB irons [61] ; 10 ppb in IIIAB irons [96] ; generally between 10 ppb and 100 ppb for IIID irons [61] ; ~100 ppb in IC irons [61] ; ~100 ppb in IIIE irons [61] ; ~100 ppb in IVA irons [96] . All these values are much lower than the mean concentration of Ir in any group of chondrites with minimal content of Ir as 336 ppb in the LL chondrites and maximum content of Ir as 3235 ppb in the CB chondrites [47] . The mean concentration of Ir in most primitive carbonaceous chondrites ranges from 472 ppb for the CI chondrites to 2635 ppb for the CB chondrites [47] ; see also Table 3. Ir content in Acapulcoites is within the range of 536 - 2125 ppb [74] that is also much higher than was observed in many iron meteorites. Interestingly, the highest concentrations of Ir are: ~1000 ppb for the IIB irons, and <~1000 ppb for the IC and IAB irons [61] . These facts clearly contradict to the late veneer hypothesis.

On the top of that, the late veneer hypothesis cannot explain many important facts and features. Among them are: presence of non-chondritic platinum-group element ratios (e.g., Pd/Ir) in the Earth's mantle [79] ; non-chondritic distribution of the highly siderophile elements found in several types of rock (e.g., sulfides) derived from the Earth's mantle [97] ; slightly supra-chondritic Ru/Ir and Pd/Ir ratios in the Primitive Upper Mantle [83] , etc.

Maier et al. [37] [85] show that the lowest PGEs contents are found in >3.5 Ga komatiites from the Barberton (South Africa) and Pilbara (Western Australia) terranes.The authors [37] [85] also show that the most of Early Archaean (3.5 - 3.2 Ga) komatiites from the Barberton greenstone belt and the Pilbara craton are depleted in PGEs relative to late Archaean (~2.9 - 2.7 Ga) and younger (2.5 - 2.0 Ga) komatiites. Maier et al. [85] proposed that the trends reflect sluggish refertilization of PGEs in the Early and Middle Archean lower mantle with the late veneer, and that the Archean mantle had heterogeneous PGEs content. Maier et al. [85] show that by ca 3.0 Ga, the lower mantle appears to have equilibrated with the late veneer. The authors show that PGEs depletion is absent in certain 2.9 Ga komatiites that is linked to equilibration of the lower mantle with the late veneer producing PGE concentration close to modern primitive mantle through gradual homogenization of the meteorite produced PGEs concentration in the entire mantle.

Sankaran [98] analyzed problems related to reflection of trends in the mantle incorporation of meteoritic PGEs by komatiites through some time intervals. He also indicates severe depletion in PGEs in the oldest (>3.5 Ga) komatiites from the Barberton GSB (South Africa) and Pilbara Craton (Western Australia). The authors [98] shows that the Pt and Pd contents were low (~2 - 5 ppb) in the Barberton GSB and Pilbara Craton, moderate (~8 ppb) in the Late Archean komatiites from Canada, Zimbabwe, South Australia and Russia, and high (~8 - 10 ppb) in the Mesozoic komatiites of Gorgona Island (Colombia).

Fiorentini et al. [99] point at a systematic linear increase of the PGE contents in komatiite since the oldest age (>3.5 Ga) in the Late Archaean group (2.7 Ga), regardless of petrogenetic type, as well as linear secular trend of increasing Pt at a given MgO content, and in Pt/Ti from 3.5 Ga to 2.7 Ga across all komatiite types. They state that the Late Archaean komatiites have higher PGE contents as a result of progressive mixing-in of late veneer material over the period between the Late Heavy Bombardment (4.3 - 3.9 Ga) and 2.7 Ga, at which time the mantle had become effectively homogeneous with respect to PGEs.

Maier et al. [37] compared the concentrations of the PGEs and gold in mantle xenoliths from more than 20 kimberlite pipes in the South Africa and Karelian craton (Fennoscandian Shield) and led to the following conclusions: the noble metal content of the primitive mantle cannot be accurately determined using cratonic mantle samples; the data indicate considerable heterogeneity within individual pipes, structural blocks within the individual cratons, and between the cratons; relatively low PGEs contents result from heterogeneous equilibration of the Early- to Mid-Archean mantle with the late veneer; the shallow mantle rocks postdate the onset of crust formation, which is inconsistent with the definition of primitive mantle reservoirs.

Prescher et al. [82] state that the Archean komatiites indicate that the PGEs content of the Earth's mantle increased from about half of their present abundances at 3.5 Ga to their present abundances at 2.9 Ga. The authors [82] proposed that this secular increase in PGE content suggests a progressive mixing of the late veneer material into the Earth's mantle and that the whole mantle was relatively well mixed by 2.9 Ga.

In contrast, Pattou et al. [79] note that the mantle is heterogeneous in its PGEs content on scales of ~100 kilometers.

The idea that PGEs contents in the upper mantle during the Early Archeanwere heterogeneous, but during Late Archean they were homogeneous seems to be ungrounded ones. How is it possible that PGEs contents from Early to Middle Archean were always low, but at Late Archean they everywhere range from moderate to high? Does it mean that significant amounts of PGEs were added to the upper mantle between the Early to Middle Archean and Late Archean? Of course not, because heavy bombardment finished right before the Archean Eon. It means that geological records do not support transformation of PGEs concentration in the upper mantle from heterogeneous during Early-Middle Archean to homogeneous during Late Archean.

Righter et al. [81] analyzed partitioning of palladium (Pd) at high pressure and temperature. They show that the results of experiments at low pressures and temperatures gave very high partition coefficients for the highly siderophile elements (in particular a coefficient for palladium $>10^4$) that was used for hypothesizing that the high mantle concentrations of the highly siderophile elements and also of volatile elements originated from the addition of chondritic material after core formation as "late veneer". However, results of determination of the partition coefficient at higher pressures and temperatures that approximate the early Earth conditions gave much lower partition coefficients for the palladium (~480) consistent with an equilibration scenario [81] . Based on this finding, the authors [81] led to the conclusion that there is no need for a late veneer to explain the terrestrial-mantle palladium content.

Tagle and Claeys [100] reported contents of PGEs in the Popigai (Siberia, Russia) impact melt caused by the Late Eocene comet or asteroid showers initiated from the Oort Cloud. Since comets are believed to be primitive bodies with a composition like that of carbonaceous chondrites [100] , it should be expected that concentrations of PGEs in comets are high. However, in contrast average contents of PGEs are: 1.75 ppb of Pd, 2.58 ppb of Pt, 1.77 ppb of Ru, 0.42 ppb of Rh and 0.96 ppb of Ir [100] . These values are very close to PGEs content in the Earth's crust and some magmatic rocks and even much lower than the PGEs content in several magmatic rocks (see Table 1).

All these facts do not support the late veneer hypothesis of the late PGEs addition to the Earth's upper layers (slabs).

Another case is the Younger Dryas impact hypothesis or Clovis comet hypothesis at about 12,900 BP [e.g., [101] [102]]. Analysis of data from 12 of ~50 sites of the impact shows that Ir concentration in bulk sediments is within the range from <0.5 to 3.8 ppb, and Ir concentration in magnetic fraction is within the range from <1 to <2 ppb in five sites and in four sites it consists of 15, 24, 51 and 117 ppb. Upon retesting aliquots of high-Ir samples, five from nine sites were confirmed, but Ir abundances were below the detection level in four retests. The Ir concentration values in bulk sediments are of same range of magnitude as in most terrestrial rocks (see Table 2), as well as in achondrites, Martian and Lunar meteorites (see Table 3), but they are more than on 2 orders of magnitude lower than the Ir concentration in the CI carbonaceous chondrites. On other hand, as it is clear from Table 1 [101] that there is no any correlation between Ir, Ni, Fe and Ti, which is very unusual for siderophile elements [101] [102] . It is known that Ir peaks were found at major geologic boundary layers with no confirmed impacts, and at least some of those Ir concentrations may have resulted from volcanism [101] . This means that any slight PGEs enrichment could be caused by volcanism rather than be delivered by meteorites. Moreover, Paquay et al. [103] state that they did not found any osmium or iridium anomalies in the Younger Dryas sediments. All these facts also not support the late veneer hypothesis, because no really high concentrations of PGEs were detected, how it should be expected from the carbonaceous chondrites and comets according the late veneer hypothesis.

In both cases of the Popigai and Younger Dryas impact structures we should expect that delivered PGEs would stay on place, because there was no magma-ocean for them to sink to deep levels or some other processes, that would make them disappear. This means that absence of any significant PGEs contents in such cases indicates against the late veneer hypothesis, because it is difficult to expect that all carbonaceous chondrites and comets involved to the heavy bombardment before ~3.8 Ga (which did not leave a trace) were extremely rich in PGEs, but later impacts of carbonaceous chondrites and comets did leave any traces of significant PGEs contents delivered to the Earth surface.

All above-mentioned means that there must have been other mechanisms of preservation and concentration of PGEs and PGMs in the Earth's upper slabs.

4. INITIAL CONCENTRATION OF PGES AND PGMS DURING THE EARLY EARTH EVOLUTION

It is known that all matter on the Earth originated from the solar nebula by infalling of the nebular material onto the evolving planet during its accretion. This means that in order to accumulate significant amounts of PGEs and PGMs near the Earth's surface, there must have been certain mechanisms of their concentration in near-surface layers, their trapping, and preservation.

Research shows that PGEs are mostly presented in rocks as inclusions of native metals, alloys, sulfides, arsenides, telurides, etc. From the early stages of accretion on the Earth, a magma-ocean with a depth of up to 1000 km has been formed and existed for millions or even hundreds of millions of years [e.g., [77]]. The magma-ocean was stratified by density and iron content, with a felsic-intermediate-mafic-ultramafic sequence of layers from the top to bottom [77] . If the magma-ocean has existed for a long-time period, there is also the possibility of magma fractionation by the main rock-forming minerals, along with the formation of the "forsterite layer" [1] [78] at a certain depth, plagioclase layer at the top, etc. A plagioclase layer is known to have been formed within the Lunar magma-ocean, representing the oldest plagioclase or anorthositic Lunar crust [104] . The presence of enormous amounts of anorthosites in the Earth also indicates the possibility of such a layer formation within the Earth's magma- ocean. In addition to that, the presence of massive (1000-s to 10,000-s · km^3) anorthosites [105] of the Mid-Proterozoic suggests the earliest Earth crust had an anorthositic composition, since the formation of magma chambers capable of otherwise collecting such enormous quantities of anorthosites within the crust would be extremely unlikely and these volumes of anorthosites could have been formed within the magma-ocean only. It is also conceivable that both the differentiation of magmatic rocks and their fractionation by main rock-forming minerals evolved in parallel, depending on the thermodynamic conditions and period of the magma-ocean's existence in different regions of the Earth (e.g., [78]).

Formation of the lithosphere has started with formation of the "forsterite layer" [1] [78] [106] at depths of ~70 - 100 km. Even though forsterite has the highest melting point among silicates (2163 K at 0.1 MPa [107] and ~2307 K at 3 GPa [108]), its density is very low in both molten (2970 kg/m^3 at 2163 K [109]) and solid (3222 kg/m^3) states and it would prevent forsterite from sinking within any melt with even small content of iron (e.g., Table 4 and Table 5).

This means that from the time of formation of the "forsterite layer", heavy compounds including refractory mineral alloys had no chance to sink below the layer (slab). They were trapped above the layer and their enrichment was proportional to time from formation of the layer to at least starting of solidification of the magma-ocean upper section. On other hand, geothermal gradient within the molten planet is the adiabatic geothermal gradient that is extremely small (e.g., [78] [120] [121]). The estimated value of the adiabatic geothermal gradient is as low as 0.5 K/km [120] and 0.3 K/km [121] . This means that to keep temperature as high as ~2307 K at depths corresponding to pressure ~3 GPa the surface temperature of Earth at the time of accretion and formation of magma-ocean must have been in excess of 2260 K which is much higher than temperature of the magma-ocean start solidification at surface temperature ~1300 - 1500 K (e.g., [1] [77]). Such a high temperature of the Earth's surface was unbearable to keep for a long time. This means that solidification of forsterite with formation of the "forsterite layer" could have started at very early stage of the Earth formation, probably during the time of its accretion. It was shown earlier [77] [78] that formation of the "forsterite layer" could have started from solid forsterite, if temperature within the upper layers of forming Earth was not high enough to melt forsterite [77] [78] . All this indicates that formation of solid layer at relatively shallow depth within the forming Earth

Table 4. Density of molten forsterite, fayalite and olivine Fo_{93} (modified after [78]).

Mineral	Temperature, K	Density, kg/m^3	Reference
Forsterite	2163	2970	[109]
Fayalite	1573	3757	[110]
	1673	3725	[110]
	3000	3529	[111]*
	2163	3810	[112]
	1773	3747	[113]
	1573	3750	[73]
	1673	3766	[73]
	1773	3688	[73]
	1300	~3875	[114]
	1500	~3861	[114]
	1600	3850	[114]
	1973	3593	[115]
	2023	3586	[115]
	2073	3587	[115]
	2123	3560	[115]
Olivine Fo_{93}	-	2865	[116]

*Calculated using data of [111].

Table 5. Density of some ultramafic rocks and chromites.

Rock	Density, kg/m^3	Reference
Cobalt chromite, $CoCr_2O_4$	5140	[22]
Copper chromite, $CuCr_2O_4$	5400	[22]
Iron chromite, $FeCr_2O_4$	5000	[22]
Zinc chromite, $ZnCr_2O_4$	5290	[22]
Peridotite	3230	[117]
Dunite	3280	[117]
Dunite	3160	[118]
Dunite xenoliths from basalts	3244 - 3326	[119]

could have started during the accretion or at the earliest stages of the magma-ocean evolution. This means that such solid layer could have formed a barrier for all heavy species, including siderophile elements and, what important here, PGEs and PGMs, which would prevent them from sinking to the depths of core or lower mantle. This fact could be a cause of preservation of PGEs and other siderophile elements at shallow depths (above 70 - 100 km) within the forming Earth. Among next to the "forsterite layer" in the upper mantle part of the forming early Earth's lithosphere according melting points at 0.1 MPa solidification process pass: chromites of different composition with the melting points within a range of temperatures from 1908 K for incongruent melting [122] to 2473 - 2673 K [123] ; dunite with the melting point of 2073 K and olivine composition of $Fo_{90}Fa_{10}$ [124] ; harzburgites with the melting point of 1633 - 1773 K at pressure of 1.0 - 1.5 GPa [125] ; and lherzolites with the melting point of 1543 - 1693 K at pressure of 1.0 - 1.5 GPa [125] . These data show that some chromites have melting point the same or close to that of dunites, even though most of chromites have much higher melting points than dunites, but melting points of chromites much higher than that of harzburgites and lherzolites with a significant gap. This testifies that some of chromites were solidifying with dunites at the same time or some time earlier. It is clear that significant amount of chromites (Table 6) has melting points greater than that of forsterite and it indicates that those chromites, taking into account their high density (5000 - 5400 kg/m^3), were mostly plunged to the great depths within the upper mantle before the "forsterite layer" formation.

Table 6. Melting point of some ultramafic rocks and chromites.

Rock	Melting point, K	Reference
dunite	2023	[126]
dunite	2073	[124]
harzburgite (at 1 GPa)	1633 - 1743	[125]
harzburgite (at 1.5 GPa)	1693 - 1773	[125]
harzburgite (at >5 GPa)	1723 - 1873	[127]
lherzolite (at 1 GPa)	1543 - 1633	[125]
lherzolite (at 1.5 GPa)	1633 - 1693	[125]
peridotite (at 4.5 - 5.0 GPa)	1723	[127]
chromites (different)	2123 - 2473	[128]
chromite (monocalcium)	2443	[129]
chromites (different)	2473 - 2673	[123]
chromites (Lanthanum)	≥2273	[130]
chromite (incongruent melting)	1908	[122]
(Fe, Mg) Cr_2O_4 (spinel)	~1873	[131]
$MgCr_2O_4$	2623	[132]
chromite	2453	[133]
chromites (Slag liquidus)	1873 - 2273	[134]
$MgCr_2O_4$ (sintering)	~2023	[132]
$NiCr_2O_4$ (sintering)	~2096	[135]
$MnCr_2O_4$ (sintering)	~1828	[136]

However, some chromites, which did not sink to the great depths by some reason or appeared within the magma-ocean along with other matter of continuing infalling of cosmic objects, would be blocked from sinking by the "forsterite layer". It is in agreement with the fact of presence of significant amounts of PGEs associated with chromites/chromitites usually enveloped by dunites found in regions of obduction, formation of ophiolites and serpentinites. For example, PGEs rich chromitite occurs as centimetric veins, pods and schlieren enveloped by dunite at the Middle Arm Brook, Central Advocate ophiolite complex (Baie Verte Peninsula, Newfoundland, Canada) [137] . Analysis of olivine inclusions in the chromites/chromitites found in regions of ophiolites formation demonstrates that those inclusions contain olivines with forsterite content up to Fo_{98-99} in ophiolites: Kempirsai, Kazakhstan [138] ; Shangla, Pakistan [139] ; Jungar, China [140] ; Egypt [141] ; and generally with forsterite content > Fo_{92}. This testifies that at the time of formation of chromites in different regions, olivine inclusions within chromites/chromitites were already solidified and it places constraint on temperature of solidification of those chromites.

Estimations of temperature solidification of chromites found in mentioned ophiolites gave temperatures of <2108 - 2156 K. At the same time, the fact of enveloping chromites by dunites shows that such association was formed either when dunite was at the early stages of its solidification, or dunitic magmas rising from below the "forsterite layer" were enveloping by solid chromites above the layer. This is in agreement with a conclusion made by Pushkarev et al. [142] that chromitites in the Ural-Alaskan type mafic-ultramafic complexes start to form when dunite was almost solid at the end of plastic deformation process. Such a model of formation of chromite-dunite complexes is supported by the fact that content of forsterite in dunites is always lower than in the corresponding chromites. For example, within the Alaskan-type Tulameen complex (British Columbia) olivine inclusions in chromites contain olivines with Fo_{92}-Fo_{95} and olivines in dunites have Fo_{88}-Fo_{91} [143] ; in the Logar Ophiolite Complex (Afghanistan) olivines in chromites and dunites contain Fo_{94}-Fo_{96} and < Fo_{92}-Fo_{94}, respectively [144] ; in the Mawat ophiolite (Iraq) olivines in chromites and dunites contain Fo_{91}-Fo_{95} and Fo_{89}-Fo_{92}, respectively [145] ; in Jungar ophiolite (China) olivines in chromites and dunites contain $Fo_{96.1}$-$Fo_{98.3}$ and Fo_{90}-Fo_{94}, respectively [140] . Interestingly, olivine composition in chromitites and dunites in PGE deposits of the Singhbhum Craton (Indian Shield) contain Fo_{96}-Fo_{98} and Fo_{92}-Fo_{95}, respectively. The temperatures of solidification of dunites in the referencing regions range between 2081 K (for Fo_{88}) and 2129 K (for Fo_{95}). All these temperatures are obtained for the pressure of 0.1 MPa and at different depths they were higher depending on the real pressure, but it is clear that a sequence of solidification was forsterite-chromite-dunite with small differences in temperature that makes timing of their formation pretty close on geologic time scale. Obviously, solid "forsterite layer" has trapped chromites with PGEs above itself, and then solid chromites with PGEs were enveloped by solidifying dunite. However, it is known that in numerous cases the PGE-rich chromites are enveloped by dunites that indicates chromites were already solid while dunites still be partly molten at the time of formation of such complexes. If dunites were molten, completely denser PGE-rich chromites would sink through them. At the same time, it means that forsterite and olivine with Fo content ≥94% - 95% were already solid and solid "forsterite layer" was already in the place preventing chromite and PGEs from sinking within magma-ocean. However, presence in chromites olivines with Fo content ≥94% - 95% means that solidification of those chromites took place when inclusions of Fo-high olivines were already solid. It places some constraints on thermal conditions of formation of PGE-rich chromites, as well as constraints on formation of chromite-containing dunite complexes. Such a model of formation of chromite-dunite complexes is supported by the fact that content of forsterite in dunites always lower than it is in the corresponding chromites.

Solidification of the crustal part of lithosphere most likely started with the solidification of quartz (christobalite), leucite, tridymite, anorthite and β-quartz having melting points 1996 K, 1958 K, 1943 K, 1830 K and 1823 K, respectively. Since both high-temperature quartz and leucite have lowest densities among the main rock-forming minerals, within magma-ocean they would most likely pre- sent within the topmost layers of the magma-ocean and they were first to solidify within the forming crust. It should be stated that any new solid layer forming during the early Earth evolution would form a barrier for heavy species to sink through the molten upper layers of forming Earth to its deep levels (e.g., crust, lower mantle). Any such layer would be formed similarly with the "forsterite layer" under strict control of the density of layer composing rocks at corresponding depths.

For characteristics of pristine lithosphere composition, a simplified model of the composition was compiled (Table 7).

All the afore-mentioned shows that there were mechanisms available for trapping PGEs and other siderophile elements at different levels of the forming Earth above 70 - 100 km depths from the early Earth evolution stages as old as the accretion process and during the evolution and solidification of magma-ocean.

Since the magma-ocean was represented by viscous liquid magmas, entities infalling from the accretion disc were also separated by density within it. Such highly dense metals as PGEs would certainly sink in the magma. Viscosity of magmas in the magma-ocean was strongly depend on the magma composition and its temperature, and viscosity controlled processes of differentiation of matter within the magma-ocean and terminal velocity of sinking species continuing to infall into the magma-ocean from the cosmic space. The importance of the thermal characteristics for rate of differentiation of matter is related to the fact that increase of temperature leads to significant decrease of viscosity of molten rocks [e.g., [146]] that would speed up the differentiation at high temperature conditions and slow down the differentiation on cooling or at low temperature conditions. It would work for sinking both liquids and solids within less dense liquid. For sinking solids, terminal velocity of an object sinking into viscous liquid depends on its size, density, viscosity of liquid, and it can be described by the Stokes equation (e.g., [147]):

$$v = \frac{2r^2 g(\sigma - \sigma_M)}{9\eta} \quad , (1)$$

Table 7. Simplified model of pristine lithosphere formed during solidification of magma-ocean.

Layer number	Layer density, kg/m³	Melting point (in K) at 0.1 MPa,	Layer composition
1	<2500	1823 - 1986	SiO_2-rich layer
2	2500 - 2600	T_m > 1391 - 1423	SiO_2 + K-Fsp + Ab-rich ± Olg
4	2620 - 2660	1400 < T_m < 1555	Olg + Qtz ± And-rich
6	2660 - 2690	1460 < T_m < 1645	And ± Lab-rich
9	2690 - 2720	1555 < T_m < 1763	Lab ± Byt-rich
11	2720 - 2760	1645 < T_m < 1830	Byt ± An-rich
13	2760 - 2900	1830	An-rich (Anorthite layer)
15	3160 - 3280+	2023 - 2073+	Dun ± Chr
16	3222	2163	Fo-rich (forsterite layer)

SiO_2, amorphous silica; K-Fsp, potassium feldspar; Ab, albite; Olg, oligoclase; Qtz, quartz; and, andesine; Lab, labradorite; Byt, bytownite; An, anorthite; Dun, dunite; Chr, chromite; Fo, forsterite.

where v is the terminal velocity of a sphere with radius r, σ is the density of a sphere, g is the gravity acceleration, η is the viscosity of the medium, and σ_M is the density of the medium.

It is clear from Equation (1) that terminal velocity of the sinking species strongly relates to the radius (proportional to r^2) of species and it would be high for the species with the large radius and it would be low for the species with the small radius. The terminal velocity is inversely proportional to viscosity (η) and proportional to density difference ($\sigma - \sigma_M$) between the sinking species (σ) and the melt (σ_M). This means that the terminal velocity for PGEs by density would be highest for osmium (Os) and lowest for palladium (Pd).

Performed analysis shows that viscosity of rocks (depending on temperature and type of melt) changes from 0.01 - 10 Pa・s for komatiites to ~10^{11} Pa・s for anhydrous granite and dacite. Forsterite has viscosity valuewhich is very close to that of komatiites.

It is well known that addition of water to rocks and minerals significantly reduces their viscosity [e.g., [148] [149]], but before formation of the water-ocean (at about 3.42 - 3.26 Ga) the rocks composingthe Earth crust and mantle were essentially dry. Results of research on viscosity of melts testify that viscosity is always significantly increasing with increase of silica content along the join SiO_2- $NaAlSiO_4$ [150] and diopside-albite binary system [151] . This means that the silica-rich felsic rocks would have highest viscosity among the typical magmatic rocks.

The lowest densities in both solid and molten caused forming high temperature silica polymorphs (following in increasing order of density by SiO_2 + Al_2O_3 binary compounds (Al_2O_3 content less than 22.4 wt.%)): K-feldspars, albite, felsic magmatic rocks, intermediate magmatic rocks, basic magmatic rocks and ultramafic magmatic rocks. This means that topmost layers within the magma-ocean were represented by quartz and/or quartz-rich compounds, K-feldspars and/or K-feldspar-rich compounds and albite and/or albite-rich compounds, and so on. At the same time, increasingthe period of magma-ocean evolution ledto closeness of those layers' content by pure mineral composition due to fractionation (e.g., Table 7). It is also obvious that on contacts of the layers' such binary system as quartz-K-feldspars, K-feldspars-albite, etc., could have been developing.

Comparison of above data with data shown in Table 7 displays that within the pristine lithosphere formed during solidification of magma-ocean silica, K- feldspar, and different plagioclase (different contents of albite (Ab) and anorthite (An)) layers would form main crustal layers positioned according to their density and solidified permitting their melting points. Depending on the plagioclase composition, the plagioclase layer was characterized by a viscosity within the range of 4.5 Pa s for anorthite (An) to 10^7 Pa s for albite (Ab) at their liquidus temperatures [152] [153] ; its melting point would have been within the range of 1391 K for Ab to 1830 K for An [154] ; and density within the range of 2620 kg/m^3 for Ab to 2760 kg/m^3 for An [153] . For instance, for the Earth anorthosites with a usual composition of 40% - 80% of An [155] , the viscosity is in the range of 5.5×10^3 Pas (for An_{40}) to 1.7×10^2 Pas (for An_{80}) [153] at liquidus temperatures, and the melting point is in the range [154] of 1567 K (for An_{40}) to 1742 K (for An_{80}). For Archean anorthosites with An content of ~90% [155] , the melting point of the plagioclase is 1786 K. Viscosity of plagioclase melts at liquidus temperatures calculated using data of [152] [153] is about: 4.5 - 8.5 Pa s for anorthite (An_{90-100}); 8.5 - 35 Pa s for bytownite (An_{70-90}); 35 - 200 Pa s for labradorite (An_{50-70}); 2×10^2 - 2×10^3 Pa • s for andesine (An_{30-50}); 2×10^3 - 10^5 Pa • s for oligoclase (An_{10-30}); 10^5 - 10^7 Pa s for albite (An_{0-10}). Data of Shaw [156] indicate that even at high temperatures viscosity of albite is high: ~$10^{4.2}$ Pa • s at ~1667 K, ~$10^{4.25}$ Pa • s at ~1587 K, and ~10^5 Pa • s at ~1537 K. At the same time, viscosity of K-feldspar is much higher than viscosity of albite at the same temperatures [156] . This means that albite and K-feldspar rich rocks are suitable for trapping of heavy microscopic entities settling within the melt.

One of the most important parameters, determining viscosity, is the rate of cooling of the magma-ocean layers. The liquidus temperatures of plagioclase are very high for some of its compositions, and the difference between liquidus and solidus temperatures for its intermediate compositions like labradorite and andesine can reach ~200 K [154] . This means that if the temperature of the cooling layer dropped below its liquidus, the plagioclase could still be in viscous liquid condition, though its viscosity would increase dramatically. On top of that, Öhman and Kring [157] show that depending on the content of clasts in the melt, the viscosity of norite within the temperature range of 1500 - 1973 K, can increase by an order of up to 2 - 2.5 for 30% and 3.5 - 4 for 50% clasts content. For example, viscosity of norites with 50% clasts is: $\sim 10^{4.9}$ - $10^{7.5}$ Pa · s at ~1500 K, $10^{3.1}$ - 10^{5} Pa · s at ~1600 K, and $\sim 10^{1.5}$ - 10^{2} Pa · s at 1973 K [157] . This means that the infalling of objects composed of minerals of refractory elements and other compounds with high melting points would create conditions of significantly increased viscosity within the magma-ocean, depending on the clasts content of the original molten layer.

Hiemstra [158] employed the Stokes equation (Equation (1)) to explain the settling of PGMs during the formation of PGE deposits in the Merensky and UG-2 reefs of the Bushveld Complex. Pilchin [159] utilized this equation to clarify the exhumation of peridotites and metamorphic blocks during the serpentinization process. Matveev and Ballhaus [160] also show that the large-scale concentration of chromite in ore bodies is driven by density contrasts between the chromite-laden fluid pools and silicate melt following from the Stokes law. Interestingly that Hiemstra [158] simply dismissed the settling of small grains of PGMs on their own, and stated that larger chromite grains carried them down. However, in reality small grains would have a much slower terminal velocity and it would take a much longer time for them to settle, which represent the perfect conditions to trap them in highly viscous melts.

Even though PGEs and PGMs mineralization in many cases are associated with chromites/chromitites, there is evidence that the PGMs are rarely enclosed by chromite ([161] [162] [163]). Page et al. [163] investigated presence of IPGEs (Ir-group) in natural chromite and show that all these elements are below the detection level, except for chromites from komatiites which contain in average 268 ± 26 ppb of Ru. The authors [163] have shown that all of the plutonic chromites and almost all chromites from the lavas do not contain any IPGE in solid solution, at least at the resolution of the detection limits; on average IPGE are not controlled by crustal plutonic chromite, mantle chromite, chromite in boninites, or chromite in MORB. It was most likely caused by the fact that all chromites and PGEs were already solid at the time of magma formation or that some chromites could be formed from melt at temperatures about or below melting point of komatiites. Obviously, this is the reason why chromite can contain solid inclusions of IPGEs, but absence of IPGEs in regular chromites point at the fact that temperature within the magma ocean was not high enough for melting all chromites from the beginning or it was high only for a short period of time. On other hand, high content of PGEs in chromitites and low content of IPGEs in chromites mean that main concentration of PGEs came from plagioclase part of chromitites, from remnants of magma-ocean having high content of anorthite.

Analysis of data on the sizes of grains and inclusions of PGEs and PGMs from a vast amount of publications shows that their sizes are generally <10 μm for PGE alloys and <15 μm for PGMs. There are also numerous reports about the size of PGE-alloys and Au inclusions being less than 0.1 - 1 μm (e.g., [164] [165] [166]), and even "invisible" gold found in pyritic Cu-Zn ore from Uzelginsk (southern Urals, Russia) [43] . Microinclusions of noble metals with size <0.1 μm were found within the Merensky Reef (Bushveld Complex) [164] . Submicroscopic gold in coarse-grained pyrite of the main mineralization stage (mean concentrations up to 1.01 ppm in the Pindos ores and 0.22 ppm in the Mirdita ores) was discovered in ophiolites of the Balkan Peninsula [165] .

The terminal velocities of settling PGEs for grain sizes with a radius of 1 μm, 10 μm and 100 μm within the different kinds of magmatic rocks are presented in Table 8. Analysis of these data indicates that the lowest and highest values for terminal velocity relate to Pd and Os, respectively.

It is clear from Table 8 that the terminal velocity of even micro-grains of PGMs sinking in mafic and ultramafic magmas is very high and would cause them to quickly descend to deep levels within the magma-ocean during the first few millions of years of its existence, and only submicron-sized grains had a chance of staying in the magma at the time it solidified within the magma-ocean. In contrast, grains measuring ≤10 μm sink much slower in andesite magma and would be mostly preserved within such a layer. The terminal velocity in felsic magmas is very slow and grains with sizes ≤10 μm would almost certainly be preserved within the magma; even grains measuring up to 100 μm were in some cases able to remain. Depending on the viscosity and span of the existence of the magma-ocean, grains ≥100 μm in size could have been preserved only in certain cases within felsic magmas. This is in agreement with the fact that within the Ural-Alaskan type of PGE deposits, size of PGE grains and microinclusions (e.g., 0.5 - 2.0 mm in Yukon; [171]), is much greater than in ALMI PGE deposits, because huge PGE grains sank within the magma-ocean to greater depths, and later

Table 8. Terminal velocity of PGE inclusions settling in different magma types.

Magma type	Temperature of magma, K	Viscosity of melt, Pa s	Terminal velocity (m/year) of inclusions with radius of		
			1 μm	10 μm	100 μm
Basalt	1473	10^1 - 10^2	$(0.64 - 13.7) \times 10^{-3}$	$(0.64 - 13.7) \times 10^{-1}$	6.4 - 137
Andesite	1473	10^3 - 10^4	$(0.65 - 13.8) \times 10^{-5}$	$(0.65 - 13.8) \times 10^{-3}$	$(0.65 - 13.8) \times 10^{-1}$
Rhyolite	1473	10^5 - 10^6	$(0.66 - 13.8) \times 10^{-7}$	$(0.66 - 13.8) \times 10^{-5}$	$(0.66 - 13.8) \times 10^{-3}$
Komatiite	1873 - 2073	0.1 - 10	$(0.64 - 13.6) \times 10^{-1}$	6.4 - 136	$(0.64 - 13.6) \times 10^{3}$
Forsterite + enstatite	2043 - 2523	0.02 - 0.13	0.031 - 1.04	$(0.031 - 1.04) \times 10^{2}$	$(0.031 - 1.04) \times 10^{4}$
Forsterite	2600	0.01 - 0.1	0.062 - 1.35	$(0.062 - 1.35) \times 10^{2}$	$(0.062 - 1.35) \times 10^{4}$
Molten peridotite	near liquidus	0.1	0.62 - 1.35	$(0.64 - 1.35) \times 10^{2}$	$(0.64 - 1.35) \times 10^{4}$
Granite, anhydrous	1073	$\sim 10^{11}$	$(0.66 - 1.38) \times 10^{-12}$	$(0.66 - 1.38) \times 10^{-10}$	$(0.66 - 1.38) \times 10^{-8}$
Granite with P_{H2O} = 500 MPa	1073	2.6×10^5	$(1.72 - 3.59) \times 10^{-6}$	$(1.72 - 3.59) \times 10^{-4}$	$(1.72 - 3.59) \times 10^{-2}$
Granite with P_{H2O} = 2000 MPa	1473	0.5×10^2	$(3.3 - 6.9) \times 10^{-4}$	$(3.3 - 6.9) \times 10^{-2}$	$(3.3 - 6.9) \times 10$
Basalt lava	in eruption	10^2 - 10^4	$(0.64 - 137) \times 10^{-5}$	$(0.64 - 137) \times 10^{-3}$	$(0.64 - 137) \times 10^{-1}$
Andesite lava	in eruption	10^4 - 10^6	$(0.65 - 138) \times 10^{-7}$	$(0.65 - 138) \times 10^{-5}$	$(0.65 - 138) \times 10^{-3}$
Dacite lava	in eruption	$\sim 10^{11}$	$(0.65 - 1.38) \times 10^{-11}$	$(0.65 - 1.38) \times 10^{-9}$	$(0.65 - 1.38) \times 10^{-7}$

Viscosities of magmatic rocks are after: [117] [167] [168] [169] [170].

these grains were tectonically delivered to surface by obduction and other tectonic processes. By other words, small grains of PGEs at the time of the magma-ocean were mostly concentrated within its top felsic layer, with some PGEs enrichment possibly forming within the intermediate magma layer. Meanwhile, grains of PGMs reaching mafic and ultramafic magma layers would sink through them relatively quickly. This is the main reason of very low PGE contents in mafic and ultramafic magmas (see Table 2).

It must also be taken into account that the settling of PGM grains within a layer (slab) can take place only before the solidification of that layer and eventually of the entire magma-ocean at all levels trapped the settling grains in some place within the layers. Given that various rocks and minerals have different melting points we can expect that some of them would begin solidifying before others. This would lead to the formation of solid layers within the magma-ocean at different depths prior to its complete solidification (e.g., the "forsterite layer" formation; [1] [77]). The high melting point of An-rich plagioclase also points to the possibility of the formation of the Anorthite and/or Bytownite layer within the magma-ocean. If such a layer had to form, it would prevent the further sinking of PGM grains, and they would be collected on the top of this solid layer (slab) within the forming crust. This brings forward the possibility of the existence of massive-type ore deposits formed above such a layer within the crust. Otherwise, with gradual solidification of the molten layer, the distribution of settling grains would more closely resemble to disseminated mineralization. At the same time, the melting point of plagioclase is much greater than that of felsic magmatic rocks, which would continue to be molten much longer than the plagioclase layer. Since the molten felsic rocks have lower density comparing with the solid plagioclase, their position would therefore also be above the plagioclase layer, continuing to accumulate the infalling grains.

Let us analyze the above facts in comparing with the available geological data. Examination of numerous data about the PGE ore deposits indicate that PGEs deposits and mineralization were found in anorthosite-bearing layered mafic intrusions (ALMIs) such as the Bushveld Complex (South Africa), the Stillwater complex (Montana, USA), the Duluth Complex (Minnesota, USA), and many others (e.g., [42]) known to contain the largest PGE deposits in the Earth.

The Bushveld complex includes two largest PGEs containing horizons (UG2 chromitites and the Merensky Reef) [42] [172] . Economic and high contents of PGEs in the anorthosite-bearing complexes were found in layers of chromitites, gabbroid cumulates (gabbros, leucogabbros, gabbronorites, norites, leuconorites, etc.) and anorthosites. The most famous PGE bearing chromitites are those of UG2 chromitite of the Early Proterozoic Bushveld Complex [42] [172] . Chromitites (from less than 1 cm to more than 2 m thick) in the Eastern Bushveld Complex are associated with anorthosites, norites, gabbros, and mostly with feldspatic bronzitites in the Upper and Lower Critical Zones [173] . PGE mineralization in the Aurora of Bushveld complex is associated with gabbros and leucogabbros [174] .

The principal PGEs deposit of the Archean Stillwater complex, the J-M reef, is located within the Olivine-Bearing zone I (OB I) and composed of troctolite, dunite, gabbronorite and anorthosite [175] . The Picket Pin Pt/Pd deposit is a zone of disseminated PGEs-bearing sulfide which occurs in the upper 150 m of Anorthosite subzone II (AN II) of the Stillwater Complex [176] .

PGEs content of troctolitic series in the Mid-Proterozoic Duluth Complex contains up to 14 ppm of Pt + Pd [177] .

In the Archean Lac-des-Iles complex of Ontario (Canada) the highest precious metals mineralization was revealed in the Roby zone of the western gabbro consisting of interlayered cumulates of gabbroic (70%), noritic (20%), pyroxenitic (10%) and minor anorthositic rocks [178] .

Enrichment of PGEs in the Late Archean Legris Lake Complex, Wabigoon Subprovince (Canada) occurs in the leucogabbro breccia and mottled anorthosite [179] .

PGE deposits were discovered in Late Archean-Early Proterozoic in Federovo Pansky ALMI (Kola Peninsula, Russia): Main Gabbronorite Zone consisting of contrasting alteration of gabbronorite, norite, pyroxenite and interlayers of leucocratic gabbro and anorthosite, and Upper Layered Horizon consisting of olivine-bearing troctolite, norite, gabbronorite and anorthosite [180] .

It should be stated that even though chromitites are not gabbroic cumulates, they have strong relationships with the anorthosite-bearing gabbroic cumulates and anorthosites. Cameron [173] distinguished three types of chromitite-silicate rock series based on variations in modal proportions and presented in the Critical Zone of the Bushveld Complex: chromitite-bronzitite, chromitite-dunite, and chromitite-anorthosite.

There is evidence about direct presence of PGEs mineralization in anorthosites. PGEs-bearing anorthosites are known in: the Archean Stillwater Complex [176] ; the Early Proterozoic Bushveld Complex [172] ; the Merensky Reef at Impala Platinum, Bushveld Complex [161] ; the Middle Proterozoic Duluth Complex [177] ; the Archean Lac-des-Iles complex of northwestern Ontario (Canada) [178] ; the Late Proterozoic Ioko-Dovyren layered massif (Northern Transbaikalia, Russia) [181] ; the Early Proterozoic Dzhugdzhur massif-type anorthosite (Aldan Shield, Russia) [182] ; the Late Archean Legris Lake Complex (Wabigoon Subprovince, Canada) [179] ; the Early Proterozoic Nadezhda sill-like gabbronorite body (Lukkulaisvaara layered intrusion, Fennoscandian Shield, Russia) [183] ; the Sittampundi anorthosite Complex (Tamil Nadu, India) [184] ; the Geransky massif (Far East, Russia) [185] ; the Mid-Proterozoic Steel Mountain anorthosite (Canadian Appalachians, Newfoundland) [186] ; Indian Head anorthosite (Newfoundland) [187] (Pilchin, 2014). Orsoev et al. [181] state that PGE mineralization occurs mainly in anorthosites in the Ioko-Dovyren dunite-troc- tolite- gabbro layered massif, Northern Transbaikalia (Russia).

It is also clear that anorthosites and especially An-rich anorthosites can be perfect traps and keepers for PGMs, because there are not so many magmas that able to re-melt them and release PGMs. Plagioclase of intermediate composition An_{40-60} has melting points in the range of ~1700 - 1750 K. However, plagioclase of intermediate composition has difference between liquidus and solidus temperatures of ~200 K [154] . This means that none regular basaltic magma will be capable to re-melting such a plagioclase. Even pure albite, having melting point of 1391 K, can be remelted only by some, mostly dry, basaltic magmas. It is unrealistic to suppose that any basaltic magma of typical composition would be capable of melting anorthite which has melting point of 1830 K. Only komatiites and some picrites could be able to melt anorthosites to release PGMs. At the same time, elemental sulfur could get into anorthosites as fluid through porous and fractures and react with metals forming sulfides, which would be also trapped within anorthosites.

In a continuous borehole core (BV-1) drilled through the uppermost 2.8 km of the intrusion in the northern lobe of the Bushveld complex, typical rocks range upwards from troctolite, through gabbronorite and ferrogabbronorite to ferrodiorite, with extreme examples of anorthosite, magnetitite and feldspathic pyroxenite [188] . Most of the rocks contain at least 50% plagioclase, and plagioclase contents average to 60% for all samples analyzed. Within the cross-sec- tion of the borehole forty-five anorthosite layers, ranging from 1 to 23 m thick, and twenty layers of magnetite have been identified. The mentioned authors find no evidence supporting flotation or prolonged suspension of plagioclase. The entire length of core discussed shows a change in upward plagioclase composition from An_{77} to An_{36}.

PGMs were found in the Duluth Complex as inclusions in plagioclase and sulfide halos within plagioclase [189] , and the highest concentration of PGEs in anorthositic series rocks found to date is 163 ppb Pt. Marma [190] also reported about PGEs occurring in the poikilitic anorthite-rich plagioclase (An_{75}-An_{95}).

In numerous cases PGE mineralization and ore deposits are found in gabbroic and noritic rocks which are gabbroic cumulates (plagioclase rich rocks). Moreover, chromitite layers of up to 20 m thick are known to concentrate in the Anorthosite unit [191] and at the top of the Upper Leucogabbro unit of the Middle Archean (~2.97 Ga) Fiskenæsset Complex, SW Greenland [192] . Anorthosite bearing plutonic suite of Middle Archean (~3.1 Ga) containing chromitite bodies and chromite deposits is known in the Nuggihalli greenstone belt, Dharwar craton (India) [193] . These facts clearly point at trapping and concentration of chromites within anorthosites as well.

The various properties of the PGE- and gold-PGE-bearing formations within the conglomerate-quartzite formation of the Kursk Magnetic Anomaly of the Voronezh Crystalline Massif (Russia) are presented in [16] . This points at the possibility of trapping some PGEs in the silica-rich layers.

An excellent example of the settling of PGMs and minerals of some transitional elements within the komatiitic basalt lava lake in the Vetreny Belt (Baltic Shield) is described in [194] , where authors clearly illustrated the distribution of Os, Ir, Ru, Cr and Ni with a depth. Data presented in Table 3 of [194] display an evident pattern of enrichment in Cr, Ni, Os, Ir and Ru with a depth, clearly supporting concentration through the settling of heavy materials within the komatiitic magma. Certain elevated values of these elements somewhat close to their average values throughout the entire topmost layer of the lava can be explained by a relatively quick cooling and consequent sharp increase in the viscosity of the layer. However, the significant accumulation of these elements in the two bottom layers, in particular Os, Ir and Ru, is easily explained by their settling within the magma. In the same token, the enrichment observed for Cr and Ni in the second layer from the bottom compared to the bottom layer can be accounted by the relatively low density of these elements compared to PGEs, which would result in settling slower and solidifying magma before most of them could reach the bottom.

It should also be stated that very low concentrations of PGEs and PGMs in typical terrestrial magmatic rocks (see Table 2) and achondrites (see Table 3) clearly show that the typical mafic magmatic rocks are not able to trap PGEs and PGMs which would settle through them very fast. This is also the reason why typical mafic magmatic rocks cannot be either source or carrier of PGEs and PGMs for formation of PGEs ore deposits.

Everyone aforementioned clearly point at strong relationships between the anorthosites and other gabbroic cumulates (troctolites, gabbros, norites, etc.) with PGE mineralization and PGE deposits. Since, anorthosites represent the pristine continental crust, it is likely that they have been collected PGEs and PGMs at the time of the magma-ocean before their solidification. It is likely that they also were the initial containers of PGEs and PGMs which were found themselves within anorthosites and stack within them after their the solidification.

So-called gabbroic cumulates contain plagioclase of up to: 70% in troctolites, 65% in norites, and 55% in olivine gabbro and gabbronorites [195] ; 40% in chromitites [34] [161] ; 65% - 85% in leuconorites [196] ; 80% in leucogabbronorites or mottled anorthosites [197] ; 70% - 80% in anorthositic gabbro [198] , and 85% - 95% in anorthosites [199] . It is evident that all these rocks require presence of huge amounts of plagioclase which could be provided only by anorthosites. Anorthosites, norites and troctolites are also typical Moon rocks, containing extremely high volume of calcic plagioclase (~94% - 96% of anorthite): about of 90%, 60% and 60%, respectively [198] .

Interestingly, such rocks as norites, gabbros and anorthosites are known to generate families of rock associations (e.g., gabbro-norites, gabbro-anorthosites, norite-anorthosites, etc.), and troctolite which is known as olivine-rich and pyroxene-depleted relative of gabbro [200] . Both gabbros and norites contain only high-Ca plagioclases (labradorite-bytownite) and olivine, but gabbros contain clinopyroxene and norites containing mostly orthopyroxene; and all three anorthosites, gabbros and norites hold chromite as an accessory mineral [201] .

It should be stated that no troctolite corresponding in composition to a partial melt of peridotite was found [200] . This means that formation of troctolite in ALMI could take place only by interaction of the plagioclase-rich rock (e.g., anorthosite) with ultramafic melt.

The Archean Stillwater Complex contains two main anorthosite zones AN I (~370 m thick) and AN II (~630 m thick) in the Middle-Banded Series, and five additional anorthosite subzones forming 20-to 90-m-thick layers [199] . The horizon at the top of AN II, its upper 20 - 30 m [202] is the PGE-enriched Picket Pin zone. However, the Olivine-Bearing zone I (OB I), composed of troctolite, dunite, gabbronorite and anorthosite, is host of the principal platinum-group element deposit of the Stillwater complex (Bushveld Complex) [175] . The Picket Pin Zone is of lower grade than the Main Zone and the ore is notably enriched in Pt and Pd relative to the other PGE's. Czamanske and Bohlen [199] state that it seems inconceivable that introduction of any new liquid could promote in situ crystallization of 370- and 630-m thick layers of plagioclase cumulates that contain 85 to 95 vol% plagioclase and vary insignificantly in composition. Boudreau and McCallum [202] testify that the PGE-bearing anorthosites are enriched (relative to unmineralized anorthosites) in chalcophile and some incompatible trace elements, and that the sulfide-bearing zones occur within pyroxene-poor anorthosites.

The Bushveld Complex (South Africa) has been investigated in detail by various authors. At many localities in the Bushveld Complex, the sulfide mineralization is largely hosted by mela-, meso-, and leucogabbronorites, as well as norites and anorthosites [172] . The contact between the Main Zone and the underlying Platreef at Sandsloot is often defined by a mottled anorthosite [172] . The Northern Lobe mineralization is consistently Pd-rich, rather than Pt-rich, and there the Lower Zone is overlain by the Platreef and by the Grasvally norite-pyrox- enite-anorthosite (GNPA) member [174] . In some places the transition between the Upper Critical and Main Zones of the Bushveld Complex is within or above the distinctive Giant Mottled Anorthosite [203] . Eales et al. [196] draw the boundary between the Critical and Main Zones at the top of the Bastard Unit (at the upper contact of the Giant Mottled Anorthosite). Well-defined cyclicity is exhibited by a transition from pyroxenite, through melanorite, norite (35% - 65% plagioclase) and leuconorite (65% - 85% plagioclase) to anorthosite (≥85% plagioclase) in the Bastard and Merensky Units [196] . The authors [196] reported presence of abrupt transitions such as chromitite-anorthosite-chromitite, harzburgite-anorthosite-harzburgite, and leuconorite-anorthosite-leuconorite for the Western Bushveld Complex. Mitchell and Manthree [203] demonstrate that the uppermost unit of the Upper Critical Zone (an anorthosite with large intercumulus pyroxene mottles, designated unit HW5 at the Impala Platinum Mines), whilst petrographically part of the Giant Mottled Anorthosite, has some chemical attributes that link it to the Main Zone, and they attribute the hybrid geochemical signature of HW5 unit to partial melting of the anorthosite, due to an influx of the Main Zone magma above it. The Merensky Reef contains the Merensky mottled anorthosite of about 5 m thick, and the Bastard cyclic unit contains the Giant mottled anorthosite c. 50 m thick [204] . PGEs were revealed in anorthosites of the Merensky Reef at Impala Platinum (Bushveld Complex) [161] .

The Duluth Complex (Minnesota, USA) consists of an extensive cap of anorthositic rocks which are underlain and intruded by troctolitic rocks. Anorthositic series rocks are found as inclusions in the troctolitic series on a scale from centimeter-sizes to mappable units [205] . All the rocks in the anorthositic series are plagioclase cumulates and the anorthosite inclusions are presented in the olivine diabases in the North Volcanic Group lavas. Plagioclase shows a gradual decrease in An content upward in the section from An85 in the basal zone to An40 in the upper layered unit. Weiblen and Morey [205] led to conclusion that the anorthositic series is older than the troctolitic series, and possibly these two series are not genetically related.

The fact of increasing anorthite-content (and density) in plagioclase with a depth within the layered intrusions shows that it is cumulate plagioclase repre- senting the primordial plagioclase layer of the magma-ocean, like the oldest Lunar crust. Inversion of the density distribution of olivine (increase of Mg-content and decrease of Fe-content downwards) in layered intrusions indicate that olivine is not part of primordial cumulates and was delivered later by magma flows. Increasing melting points of both plagioclase and olivine with a depth the high melting point of olivine-containing magma indicate necessary to re-melt and intrude the plagioclase layer to form local magma chambers within it at different levels. To intrude the plagioclase layer, magma must have a heat energy in excess of that required to melt the plagioclase, providing the necessary latent heat of fusion. Analysis of the difference between melting points of coexisting olivine and plagioclase inclusions in the Bushveld layered intrusion shows the disparity to be ~200 - 400 K; and for carbonaceous chondrites it is ~250 - 400 K. This means that to re-melt the plagioclase and intrude the plagioclase layer, magma must have a significantly higher melting point than plagioclase (1830 K for anorthite, 1391 K for albite) or to be highly overheated. In contrast, such difference between the melting points in the Lunar meteorite Sayh al Uhaymir 300 is only ~100 - 290 K, which explains preservation of the early anorthositic Lunar crust. Magmas with greatest temperatures could intrude the plagioclase layer's lower parts, while moderate-temperature magmas could only intrude upper ones. Each such intrusion has created a magma chamber and relative cyclic unit.

It is known that gabbroic cumulates were formed later during interaction of anorthosites with melts. For example, troctolites were most likely formed during interaction of komatiites or dunites (dunite-chromites) with anorthosites. The enrichment of chromites with PGEs could have happened by simultaneous their accomplishment above a solid barrier within the magma-ocean (e.g., the "forsterite layer") or by underplating an anorthosite layer by dunite-chromite melt with remelting of the anorthosite layer and PGEs accumulation within the anorthosite layer (according to the Stokes law) into the dunite-chromite layer where PGEsare finally accumulated. Similar situation could take place with underplating of anorthosite layer by komatiites or other ultramafic melt with formation of troctolites enriched with PGEs from the anorthosite. Interaction of anorthosites with different ultramafic magmas in different proportions would have formed gabbroic cumulates which would inherit and in some cases would be enriched in PGEs. Possibilities of such underplating are in line with the presence of ALMIs (e.g., Bushveld Complex, Stillwater Complex, etc.) having ultramafic layer at the bottom of each ALMI complex, as well as ultramafic layers could be presented at other levels within the ALMI inside each cyclic unit. It is also in agreement with the facts [206] [207] showing that temperature of melt forming dunite pipes (Mooihoek, Onverwacht, Driekop and Twyfelaar) within the Bushveld complex were in excess of 1786 - 2096 K. The Driekop and Mooihoek pipes are emplaced in the Upper Critical zone, the Onverwacht and Twyfelaar pipes occur in the Lower Critical and Lower zones, respectively. The core-zones at Mooihoek and Onverwacht consist of hortonolite dunite and hortonolite wehrlite pegmatites (hortonolite is a moderately iron-rich olivine; average composition is $Fo_{50?44}$ that points at melting temperature $T_m \sim 1786$ - 1826 K). The core-zone at the Driekop pipe differs somewhat in that it consists of discrete mineralized pods and veins of iron-rich dunite and iron-rich wehrlite (average composition is $Fo_{73?72}$ with $T_m \sim 1974$ - 1981 K). Scoon and Mitchell [208] point at presence of two principal assemblages in the Onverwacht pipe with olivine $Fo_{83?80}$ ($T_m \sim 2028$ - 2049 K) and olivine $Fo_{64?44}$ ($T_m \sim 1786$ - 1921 K). Here for estimations of T_m of geological rock by olivine phenocrysts composition, a method described in [78] , was employed. We must underline that for effective re-melting of rocks in layer (slab) and underplating this layer, the temperature of magma must be in excess, because it must provide a heat energy for warming rock to its melting point overcoming heat of latent fusion and stay molten by itself. Extremely low viscosities of ultramafic melts also support possibility of underplating of crustal layers by ultramafic melts. Hall [117] described dunitic magma with unusually low viscosity from the Nordre Bumandsfjord pluton (approximately of Cambrian age) in Norway caused partial melting of gabbroic xenoliths and modified by assimilation of gabbroic and sedimentary material. It should be stated that chromite is a widespread accessory mineral in komatiites [209] , it is least abundant in highly magnesian chrome-undersaturated lavas and most abundant in strongly differentiated layered cumulate bodies; the natural range of Cr contents in olivine-chromite cumulates should be from ~5 × 10^3 to 10^4 ppm and chromite is a very widespread accessory mineral in almost all komatiitic rocks. It typically makes up a few tenths of 1% of the rock by volume, but in some cases it may be much more concentrated. The most extreme example of this is the Shurugwe (formerly Selukwe) massive chromitite deposits of Zimbabwe [210] [211] , where chromitite seams up to 10 m thick are associated with highly altered, deformed and metamorphosed 3.4 Ga serpentinites of probable komatiitic affinity.

In the Moon, where anorthosite layer was preserved and never was re-melted, temperature of melt was not in excess compare to the melting point of Lunar anorthosites. Analysis of data for Lunar meteorite Sayh al Uhaymir 300 comprising anorthositic clasts presented by Hudgins et al. [212] and shows that its matrix contains inclusions of plagioclase ($An_{94?98}$; T_m~ 1804 - 1821) and olivine ($Fo_{62?72}$; T_m~ 1907 - 1975 K), and mineral fragments of plagioclase ($An_{93?96}$; T_m~ 1799 - 1812) and olivine ($Fo_{62?73}$; T_m~ 1907 - 1982 K). This means that excess of temperature in the range 103 - 170 K was not enough for re-melting these plagioclase fragments.

Scarfe et al. [213] show that viscosity of anorthite is relatively low at temperatures of its subsolidus temperature and it changes from 2.15 Pa s at 1898 K and 4.07 Pa · s at 1830 K in liquid state to ~25.2 Pa · s at 1673 K and ~37 Pa · s at 1648 K in solid state, which are still in many orders lower than viscosity of albite. This means that if the high temperature magma underplated anorthite layer containing PGEs in it and magma would be able only heat up the anorthite to high temperature, but not melt it (e.g., in the case when magma temperature is not high enough or volume of magma is relatively small) the viscosity of heated anorthite could be low enough for fast settling within it PGEs and PGMs with their deposition within or just above the ultramafic magma complex that underplated the anorthite layer.

Analysis of appearance of PGEs in different mineralization zones and ore deposits show that in numerous cases the PGE economic concentrations are related to their association with chromites/chromitites. In case of PGE associations with chromites, sizes of combined grains were significantly larger than PGE grains and they were sinking within the magma-ocean much faster and should have sank to levels of the middle mantle and core during the time span of magma-ocean existence. However, their numerous presence in different PGE ore deposits shows that not all grains of PGEs with chromite sank to the great depths within the Earth. This fact could be explained by formation of “forsterite layer” [66] [77] which prevent them from sinking to the deep levels.

Let us analyze from this point of view increase of PGE concentrations in komatiites from the Early Archean (>3.5 Ga) to Late Archean. It is well known that melting temperature of komatiites increases with increase of MgO, MgO-content of komatiites decreases with age youngling, melting point of komatiites is also decreasing with age youngling and viscosity of melt increases with temperature decline. Thus, it is clear that lower contents of older komatiites in PGEs than in young komatiites could be explained by higher viscosity, longer time of cooling during hotter early periods of Earth evolution and higher settling speed of heavy species in higher viscous melt.

All this indicates that PGEs and PGMs during the early Earth evolution could have been prevented from sinking toward a center of the Earth by formation of solid layers (e.g., the "forsterite layer", dunite-chromite layers, anorthite layer, etc.) as well as collecting micro-grains of PGEs and PGMs within anorthosite or other silica-rich layers.

All the above-mentioned demonstrates that the initial concentration of PGEs and PGMs is related to their accumulation within the magma-ocean and they being trapped in the forming solid layers of rocks and minerals at shallow depths with high melting points and within highly viscous magmatic layers, mostly represented by plagioclase melt and felsic magmas.

It is also important to take in consideration that processes of plate tectonicsduring the Hadean and Early Archean were not operational in the Earth, and neither were operational processes of exhumation at the time of magma-ocean. There was no water-ocean before ~3.42 - 3.26 Ga [77] [78] and it means that any metal transport by aqueous fluids, the main transport tool in hydrothermal ore deposits (e.g., [214]), was not possible.

5. RELATIONSHIPS OF PGES AND PGMS ORE DEPOSITS FORMATION WITH PROCESS OF SERPENTINIZATION

Five largest known PGE deposits in the Earth are [215] : the Merensky and UG2 Chromitite Reefs of the Bushveld Complex (South Africa), the Main Sulfide Zone of the Great Dyke (Zimbabwe, South Africa), the Johns-Manville (JM) Reef of the Stillwater Complex (USA), and series of the Noril'sk-Talnakh deposits (Northern Russia).

The first four largest PGE deposits are related to large anorthosite-bearing layered mafic-ultramafic intrusions (ALMI). ALMI are rare, but host magmatic ore deposits containing most of the world's economic concentrations of PGEs. "PGE reefs" are stratabound PGE-enriched lode mineralization in mafic to ultramafic layered intrusions [38] . Contact-type Cu-Ni-PGE deposits is another relatively common type of PGE deposits [38].

PGE-enriched sulfide mineralization is also found near the contacts or margins (contact-type Cu-Ni-PGE deposits) of ALMI [216] . The largest contact-type Cu-Ni-PGE deposit is also mined in the Bushveld Complex [38].

The fifth largest and unique PGE-Cu-Ni Noril'sk and Talnakh deposits (Russia) are located in the northwest flank of the Triassic basalt trap formation of Siberia. They were formed on regional basement, which comprises Archean crystalline complexes and Lower-Middle Proterozoic volcano-sedimentary rocks and they were subjected to multiple tectono-magmatic transformations [217] . The main tectonic peculiarity of the region is presence of several extended long-lived Early Proterozoic rifting structures [217] . The authors [217] led to conclusion that ore elements of these deposits (except iron) are derived from the Earth crust rather than from the mantle.

Another usual type of PGE deposits is the Ural-Alaskan type: the lode sources of the most significant economic PGE placer resources and reserves [218] [219] . The Ural-Alaskan type complexes consist of large intrusive bodies of dunite, chromitite, wehrlite, clinopyroxenite, amphibole gabbro, and hornblendites. In contrast, Alpine type (ophiolitic lherzolite type) mafic-ultramafic complexes are known to contain only modest economic concentrations of PGE in placer deposits [219] .

Potentially economic deposits of Pt and Pd occur in a range of hydrothermal deposits without obvious direct or even indirect association with magmatism [220] . A number of PGE deposits are by-products or potential by-products from the production of such metals as uranium, gold and base-metals (e.g., orogenic gold deposits) [220] .

One of the main patterns of magma-involved PGE deposits is presence of high-temperature ultramafic magmatic rocks (mostly dunites) in all of them. Even in Noril'sk and Talnakh PGE deposits (Russia) presence of significant amounts of Mg-rich magmatic rocks was reported [217] . Presence of thick ultramafic layers is one of the main characteristics of ALMIs. Moreover, rich PGEs content in some dunite pipes (e.g., Mooihoek, Driekop, Onverwacht and Twyfelaar dunite pipes, Bushveld complex, South Africa) were discovered [206] [207] [221] . One of the main properties of ultramafic and Mg-rich rocks is their ability to be serpentinized with presence of water and temperature within the range of about 473 - 729 K (and even up to 843 K) which unavoidably would be reached during cooling of magmatic rocks and/or their later re-heating. It is in agreement with presence of the Ural-Alaska and Alpine types of PGE deposits which are always formed within the serpentinite and/or ophiolite belts in association with the serpentinized ultramafic rocks. Many PGE deposits are associated with chromites and/or chromitites which known to associate with dunites. Serpentinites are known in almost all ALMIs: Bushveld Complex [172]

[221] , the Great Dyke of Zimbabwe [222] , Stillwater Complex (Montana, USA) [223] ; Duluth complex (USA) [224] , Lac-Des-Iles Complex (Northwestern Ontario, Canada) [225] , Burakov intrusion (Fennoscandian Shield, Russia) [226] , Mt. General'skaya (Kola Peninsula, Fennoscandian Shield, Russia) [180] , the Guli Clinopyroxenite Dunite Massif (Siberian Craton, Taimyr Province, Russia) [227] , and in other numerous PGE deposits relating to ALMIs.

It should be stated that formation of PGE deposits in association with continental rifts (e.g., Noril'sk and Talnakh PGE deposits of Russia) and some PGE deposits not related to magmatic processes (e.g., PGEs associated with the orogenic gold deposits) are most likely also related to the serpentinization processes, because most continental rifts since Middle Proterozoic were formed with involvement of riftogenic ophiolites [228] [229] and the most orogenic processes have been also involved serpentinites and/or ophiolites. McKelson et al. [227] led to the conclusion that serpentinization, possibly driven by metasomatising fluids, seems to be the important process in PGE concentration. Interestingly, serpentine is one of principal minerals of carbonaceous chondrites [69] , which are usually used as the standard for PGEs content (in CI chondrites) for a comparison of PGE contents in the terrestrial rocks.

From this point of view, it is important to analyze the possible relationships of PGE and PGM in ore deposits with the serpentinization process.

Since serpentinization process is known for its association with more powerful tectonic processes (e.g., obduction, orogenic, thrusting, exhumation, etc.) and generation of overhigh pressure [77] [159] [230] [231] , it is obvious that tectonic factors play important role both in transportation and formation of the PGE deposits.

The processes of the PGEs enrichment can be related to the following processes: initial collecting of PGEs within magma-ocean above solidified ultramafic layer (forsterite, forsterite-dunite, dunite, dunite-chromite, etc.); initial trapping of PGEs within highly viscous layers of magma-ocean (felsic and plagioclase layers) with following their liberation, re-distribution and secondary enrichment by means of magmatism (re-melting of PGEs containing rocks) and/or alteration (by sulfur, water solutions, etc.). The formation of PGEs mineralization and deposits has been carried out at some depths from which they were tectonically delivered to shallow depths or earth's surface. It is in agreement with the fact of PGE abundances in the Samail Ophiolite (Oman) and the proposition that PGE composition very similar to Lorand et al. [83] estimates for the Primitive Upper Mantle of the Earth. But sulfur and hydrothermal activity are impossible within the upper mantle where forsterite, dunite and dunite + chromite layers were formed during solidification of magma-ocean and where most of PGEs were initially trapped and enriched. This means that first step of the PGEs transport to area of their mineralization and/or PGE deposits formation always was tectonic one.

PGE mineralization and deposits are usually associated with anorthosites and plagioclase-rich rocks. Therefore, it is important to mention that huge anorthosite massifs and ALMIs were emplaced tectonically to their present positions.

Really, since PGEs are extremely dense they should have sunk to the core or other great depths during the Earth accretion and evolution of magma-ocean. The traps for PGEs were formed only by either late veneer (late bombardment/ infall) or formation of solid layers within the magma-ocean at some depths (e.g., "forsterite layer"). At the same time, most of typical magmas contains low amounts of PGEs and such magmas as MORB and Alkali Basalt contain negligible amounts of PGEs (see Table 2) that makes it unrealistic or even impossible to deliver PGEs to the surface from large depths. Moreover, PGEs content in the primary mantle and bulk crust is also extremely low (see Table 2) and their simple melting would not provide sufficient amounts of PGEs. This means that PGEs were first concentrated somehow at different depths and then they were delivered to the surface. The possible processes of delivery of such concentrations to the surface are tectonic (e.g., obduction, thrusting, uplift, exhumation, etc.). All these processes are known to associate with serpentinization process [159] [230] [231] . This means that serpentinization process must have played a crucial role in the PGEs delivery to the surface and the main transport of PGEs to the final location of PGEs ore deposits was realized tectonically. It is in complete agreement with the following facts: PGE ores are presented in ophiolites and serpentinites, PGEs containing layered intrusions contain ultramafic layers and serpentinites, Ural-Alaskan type complexes are presented in regions of obvious obduction and formation of ophiolites and serpentinites.

The most important fact is that the ultramafic layer in layered intrusion containing PGE deposits does not need to be serpentinized by itself, because it would just destroy the complex. Serpentinization of layer below the layered intrusion would either destroy the intrusion or cause its vertical exhumation. The latest is possible in the Ural-Alaskan type complexes around the world, which are frequently associated with the PGE enriched placer deposits. Really, placer deposits can best be formed in the case of complete destruction of PGEs containing layers/blocks with following interaction with water flows and other leaching agents.

One of the major characteristics of the PGE deposit formation is the fact of their presence in very thin layers within huge complexes. For example: the Merensky Reef, in Impala Platinum Mines (Bushveld Complex) is only 46 cm thick and consists from bottom to top of leuconorite, anorthosite, chromitite and a very coarse-grained melanorite [34] ; the Bastard cyclic unit of the Bushveld Complex contains thin basal chromite usually <5 mm thick [204] ; chromitites in the Critical Zone (Bushveld Complex) range from less than 1 cm to more than 2 m thick [173] ; in the Bird River Sill (Manitoba, Canada) chromite layers are concentrated within the upper part of the ultramafic series, they are typically <20 cm thick (maximum of ca. 1 m thick) and have planar, regular contacts with adjacent dunite and peridotite layers [232] ; marker chromitite is a prominent 48 cm thick layer of the Lower Critical Zone (Bushveld Complex) [233] ; three massive chromitite layers, each with thicknesses between 28 and 35 cm were reported for the western Bushveld Complex [234] ; Mondal and Mathez [235] called the massive 70 cm thick UG2 chromitite layer of the Bushveld Complex; in borehole core (BV-1) drilled through the uppermost 2.8 km of the intrusion in the northern lobe of the Bushveld complex forty-five anorthosite layers, ranging from 1 to 23 m thick, have been identified [188] . All of these layers are significantly horizontally extended. Zientek [38] states that within the layered igneous intrusions the PGE mineralized intervals are thin, generally centimeters to meters thick, relative to the stratigraphic thickness of layers in an intrusion that varies from hundreds to thousands of meters, and reef-type mineralization is laterally persistent along strike, extending for the length of the intrusion, typically tens to hundreds of kilometers. Such layers as well as ALMIs could not be formed near the surface. Moreover, all those thin layers have survived during billions of years of tectonic perturbations and it is really enigmatic. It should also be mentioned that each ALMI consists of cycling units which not only survived, but also keep their position through billions of years. Presence of undisturbed small bodies of chromite-containing reefs and thin PGE plates among layers of layered intrusions signify the exhumation of layered intrusions within a single block during obduction or serpentinization of underlying ultra-basic rocks (e.g., dunites, harzburgites). This means that oldest ALMIs (e.g., Stillwater Complex, Montana and some PGE-rich complexes of the Canadian Shield) could be among the first witnesses of plate tectonics.

Evidence of presence of giant obducted plates around the world is overwhelming. Among the giant obducted plates are known: Oman Ophiolite (Arabian Peninsula), Bay of Islands Ophiolite (Canada), and Plates in Caledonian Orogens (Scandinavia). Obduction of giant plate on the Fennoscandian Shield took place in Early Proterozoic [236] . Results of detailed analysis of repeated obduction in the Appalachians [237] show that initial obduction of oceanic lithosphere within the present Appalachian region started with detachment and thrust of the entire Greenville Province onto the Canadian Shield (see also [238]). It should also be stated that repeating of obduction in the same region could cause delivery of rocks of the deep-seated layers to the surface or to very shallow depths through a few steps as plates or plate remnants [230] . Repeated obduction processes are known in the regions of Alps, Appalachians, Urals, California, Japan, Russian Far East, Norway, Mozambique, and many others (e.g., [237]). Pilchin [187] presented results of research discussing in more detail an involvement of forsterite and dunite layers of the oceanic lithosphere in obduction during formation of the Appalachians.

It was shown earlier [1] that forsterite layer in continents is usually located at depths of 70 - 120 km. Pilchin and Eppelbaum [77] proposed that dunite and chromite were solidified next after forsterite within the magma-ocean and they would form a layer right above the forsterite layer. Obviously, it is clear that presence of forsterite xenoliths as well as dunites at and near the earth's surface, is possible only if they were delivered to the surface by some geodynamic (tectonic) processes. To examine this supposition a special research was accomplished [187] .

Since unserpentinized forsterite and dunite layers' rocks are characterized by velocities of 8.5 - 8.7+ km/s (e.g., [119] [187]), analysis of seismic data at depths where such velocities were detected within the Appalachians and some other regions was accomplished. Obtained results indicate that layers with velocities observed in some regions are: 8.6 km/s dipping eastward from 50 to 60 km beneath the southeastern Greenville province (northern NY state and eastern Ontario) [239] ; about 8.5 - 8.6 km/s at ~60 km depth beneath southern Quebec (Canada) [240] ; >8.4 km/s and >8.6 km/s were reported at depths of ~40 - 60 km beneath the Flin Flon Domain and Western Interior Platform of the Trans-Hudson Orogen (Canada) [241] ; 8.7 - 8.8 km/s at 52 km depth beneath the Gaspe Peninsula (Canada) [242] ; 8.5 km/s at 40 - 50 km depths beneath the Gaspe Peninsula (Canada) [243] ; 8.4 - 8.7 km/s in the upper part of the mantle beneath the Gulf of St. Lawrence and beneath the northeastern Newfoundland [244] ; 8.5 km/s at ~46.3 km depth beneath the Gulf of St. Lawrence (Northern America) [245] ; ~8.5 km/s at depth of ~55 km beneath Newfoundland Appalachian orogen [246] ; 8.5 km/s at depths 55 - 65 km below the Appalachians in the Canadian Maritimes [247] ; 8.5 km/s at a depth of about 13 km beneath the central ridge in the Norwegian-Greenland Sea [248] ; up to 7 km thick lower crustal body with velocity of 8.5 km/s in Norwegian Caledonides [249] ; 8.5 km/s under southeastern Møre Basin (mid-Norwegian Margin) at depths of ~22 - 25 km [249] ; 8.5 km/s at depths of ~33 - 43 km within the lowest crust across the main Caledonian suture between Barentsia and Baltica (Barents Sea) [249] ; block of lower crust with 8.5 km/s at a depth of 14.5 - 18 km below the southern Vøring Basin (mid-Norwegian Margin) [249] ; 8.8 km/s apparent velocity results at a depth of 31 km under Fairbanks (Alaska) and dipping southward under the Alaska Range to ~48 km depth [242] ; 8.6 km/s at depths of ~28.5 - 30.0 km beneath the South Island

(New Zealand) [250] ; 8.6 - 8.7 km/s in the local upper mantle block between 45 and 60 km beneath the Aralsor rift (the Pre-Caspian Basin) [251] ; 8.5 - 9.0 km/s a the depth of 43 - 48 km in the South Caspian Basin [252] ; 8.7 km/s at depths of ~41 - 43 km and 8.5 km/s at depths of ~40 - 42 km beneath the Mirnyi kimberlite field (eastern Siberian craton) [242] . These data clearly show that position of forsterite layer in the Appalachians region and numerous other regions could be at relatively shallow depths. In some regions of obvious repeated obduction (e.g., Norwegian Caledonides) this layer may be presented within the Earth crust that cannot be reworked by any processes other than obduction.

Presence of huge amounts of serpentinized and unserpentinized dunites can prove that obduction process did take place in a studied region [187] . For example, huge amounts of serpentinites in the Appalachians could be found almost everywhere and huge amounts of dunites are known: massive chromite-bearing dunite in the Nadeau Ophiolitic Mélange (Quebec Appalachians, Canada) [253] ; in the Baltimore mafic complex (Maryland, USA) [254] ; alternating layers of dunite and chromite-rich dunite in the Quebec Appalachians [255] ; serpentinized dunites from the Webster-Addie ultramafic body (about 10 km long and 5 km wide) (North Carolina, USA) [256] ; Blue Ridge dunites of North Carolina [257] ; dunites of the Bay of Islands Ophiolite (Canada) [258] ; dunites of the Sept-Îles Anorthosite Complex (Quebec, Canada) [259] ; dunites of the Thetford Mines Ophiolite Complex (Canada) [260] ; chromitites of the Middle Arm Brook are enveloped by dunites (Advocate ophiolite complex, Baie Verte Peninsula, Newfoundland, Canada) [137] . These data noticeably testify that huge amounts of dunites were directly involved in obduction processes in the Appalachians region. Moreover, in most cases dunites in the Appalachian regions are present in association with chromites and they are serpentinized. Close to the region, dunites are also known in Jamaica and Tobago [261] , in the Moa-Baracoa massif (eastern ophiolitic belt of Cuba) [262] , and in some other regions.

It is known that anorthosites are usually associated with ultramafic rocks. Such a pattern was also observed in the Appalachian region: the Steel Mountain anorthosite is located within the Long Rang Fault as a part of the Baie Verte- Brompton line [263] ; Roseland anorthosite within the Blue Ridge Province [264] ; Indian Head anorthosites in the western Newfoundland [265] ; anorthosites in the northern Gulf of St. Lawrence [266] ; anorthosites in the Arden Pluton, Appalachian Piedmont (Delaware and adjacent Pennsylvania) [267] ; the Honey Brook anorthosite in Pennsylvania [268] ; anorthosites of the CRUML Belt and the central Appalachians [269] ; the Montpelier Anorthosite within the Blue Ridge Province [270] ; the Montpelier Anorthosite in Virginia [271] .

It was also shown [187] that dunites and associated with them chromites and chromitites in the Appalachian region in many cases are relatively rich in PGEs. All these facts led to the conclusion that forsterite and dunite (dunite + chromite) layers were deeply involved to obduction processes in the Appalachians region [187] .

PGE enrichments are also known to be found in relation with the olivine-rich rocks (dunites, peridotites) in numerous regions of ophiolite formations. These formations are characterized by serpentinization of Mg-rich olivine, which links the ophiolites with the forsterite layer, with dunites solidifying on top of it. Associations of chromites with serpentinites are also known in the following areas: chromitic serpentinites in Newfoundland, Canada and Baltimore Mafic Complex, USA [187] [272] ; chromite-serpentinite layers in the Newlands pit, Coobina Intrusion, Pilbara Craton [273] ; Kalung-Kalungan island of North Borneo [274] , etc. Moreover, chromitite is a part of the Thetford Mines ophiolite, Canada [275] .

For analysis of contents of PGEs in rocks associated with serpentinites, a careful analysis of published data has been performed (Table 9 and Table 10). In Table 9 only examples with maximal PGEs content of 3.2 ppm and higher were presented and in Table 10 only examples above 0.5 ppm and less than 3.0 ppm are given. Cases with the RGE contents (single PGEs and total) in rocks associated with serpentinites of a few hundreds of ppb, which are about two orders of magnitude of PGE contents in typical magmatic rocks, are numerous, but they are too low for economic PGE deposits and much lower than the PGE contents in chondrites.

Çina et al. [283] reported four types of PGE mineralization in the Albanian ophiolites with the PGE contents ranging from: 1) 160 - 350 ppb, 2) 580 - 2050 ppb, 3) 2195 - 8820 ppb, and 4) 1943 - 9730 ppb.

PGE deposits are usually associated with chromites and/or chromitites, which in turn are associated with dunites (Mg-rich end member of the peridotite family) having maximal effect during serpentinization process. This is an agreement with the fact of presence of significant amounts of PGEs associated with chromites/chromitites usually enveloped by dunites found in regions of obduction, formation of ophiolites and serpentinites. Thus, it is obvious that high contents of PGEs in chromites even in regions with absence of clear evidence of serpenti nization and ophiolite activity could also be treated as related to those, because blocks containing those chromites/chromitites could represent parts of greater blocks which were partially serpentinized. The mechanism of such thrusting is described in [159] [230] [231] . From this point of view, some PGE concentrations associated with chromites from known PGE deposits are accomplished (Table 11).

Table 9. High PGE contents in rock inclusions within ophiolites, serpentinites, peridotites and associated with serpentinites (including samples).

Area, region	Rock	ΣPGE, ppm	Reference
Cliff, Shetland ophiolites	chromite-rich dunite	≤60.0	[41]
Cliff, Shetland ophiolites	chromite-poor dunite	≤6.4	[41]
Veria, Greece	chromite	≤25.0	[276]
Pindos, Greece	chromites	≤4.22	[35]
Eastern Desert, Egypt	chromitites	≤3.2	[277]
Kytlym, Northern Urals	chromitites	≤61.0	[40]
Korydallos, Pindos ophiolite	-	≤6.864	[278]
Butirin, Northern Urals	-	22.972	[279]
Nurali complex, Urals	CHRI, CHRII	≤11.61	[280]
Ouen Island, New Caledonia	-	11.5	[281]
Al 'Ays Ophiolite, Saudi Arabia	-	≤25.5	[282]
Albanian ophiolites	-	≤11.1	[283]
Legris Lake Complex, Canada	(Pd+Pt only)	≤3.625	[179]
Lac Des Iles Mine, Canada	(Pd only)	≤7.62	[284]
Jinchuan complex, China	-	≤3.524	[285]
State Line District, Appalachians	Chromite	≤5.0	[286]
Mayari-Baracoa Ophiolite Belt, Cuba	Chromite	≤3.7	[287]
Misheguk Mountain, Alaska	Chromite (Pt only)	≤4.2	[288]
Misheguk Mountain, Alaska	Chromite (Pd only)	≤4.7	[288]
Bear Creek, Alaska	(Pt only)	≤6.5	[288]
Danielson Creek, Alaska	(Pt only)	≤10.0	[288]
Halibut Bay & Sturgeon River, Alaska	Chromite (Pd only)	≤11.52	[288]
Halibut Bay & Sturgeon River, Alaska	Chromite (Pt only)	5.453	[288]
Dust Mountain, Alaska	Chromite (Pd only)	11.936	[288]
Dust Mountain, Alaska	Chromite (Pt only)	8.918	[288]

Table 12 shows comparison of the mean PGE concentrations in different chondrite groups.

In Table 9 and Table 10, along with the mean PGEs concentrations in some PGEs mineralization zones and deposits, sample concentrations are presented as well (especially for Alaska). It was done to perform the real comparison of PGEs contents in terrestrial deposits and samples with those in meteorites (e.g., chondrites). Since some meteorite groups (especially carbonaceous chondrites) are characterized by small number of examined meteorites which is not enough for

Table 10. Moderate and elevated PGE contents in rock inclusions within ophiolites, serpentinites and peridotites (including samples).

Area, region	Rock	ΣPGE, ppm	Reference
Skiros, Greece	chromite	2.3	[35]
Milia, Pindos, Greece	chromite	≤1.059	[286]
Rhodope Massif, Bulgaria	chromitite	≤1.6	[287]
Mawat ophiolite, Iraq	chromitites	≤1.094	[145]
Oman ophiolite	chromitite	≤1.5	[274]
Kempirsai Massif, Urals	chromitite	≤1.0	[138]
Kempirsai Massif, Urals	chromite	≤1.0 - 2.0	[288]
Polar Urals	sulfide-chromite	0.7 - 0.8	[289]
Polar Urals	sulfide-bearing	1.0 - 1.5	[289]
Eastern Desert, Egypt	dunite envelopes	up to 0.738	[274]
Turkey ophiolites	Cr-PGE	≤1.305	[290]
Eskişehir, Turkey	chromitites	≤0.533	[291]
Loma Caribe, Dominican Republic	chromitites	1.82 - 2.04	[292]
West Raglan, Cape Smith Belt, Canada	-	2.54	[293]
Avalon terrane, Canada	-	≤2.4	[294]
Thetford Mines ophiolite, Canada	chromitite (Only Pt)	≤2.0	[295]
Advocate ophiolite, Newfoundland, Canada	-	≤1.028	[137]
Bird River Sill, Canada	chrome property	≤0.7	[231]
Dovirensky layered complex, Siberia	(Pd only)	≤1.4	[296]
Blue Ridge Province, North Carolina	chromite (Pt+Pd+Rh)	≤0.746	[297]
Avan Hills, Alaska (Pt + Pd)	-	≤1.029	[285]
Rabbit Creek, Alaska (Pt + Pd)	-	≤2.298	[285]
Quartz Creek, Alaska (Pt only)	serpentinite	≤1.2	[285]
Emerick prospect, Alaska (Pt + Pd)	serpentinite	≤1.966	[285]
Caribou Mountain, Alaska	chromite (Pt only)	≤1.337	[285]
Salt Chuck Mine, Alaska (Pd only)	-	1.96	[298]
Red Mountain, Alaska (Pt only)	-	2.27	[298]

Table 11. High total PGEs contents in some chromites and/or chromitites related to ultramafic rocks but not enclosed in serpentinite layers.

Area, region	ΣPGE, ppm	Reference
Lower Chromitite, Platreef, Bushveld Complex	≤6.0	[166]
Upper Chromitite, Platreef, Bushveld Complex	up to >11.0	[166]
Total UG2, Bushveld Complex	5.7	[42]
Chromites of India	≤18.0	[302]
Chromite-sulphide, Orissa, India (Pt + Pd)	11.18	[303]

Table 12. Mean PGEs Total for the different chondrite groups (calculated using data of [47]).

Chondrite group	PGEs Total, ppm
CI	3.348
CO	4.481
CM	4.109
CV	5.074
CK	5.240
CR	4.145
CH	6.347
CB	9.605
K	2.205
R	2.161
H	5.303
L	3.589
LL	2.544
L/LL	3.071
EH	4.449
EL	4.127

Table 13. Average PGE contents in some known terrestrial economic PGE deposits.

Area, region	ΣPGE, ppm	Reference
Noril'sk, Russia	10.03	[42]
Monchegorsk, Russia, (NKT orebody)	8.59	[42]
Sudbury, Ontario	1.173	[42]
Great Dyke, South Africa	5.418	[42]
Total Merensky reef, Bushveld complex	6.214	[42]
Total UG-2 reef, Bushveld complex	5.700	[42]
Total Platreef, Bushveld complex	4.121	[42]
Total Bushveld complex	5.670	[42]
Stillwater, USA	20.699	[42]
Duluth Complex, Minnesota	Pt + Pd = ~9	[190]

conventional statistical analysis, they should also be treated as samples. This is the reason why terrestrial samples with high, moderate and elevated PGEs should also be taken in consideration for comparison with PGEs concentrations in chondrites. Many cases of high PGEs contents related to sulfides in terrestrial rocks, even those which clearly associated with ultramafic rocks were not taken here in consideration, because our goal here is to analyze PGEs contents only in rocks directly associated to serpentinites/ophiolites. For comparison, average PGE contents for some known economic PGE deposits are presented in Table 13.

Comparison of data from Tables 9-12 indicates that in cases when PGEs contents in terrestrial rocks of the same values or even greater (in some cases much greater) than those in chondrites are not so rare. Even in some cases, PGE contents listed in Table 10 show that some of those amounts are pretty close to those of chondrites with low PGEs concentration. It is also clear that cases of PGEs contents listed in Tables 11-13 in general higher and in some cases much higher than those in chondrites (see Table 12). This simply means that if all economic deposits listed in Table 13 were entirely composed of CI chondrites they would not have medium PGEs contents which now recognized in the deposits.

Analysis of presence of chromite in carbonaceous and ordinary chondrites shows that: the Ivuna (CI1) carbonaceous chondrite contain chromite and olivine [304] and serpentine [305] ; chromite and olivine and chromite-olivine pairs are presented in type II chondrules in CM, CO, CV, CR, H, L, LL, and two ungrouped chondrites [306] ; the Mokoia CV matrix contains chromite and chromites presented in partially characterized phases [307] ; dissolution of samples from two Ivuna phases left behind a tiny acid-resistant residue highly enriched in Cr, most likely a chromite-spinel phase [308] ; chromites were found in the Murchison CM2 chondrite [309] ; chromite was found in powdered meteoritic material of carbonaceous chondrites [310] ; in the Dhofar 225 CM chondrite olivine aggregates and chondrule-like olivine objects are common and the chondrules consist of olivine with minor chromite and sulfides [311] ; chromite is presented in the carbonaceous chondrites but is considerably smaller than chromite in the ordinary chondrites (~25 μm compared to ~270 μm) [312] ; Ramdohr [313] has investigated chromite in the Jilin chondrite which shows mostly coarse chromite (exsolution chromite), but with some grains too small to be included in the study; in NWA 4428 (CM2) chondrite grains as small as 11 μm, 16 - 26 μm were reported [312] ; all chromite with a diameter larger than 10 μm in the Allende (CV3) chondrite is situated in porphyritic olivine chondrules, and Chromite grains smaller than 10 μm (typically 1 - 3 μm) are also occurred in the porphyritic olivine chondrules [312] ; the maximum diameter of chromite in the L3 chondrites is 34 - 50 μm, in the L4- 84 - 150 μm, in the L5- 76 - 158 μm and in the L6- 253 - 638 μm [314] ; olivine and chromite are presented in the Tafassasset (CR?) chondrite [315] ; the H chondrites contain ~0.22 wt% chromite and the L chondrites contain about ~0.27 wt% chromite [314] ; chromite is the most common spinel group mineral in the ordinary chondrites [316] ; chromite is more common in type 4 - 6 ordinary chondrites than in type 3 ordinary chondrites [317] .

These data clearly show that both chromite and olivine are presented in carbonaceous and ordinary chondrites and in carbonaceous chondrites chromite grains have limited size of microinclusions pretty similar to that of chromites and PGEs in the most of terrestrial deposits. Relatively huge size of some microinclusions in ordinary chondrites are of the same range or even smaller than size of grains of PGEs and chromites found in most of the Ural-Alaskan type mafic-ultramafic complexes (e.g., 0.5 - 2.0 mm in Yukon; [171]). Even presence of some extra carbonate in carbonaceous chondrites is not accidental, because formation of carbonate platforms is known in association with serpentinite and ophiolite belts [318] .

There is some evidence of similarity of dunites from certain ophiolite belts with carbonaceous chondrites. Huang et al. [319] show that definite some characteristics of chromites (e.g., $^{187}Os/^{188}Os$) from the Nagqu ophiolite (Tibet) suggest that the mantle source for the dunite is similar to that of the carbonaceous chondrites. Yang et al. [320] reported analysis of magnesium isotopic composition ($\delta^{26}Mg$) of mantle peridotite xenoliths from Sanyitang and Beiyan (North China craton) which are similar to $\delta^{26}Mg$ values of two additional dunite standards (DTS-1 and DTS-2) and two carbonaceous chondrites (Allende and Murchison) analyzed in this study as well as $\delta^{26}Mg$

of all published chondrites. Büchl et al. [321] presented results of research on evolution of Os isotope heterogeneities ($^{187}Os/^{188}Os$) in the mantle sequence of the Troodos ophiolite, Cyprus. The sequence of the Troodos ophiolite was divided into eastern (Unit 1) and western part (Unit 2) [322] . The results show that most of the $^{187}Os/^{188}Os$ ratios of Unit 1 lherzolites and harzburgites (with some dunites) are chondritic to subchondritic and most harzburgites and dunites of Unit 2 have suprachondritic $^{187}Os/^{188}Os$ ratios [321] . The spinel-lherzolites of Unit 1 have $^{187}Os/^{188}Os$ ratios (0.1255 - 0.1276) similar to carbonaceous chondrites (0.127) and most of the harzburgites and dunites of Unit 2 have carbonaceous chondritic to suprachondritic $^{187}Os/^{188}Os$ ratios [321] .

6. CONCLUSION

It is clear that there are too many similarities between the terrestrial rocks with PGE mineralization and deposits formed in association with serpentinites and/or ophiolites, and carbonaceous chondrites to merely be coincidental. All of the discussed above facts about the terrestrial PGE mineralization and deposits associated with serpentinites and/or ophiolites confirm that serpentine and olivine are the principal minerals in carbonaceous chondrites. Therefore, we can conclude that carbonaceous chondrites were formed from the ophiolite/serpen- tinite massifs of their parent objects. It should also be stated that since serpentinite and ophiolite massifs cannot represent all the different terrestrial rocks and rock associations, carbonaceous chondrites (and especially extremely rare CI chondrites) should not be used as a standard for the PGE contents in terrestrial rocks. The reasons why the extremely dense and heavy PGE did not sink to great depths at the time of the magma-ocean were discussed in detail. For example, the "late veneer" hypothesis of PGEs enrichment has been comprehensively examined. A significant role of ancient meteoritic bombardment in the creation of PGE deposits has been emphasized. Obviously, this last fact could assist in the development of new enhanced strategies for combined geological-geophysical prospecting for PGE deposits.

CITE THIS PAPER

Pilchin, A. and Eppelbaum, L. (2017) Concentration of Platinum Group Elements during the Early Earth Evolution: A Review. Natural Resources, 8, 172-233. https://doi.org/10.4236/nr.2017.83012

REFERENCES

1. Pilchin, A.N. and Eppelbaum, L.V. (2009) The Early Earth and Formation of the Lithosphere. In: Anderson, J.E. and Coates, R.W., Eds., The Lithosphere: Geochemistry, Geology and Geophysics, Nova Science Publication, New York, Chapter 1, 1-68.
2. Maier, W.D. (2005) Platinum-Group Element (PGE) Deposits and Occurrences: Mineralization Styles, Genetic Concepts, and Exploration Criteria. Journal of African Earth Sciences, 41, 165-191.
3. Mungall, J.E. and Naldrett, A.J. (2008) Ore Deposits of the Platinum-Group Elements. Elements, 4, 253-258.
4. Wyborn, L.A.I., Heinrich, C.A. and Jaques, A.L. (1994) Australian Proterozoic Mineral Systems: Essential Ingredients and Mappable Criteria. Proceedings of the Australian Institute of Miningand Metallurgy Annual Conference, Melbourne, 109-115.
5. McQueen, K.G. (2005) Ore Deposit Types and their Primary Expressions. In: Cornelius, S.M. and Robertson, I.D.M., Eds., Regolith Expression of Australian Ore Systems: A Compilation of Exploration Case Histories with Conceptual Dispersion, Process and Exploration Models, CRC LEME, Australia, 1-14.
6. Naldrett, A.J. (2004) Magmatic Sulfide Deposits: Geology, Geochemistry, and Exploration. Springer, Berlin.
7. Hoatson, D.M. (1998) Platinum-Group Element Mineralization in Australian Precambrian Layered Mafic-Ultramafic Intrusions. AGSO Journal of Australian Geology and Geophysics, 17, 139-151.
8. Naldrett, A.J. and Von Gruenewaldt, G. (1989) Association of Platinum-Group Elements with Chromitites in Layered Intrusions and Ophiolite Complexes. Economic Geology, 84, 180-187.
9. Garuti, G., Pushkarev, V.E., Zaccarini, F., Cabella, R. and Anikina, E. (2002) Chromite-PGE Mineralization in the Uktus Alaskan-Type Complex (Central Urals, Russia). Transactions of the 9th International Platinum Symposium, 149-152.
10. Foose, M. and Weiblen, P. (1986) The Physical and Petrologic Setting and Textural and Compositional Characteristics of Sulfides from the South Kawishiwi Intrusion, Duluth Complex, Minnesota, USA. In: Friedrich, G.H., Genkin, A.D., Naldrett, A.J., Ridge, J.D., Sillitoe, R.H. and Vokes, F.M., Eds., Geology and Metallogeny of Copper Deposits, Special Publication No. 4 of the Socistu for Geology Applied to Min. Deposits. Springer, Berlin, 4, 8-24.
11. Augé, T., Bailly, L., Cocherie, A., Genna, A., Guerrot, C., Lerouge, C., Mukherjee, M.M. and Patra, R.N. (2002) Magmatic and Hydrothermal Platinum-Group Element Mineralization in the Baula Area, Orissa, India. Transactions of the 9th International Platinum Symposium, Billins, Montana, 21-24.
12. Thakurta, J., Ripley, E.M. and Li, C. (2008) Pre-Requisites for Sulphide-Poor PGE and Sulphide-rich Cu-Ni-PGE Mineralization in Alaskan-Type Complexes. Journal of the Geological Society of India, 72, 611-622.
13. Nemerov, V.K., Razvozzhaeva, E.A., Spiridonov, A.M. and Granina, E.M. (2008) By-Product Platinum Mineralization of Gold Ore Deposit Sukhoy Log (Eastern Siberia). In: Deposits of Natural and Technogenic Raw Materials: Geology, Geochemistry and Geophysical Methods and Ecologic Geology, Voronezh State University, 159-160. (In Russian)
14. Vasilyeva, I.E., Pozhidaev, Yu.N., Vlasova, N.N., Voronkov, M.G. and Philipchenko, Yu.A. (2010) Sorption-Atomic-Emission Determination of Gold, Platinum and Palladium in Rocks and Ores Using Sorbent PSTM-3T. Analytical Controls, 14, 16-24. (In Russian)
15. Chernyshov, N.M. (2004) Kshenskiy Type Gold-Platinum-Bearing Mineralization (Central Russia). Voronezh State University Bulletin, Geology, No. 1, 104-116. (In Russian)

16. Chernyshov, N.M. and Chernyshova, M.N. (2012) Platinum-Bearing and Gold-Platinum-Bearing Formations of Voronezh Crystalline Massif at Different Geodynamic Regimes of Formation Precambrian Lithosphere. Lithosphere, No. 1, 122-146. (In Russian)
17. Houlé, M.G. and Lesher, C.M. (2011) Komatiite-Associated Ni-Cu-(PGE) Deposits, Abitibi Greenstone Belt, Superior Province, Canada. Reviews in Economic Geology, 17, 89-121.
18. Cawthorn, R.G. (2010) The Platinum Group Element Deposits of the Bushveld Complex in South Africa. Platinum Metals Review, 54, 205-215.
19. Crowson, P. (2001) Minerals Handbook 2000-01: Statistics and Analyses of the World's Minerals Industry. Mining Journal Books Ltd., Edenbridge, Kent, UK.
20. USGS (2005) Mineral Commodity Summaries. US Government Printing Office, Washington DC.
21. Hedenquist, J.W. and Lowenstern, J.B. (1994) The Role of Magmas in the Formation of Hydrothermal Ore Deposits. Nature, 370, 519-527.
22. Lide, D.R., Ed. (2004) Handbook of Chemistry and Physics. 85th Edition, CRC Press, Boca Raton.
23. Lodders, K., Palme H. and Gail, H.P. (2009) Abundances of the Elements in the Solar System. In: Trumper, J.E., Ed., Astronomy and Astrophysics, Landolt-Bornstein, New Series, Vol. VI/4B, Chap. 4.4, Berlin, Springer-Verlag, 560-630.
24. Wang, Yu., Wang, J., Wang, L., Long, L., Liao, Z., Zhang, H. and Tang, P. (2011) Problems of PGE Metallogenesis Related to Mafic-Ultramafic Complexes in North Xinjiang, China. Geoscience Frontiers, 2, 187-198.
25. Pilchin, A.N. and Eppelbaum, L.V. (2006) Iron and Its Unique Role in Earth Evolution. Monograph, Mexican Geophys. Soc. 9, National University of Mexico.
26. MacPherson, G.J. and Boss, A. (2011) Cosmochemical Evidence for Astrophysical Processes during the Formation of Our Solar System. PNAS, 108, 19152-19158.
27. MacPherson, G.J., Simon, S.B., Davis, A.M., Grossman, L. and Krot, A.N. (2005) Calcium-Aluminum-Rich Inclusions: Major Unanswered Questions. In: Krot, A.N., Scott, E.R.D. and Reipurth, B., Eds., Chondrites and the Protoplanetary Disk, ASP Conf. Series, 341, 225-250.
28. Weiser, T.W. and Bachmann, H.-G. (1999) Platinum Group Minerals from the Aikora River Area, Papua New Guinea. Canadian Mineralogist, 37, 1131-1145.
29. Barkov, A.Y., Martin, R.F., LeBarge, W. and Fedortchouk, Y. (2008) Grains of Pt-Fe Alloy and Inclusions in a Pt-Fe Alloy from Florence Creek, Yukon, Canada: Evidence for Mobility of Os in a Na-H2O-Cl-Rich Fluid. Canadian Mineralogist, 46, 343- 360.
30. Bogdanov, K., Zaccarini, F., Garuti, G., Tzintsov, Z., and Ianakieva, N. (2012) Platinum Group Minerals (PGM) from Novoselsti Placers (Eastern Bulgaria): Evidence for a Ural-Alaskan-Type Intrusion? Transactions of the European Mineralogical Conference, 1, EMC2012-164.
31. McDonough, W.F. and Sun, S.-S. (1995) The Composition of the Earth. Chemical Geology, 120, 223-253.
32. Rudnick, R.L. and Gao, S. (2003) Composition of the Continental Crust. In: Rudnick, R.L., Ed., The Crust, 3, Ch. 3.01, Elsevier, 1-64.
33. Morgan, J.W., Walker, R.J., Brandon, A.D. and Horan, M.F. (2001) Siderophile Elements in Earth's Upper Mantle and Lunar Breccias: Data Synthesis Suggests Manifestations of the Same Late Influx. Meteoritics & Planetary Science, 36, 1257-1276.
34. Barnes, S-J. and Maier, W.D. (2002) Platinum-Group Element Distributions in the Rustenberg Layered Suite of the Bushveld Complex, South Africa. In: Cabri, L.J., Ed., The Geology, Geochemistry, Mineralogy and Mineral Beneficiation of Platinum-Group Elements. Canad. Inst. of Mining, Metallurgy and Petroleum, Special Vol. 54, 431-458.
35. Economou-Eliopoulos, M. (2011) Platinum-Group Elements (PGE) in Various Geotectonic Settings: Opportunities and Risk. Hellenic Journal of Geosciences, 45, 65-82.

36. Hauri, E.H. and Hart, S.R. (1997) Rhenium Abundances and Systematics in Oceanic Basalts. Chemical Geology, 139, 185-205.
37. Maier, W.D., Peltonen, P., McDonald, I., Barnes, S.J., Barnes, S.-J., Hatton, C. and Viljoen, F. (2012) The Concentration of Platinum-Group Elements and Gold in Southern African and Karelian Kimberlite-Hosted Mantle Xenoliths: Implications for the Noble Metal Content of the Earth's Mantle. Chemical Geology, 302–303, 119-135.
38. Zientek, M.L. (2012) Magmatic ore Deposits in layered Intrusions-Descriptive Model for Reef-Type PGE and Contact-Type Cu-Ni-PGE Deposits. Open-File Report 2012-1010, United States Geological Survey, U.S. Department of the Interior.
39. Voytekhovsky, Y.L. and Neradovsky, Y.N. (2008) The Cu-Ni-PGE and Cr Deposits of the Monchegorsk Area, the Kola Peninsula, Russia. 33 IGC Excursion No. 48, 15-21 August 2008, Oslo.
40. Zaccarini, F., Garuti, G. and Pushkarev, E.V. (2010) Unusual PGE-Rich Chromitite in the "Butyrin-Veins" of the Kytlym Ural-Alaskan Complex (Northern Urals) Transactions of the 11th INternational Platinum Symposium "PGE in the 21st Century, Sudbury, Ontario, Canada.
41. Prichard, H.M. and Lord, R.A. (1993) An Overview of the PGE Concentrations in the Shetland Ophiolite Complex. Geological Society, London, 76, 273-294.
42. Naldrett, A.J. (2010) Secular Variation of Magmatic Sulfide Deposits and Their Source Magmas. Economic Geology, 105, 669-688.
43. Vikentyev, I.V., Yudovskaya, M.A., Mokhov, A.V., Kerzin, A.L. and Tsepin, A.I. (2004) Gold and PGE in Massive Sulfide Ore of the Uzalginsk Deposit, Southern Urals, Russia.Canadian Mineralogist, 42, 651-665.
44. Crocket, J.H. (1979) Platinum-Group Elements in Mafic and Ultramafic Rocks: A Survey. Canadian Mineralogist, 17, 391-402.
45. Naldrett, A.J., Wilson, A., Kinnaird, J. and Chunnett, G. (2009) PGE Tenor and Metal Ratios within and below the Merensky Reef, Bushveld Complex: Implications for Its Genesis. Journal of Petrology, 50, 625-659.
46. Barnes, S.-J., Maier, W.D. and Ashwal, L.D. (2004) Platinum-Group Element Distribution in the Main Zone and Upper Zone of the Bushveld Complex, South Africa. Chemical Geology, 208, 293-317.
47. Tagle, R. and Berlin, J. (2008) A Database of Chondrite Analyses Including Platinum Group Elements, Ni, Co, Au, and Cr: Implications for the Identification of Chondritic Projectiles. Meteoritics & Planetary Science, 43, 541-559.
48. McDonald, I. and Russell, S.S. (2001) Platinum-Group Elements in Enstatite Chondrites and Enstatite Achondrites. Meteoritics and Plan. Science, 36, A128.
49. Kitts, K. and Lodders, K. (1998) Survey and Evaluation of Eucrite Bulk Compositions. Meteoritics & Planetary Science, 33, A197-A213.
50. Neal, C.R., Taylor, L.A., Ely, J.C., Jain, J.C. and Nazarov, M.A. (2001) Detailed Geochemistry of New Shergottite, Dhofar 019 (Abstract). Lunar and Planetary Science, 32, #1671, Lunar and Planetary Institute, Houston, Texas, USA (CD- ROM).
51. McSween Jr., H.Y. (2008) Martian Meteorites as Crustal Samples. In: Bell, J., Ed., The Martian Surface: Composition, Mineralogy and Physical Properties, Cambridge University Press, Cambridge, 381-396.
52. Anand, M., Taylor, L.A., Neal, C.R., Snyder, G.A., Patchen, A., Sano, Y. and Terada, K. (2003) Petrogenesis of Lunar Meteorite EET 96008. Geochimica et Cosmochimica Acta, 67, 3499-3518.
53. Snyder, G.A., Neal, C.R., Ruzicka, A.M. and Taylor, L.A. (1999) Lunar Meteorite EET 96008, PART II. Whole-Rock Trace-Element and PGE Chemistry, and Pairing with EET 87521. Lunar and Planetary Science, XXX, Abstract No. 1705.
54. Mason, B. and Taylor, S.R. (1982) Inclusions in the Allende Meteorite. Smithsonian Contributions to the Earth Science, No. 25. Smithsonian Institution Press, Washington DC.
55. Grossman, L. and Ganapathy, R. (1976) Trace Elements in the Allende Meteorite-I. Coarse-Grained, Ca-Rich Inclusions. Geochimica et Cosmochimica Acta, 40, 331- 344.

56. Connolly Jr., H.C., Huss, G.R. and Wasserburg, G.J. (2001) On the Formation of Fe-Ni Metal in Renazzo-Like Carbonaceous Chondrites. Geochimica et Cosmochimica Acta, 65, 4567-4588.
57. Weisberg, M.K., Smith, C., Benedix, G., Folco, L., Righter, K., Zipfel, J., Yamaguchi, A. and Chennaoui Aoudjehane, H. (2009) The Meteoritical Bulletin No. 95. Meteoritics and Planetary Science, 44, 429-462.
58. Lodders, K. (2003) Solar System Abundances and Condensation Temperatures of the Elements. The Astrophysical Journal, 591, 1220-1247.
59. Palme, H. and Jones, A. (2005) Solar System Abundances of the Elements. In: Davis, A.M., Ed., Meteorites, Comets, and Planets. Treatise on Geochemistry 1, Elsevier, 41-62.
60. McSween Jr., H.Y. (1999) Meteorites and Their Parent Planets. 2nd Edition, Cambridge University Press, Cambridge.
61. Weisberg, M.K., McCoy, T.J. and Krot, A.N. (2006) Systematics and Evaluation of Meteorite Classification. In: Lauretta, D.S. and McSween Jr., H.Y., Eds., Meteorites and the Early Solar System II, The University of Arizona Press, 19-52.
62. Zolensky M.E. and McSween, H.Y. (1988) Aqueous Alteration. In: Kerridge, J.F. and Matthews, M.S., Eds., Meteorites and the Early Solar System, University of Arizona Press, 114-143.
63. Krot, A.N., Hutcheon, I.D., Brearley, A.J., Pravdivtseva, O.V., Petaev, M.I. and Hohenberg, C.M. (2006) Timescales and Settings for Alteration of Chondritic Meteorites. In: Lauretta, D.S. and McSween, Jr., H.Y., Eds., Meteorites and the Early Solar System II. Univ. of Arizona Press, Tucson, 525-553.
64. Drake, M.J. and Righter, K. (2002) Determining the Composition of the Earth. Nature, 416, 39-44.
65. Anderson, D.L. (2007) New Theory of the Earth. Cambridge University Press, Cambridge.
66. Pilchin, A.N. (2011) Magnetite: The Story of the Mineral's Formation and Stability. In: Angrove, D.M., Ed., Magnetite: Structure, Properties and Applications, Nova Science Publication, New York, Chapter 1, 1-99.
67. Norton, O.R. (2002) The Cambridge Encyclopedia of Meteorites. Cambridge University Press, Cambridge.
68. Gupta, G. and Sahijpal, S. (2010) Differentiation of Vesta and the Parent Bodies of Other Achondrites. Journal of Geophysical Research: Planets, 115, E08001, 1-15.
69. Mason, B. (1972) The Mineralogy of Meteorites. Meteoritics, 7, 309-326.
70. Marvin, U.B. (2006) Meteorites in History: An Overview from the Renaissance to the 20th Century. In: McCall, G.J.H., Bowden, A.J. and Howarth, R.J., Eds., The History of Meteoritics and Key Meteorite Collections: Fireballs, Falls and Finds. Geological Society, Special Publication, London, 256, 15-71.
71. Grossman, L. (1973) Refractory Trace Elements in Ca-Al-Rich Inclusions in the Allende Meteorite. Geochimica et Cosmochimica Acta, 37, 1119-1140.
72. Haack, H. and McCoy, T.J. (2005) Iron and Stony-Iron Meteorites. In: Holland, H.D. and Turekian, K.K., Eds., Treatise on Geochemistry, Elsevier, Amsterdam, 325- 345.
73. Chen, J.H., Papanastassiou, D.A. and Wasserburg, G.J. (2002) Re-Os and Pd-Ag Systematics in Group IIIAB Irons and in Pallasites. Geochimica et Cosmochimica Acta, 66, 3793-3810.
74. Mittlefehldt, D.W. (2005) Achondrites. In: Holland, H.D. and Turekian, K.K., Eds., Treatise on Geochemistry, Elsevier, Amsterdam, 291-324.
75. Palme, H., Hutcheon, I.D. and Spettel, B. (1994) Composition and Origin of Refractory-Metal-Richassemblages in a Ca, Al-Rich Allende Inclusion. Geochimica et Cosmochimica Acta, 58, 495-513.
76. Campbell, A.J., Simon, S.B., Humayun, M. and Grossman, L. (2003) Chemical Evolution of Metal in Refractory Inclusions in CV3 Chondrites. Geochimica et Cosmochimica Acta, 67, 3119-3134.

77. Pilchin, A.N. and Eppelbaum, L.V. (2012) The Early Earth, Formation and Evolution of the Lithosphere in the Hadean-Middle Archean. In: Sato, F. and Nakamura, Sh., Eds., Encyclopedia of Earth Science Research, 1, Chapter 1, 1-93.
78. Eppelbaum, L.V., Kutasov, I.M. and Pilchin, A.N. (2014) Applied Geothermics. Springer, Berlin-Heidelberg.
79. Pattou, L., Lorand, J.P. and Gros, M. (1996) Non-Chondritic Platinum-Group Element Ratios in the Earth's Mantle. Nature, 379, 712-715.
80. Ringwood, A.E. (1966) The Chemical Composition and Origin of the Earth. In: Hurley, P.M., Ed., Advances in Earth Science, M.I.T. Press, Cambridge, MA, 287- 356.
81. Righter, K., Humayun, M. and Danielson, L. (2008) Partitioning of Palladium at High Pressures and Temperatures during Core Formation. Nature Geoscience, 1, 321-323.
82. Prescher, C., Cernok, A., Allu Peddinti, D., Bell, E.A., Wielicki, M.M., Bello, L., Ghosh, N., Tucker, J. and Zahnle, K.J. (2012) Origin and Mixing Timescale of Earth's Late Veneer. Transactions of the AGU Annual Fall Meeting, Abstract V51B- 2775.
83. Lorand, J.-P., Alard, O., and Godard, M. (2009) Platinum-Group Element Signature of the Primitive Mantle Rejuvenated by Melt-Rock Reactions: Evidence from Sumail Peridotites (Oman Ophiolite). Terra Nova, 21, 35-40.
84. Dauphas, N. and Marty, B. (2002) Inference on the Nature and the Mass of Earth's Late Veneer from Noble Metals and Gases. Journal of Geophysical Research, 107, 121-127.
85. Maier, W.D., Barnes, S.J., Campbell, I.H., Fiorentini, M.L., Peltonen, P., Barnes, S.-J. and Smithies, R.H. (2009) Progressive Mixing of Meteoritic Veneer into the Early Earth's Deep Mantle. Nature, 460, 620-623.
86. Eppelbaum, L.V. (2007) Localization of Ring Structures in Earth's Environments. Journal of the Archaeological Society of the Slovakian Academy of Science, Special Issue: Archaeological Prospection, XLI, 145-148.
87. Cooper, G.R.J. (2010) Enhancing Circular Features in Potential Field Data. Exploration Geophysics, 41, 174-177.
88. Eppelbaum, L.V. and Khesin, B.E. (2012) Geophysical Studies in the Caucasus. Springer, Heidelberg, New York.
89. Elbeshausen, D., Wünnemann, K. and Collins, G.S. (2013) The Transition from Circular to Elliptical Impact Craters. Journal of Geophysical Research, 118, 1-15.
90. Koeberl, Ch. (2006) Impact Processes on the Early Earth. Elements, 2, 211-216.
91. Safronov, V.S. (1978) The Heating of the Earth during Its Formation. Icarus, 33, 3-12.
92. Vacquier, V. (1998) A Theory of the Origin of the Earth's Internal Heat. Tectonophysics, 291, 1-7.
93. Abe, Y. (1997) Thermal and Chemical Evolution of the Terrestrial Magma Ocean. Physics of the Earth and Planetary Interiors, 100, 27-39.
94. https://doi.org/10.1016/S0031-9201(96)03229-3
95. Pollack, H.N. (1997) Thermal characteristics of the Archaean. In: de Wit, M.J. and Ashwal, M.D., Eds., Greenstone Belts, Clarendon Press, Oxford, 223-232.
96. Solomatov, V.S. (2000) Fluid Dynamics of a Terrestrial Magma ocean, In: Canup, R. and Righter, K., Eds., Origin of the Earth and Moon. Univ. of Arizona Press, Tucson, Arizona, 323-338.
97. Goldstein, J.I., Scott, E.R.D. and Chabot, N.L. (2009) Iron Meteorites: Crystallization, Thermal History, Parent Bodies, and Origin. Chemie der Erde, 69, 293-325.
98. Alard, O., Griffin, W.L., Lorand, J.P., Jackson, S.E. and O'Reilly, S.Y. (2000) Non-Chondritic Distribution of the Highly Siderophile Elements in Mantle Sulphides. Nature, 407, 891-894.
99. Sankaran, A.V. (2009) Komatiites through Time Reflect Trends in Mantle Incorporation of Meteoritic Platinum Group Elements. Current Science, 97, 1280-1281.
100. Fiorentini, M.L., Barnes, S.J., Maier, W.D., Burnham, O.M. and Heggie, G. (2011) Global Variability in the Platinum-Group Element Contents of Komatiites. Journal of Petrology, 52, 83-112.

101. Tagle, R. and Claeys, Ph. (2005) An Ordinary Chondrite Impactor for the Popigai crater, Siberia. Geochimica et Cosmochimica Acta, 69, 2877-2889.
102. Firestone, R.B., West, A., Kennett, J.P., Becker, L., Bunch, T.E., Revay, Z.S., Schultz, P.H., Belgya, T., Kennett, D.J., Erlandson, J.M., Dickenson, O.J., Goodyear, A.C., Harris, R.S., Howard, G.A., Loosterman, J.B., Lechler, P., Mayewski, P.A., Montgomery, J., Poreda, R., Darrah, T., Que Hee, S.S., Smith, A.R., Stich, A., Topping, W., Wittke, J.H. and Wolbach, W.S. (2007) Evidence for an Extraterrestrial Impact 12,900 Years Ago That Contributed to the Megafaunal Extinctions and the Younger Dryas cooling. PNAS, 104, 16016-16021.
103. Firestone, R.B. (2009) The Case for the Younger Dryas Extraterrestrial Impact Event: Mammoth, Megafauna, and Clovis Extinction, 12,900 Years Ago. Journal of Cosmology, 2, 256-285.
104. Paquay, F.S., Goderis, S., Ravizza, G., Vanhaeck, F., Boyd, M., Surovell, T.A., Holliday, V.T., Haynes Jr., C.V. and Claeys, Ph. (2009) Absence of Geochemical Evidence for an Impact Event at the Bølling-Allerød/Younger Dryas Transition. PNAS, 106, 21505-21510.
105. Shearer, C.K., Hess, P.C., Wieczorek, M.A., Pritchard, M.E., Parmentier, E.M., Borg, L.E., Longhi, J., Elkins-Tanton, L.T., Neal, C.R., Antonenko, I., Canup, R.M., Halliday, A.N., Grove, T.L., Hager, B.H., Lee, D.-C. and Wiechert, U. (2006) Thermal and Magmatic Evolution of the Moon. Reviews in Mineralogy and Geochemistry, 60, 365-518.
106. Scoates, J.S. (2000) The Plagioclase-Magma Density Paradox Re-Examined and the Crystallization of Proterozoic Anorthosites. Journal of Petrology, 41, 627-649.
107. Pilchin, A.N. (2010) The Forsterite Layer and Density of Olivine under Lithospheric and Asthenospheric P-T-Conditions. Transations of the Annual Meeting GeoCanada, Calgary, Alberta (Abstract)
108. Hewins, R.H., Jones, R.H. and Scott, E.R.D., Eds. (1996) Chondrules and the Protoplanetary Disk. Cambridge University Press, Cambridge.
109. Davis, B.T.C. and England, J.L. (1964) The Melting of Forsterite up to 50 Kilobars. Journal of Geophysical Research, 69, 1113-1116.
110. Coombs, M.L. and Gardner, J.E. (2004) Reaction Rim Growth on Olivine in Silicic Melts—Implications for Magma Mixing. American Mineralogist, 89, 748-759.
111. Shiraishi, Y., Ikeda, K., Tamura, A. and Saito, T. (1978) On the Viscosity and Density of the Molten FeO-SiO2 System. Transactions of the Japan Institute of Metals and Materials, 19, 264-274.
112. de Koker, N.P., Stixrude, L. and Karki, B.B. (2008) Thermodynamics, Structure, Dynamics, and Freezing of Mg2SiO4 Liquid at High Pressure. Geochimica et Cosmochimica Acta, 72, 1427-1441.
113. Lange, R.A. (1997) A Revised Model for the Density and Thermal Expansivity of K2O-Na2O-CaO-MgO-Al2O3-SiO2 Liquids from 700 to 1900 K: Extension to Crustal Magmatic Temperatures. Contributions to Mineralogy and Petrology, 130, 1-11.
114. Knittle, E. (1995) Static Compression Measurements of Equations of State. In: Ahrens, T.J., Ed., Mineral Physics and Crystallography: A Handbook of Physical Constants. American Geophysical Union, 98-142.
115. Ghiorso, M. (2004) An Equation of State for Silicate Melts. Parts I, III, IV. American Journal of Science, 304, 637-678, 752-810, 811-838.
116. Courtial, Ph., Ohtani, E. and Dingwell, D.R. (1997) High-Temperature Densities of Some Mantle Melts. Geochimica et Cosmochimica Acta, 61, 3111-3119.
117. Walker, D., Agee, C.B. and Zhang, Y. (1988) Fusion Curve Slope and Crystal/Liquid Buoyancy. Journal of Geophysical Research, 93, 313-323.
118. Hall, A. (1995) Igneous Petrology. Longman Scientific & Technical, Singapore.
119. Carmichael, R.S. Ed. (1982) Handbook of Physical Properties of Rocks, Vol. I. CRC Press, Baca Roton.
120. Clark Jr., S.P., Ed. (1966) Handbook of Physical Constants. Revised Edition, Geological Society of America Memoirs 97, Washington DC.
121. Adams, L.X. (1924) Temperatures at Moderate Depths within the Earth. Journal of the Washington Academy of Sciences, 14, 459-472.

122. Clauser, Ch. (2009) Heat Transport Processes in the Earth's Crust. Surveys in Geophysics, 30, 163-191.
123. Rubin, A.E. (2003) Chromite-Plagioclase Assemblages as a New Shock Indicator; Implications for the Shock and Thermal Histories of Ordinary Chondrites. Geochimica et Cosmochimica Acta, 67, 2695-2709.
124. Alexandrov, V.I., Gordon, V.G., Muraviev, E.N., Osiko, V.V., Prokhorov, A.M., Spiridonov, E.G. and Tatarintsev, V.M. (1977) Method of Synthesis of Metal Chromites. United States Patent 4035266.
125. Bowen, N.L. and Tuttle, O.F. (1949) The System MgO-SiO2-H2O. Geological Society of America Bulletin, 60, 439-460.
126. Falloon, T.J., Green, D.H., Danyushevsky, L.V. and McNeill, A.W. (2008) The Composition of Near-Solidus Partial Melts of Fertile Peridotite at 1 and 1•5 GPa: Implications for the Petrogenesis of MORB. Journal of Petrology, 49, 591-613.
127. Azmon, E. (1965) Melting Temperature of Complex Silicates. Annals of the New York Academy of Sciences, 123, 481-494.
128. Canil, D. (1991) Experimental Evidence for the Exsolution of Cratonic Peridotite from High-Temperature Harzburgite. Earth and Planetary Science Letters, 106, 64-72.
129. Opta Minerals Inc. (2012) Chromite.
130. http://www.optaminerals.com/Foundry/Chromite.html
131. El-Rafei, E.A. (1971) Reversible Phase Transformation in the System Calcium Oxide-Chromium Sesquioxide-Oxygen. Jour. of Applied Chemistry and Biotechnology, 21, 261-265.
132. Mori, M., Yamamoto, T., Ichikawa, T. and Takeda, Y. (2002) Dense Sintered Conditions and Sintering Mechanisms for Alkaline Earth Metal (Mg, Ca and Sr)-Doped LaCrO3 Perovskites under Reducing Atmosphere. Solid State Ionics, 148, 93-101.
133. Crundwell, F., Moats, M., Ramachandran, V., Robinson, T. and Davenport, W.G. (2011) Extractive Metallurgy of Nickel, Cobalt and Platinum Group Metals. Elsevier.
134. Morozova, L.V. and Popov, V.P. (2010) Synthesis and Investigation of Magnesium Chromium Spinel. Glass Physics and Chemistry, 36, 86-91.
135. Oo, W.W.K.K., Aye, Sh.W.H. and Lwin, K.T. (2008) Study on the Production of Chromite Refractory Brick from Local Chromite Ore. World Academy of Science, Engineering and Technology, 46, 569-574.
136. Hockaday, S.A.C. and Bisaka, K. (2010) Sone Aspects of the Production of Ferrochrome Alloys in Pilot DC Arc Furnaces at Mintek. Proceedingas of the 12th International Ferroalloys Congress, Helsinki, Finland, 367-376.
137. Augustion, C.O., Prabhakaran, D. and Srinivasan, L.K. (1993) Fabrication and Characterization of NiCr2O4 Spinel. Journal of Materials Science Letters, 12, 383-386.
138. Song, S.-H., Yuan, Z.-X. and Xiao, P. (2003) Electrical Properties of MnCr2O4 Spinel. Journal of Materials Science Letters, 22, 755-757.
139. Escayola, M., Garuti, G., Zaccarini, F., Proenza, J.A., Bédard, J.H. and Van Staal, C. (2011) Chromitite and Platinum-Group-Element Mineralization at Middle Arm Brook, Central Advocate Ophiolite Complex, Baie Verte Peninsula, Newfoundland, Canada. Canadian Mineralogist, 49, 1523-1547.
140. Melcher, F., Grum, W., Simon, G., Thalhammer, T.V. and Stumpfl, E.F. (1997) Petrogenesis of the Ophiolitic Giant Chromite Deposits of Kempirsai, Kazakhstan: A Study of Solid and Fluid Inclusions in Chromite. Journal of Petrology, 38, 1419-1458.
141. Arif, M. and Qasim, J.M. (1993) Chemistry of Chromite and Associated Phases from the Shangla Ultramafic Body in the Indus Suture Zone of Pakistan. Geological Society, London, Special Publications, 74, 101-112.
142. Ligui, P. (1987) A Study on Inclusion in Chromite of Alpine-Type Ultrabasic Rocks in Western Jungar Area of Xinjiang, China. Acta Geosicientia Sinica, 9, 103-119.

143. Ahmed, A.H., Arai, S. and Attia, A.K. (2001) Petrological Characteristics of Podiform Chromitites and Associated Peridotites of the Pan African Ophiolite Complexes of Egypt. Mineralium Deposita, 36, 72-84.
144. Pushkarev, E.V., Anikina, E.V., Garuti, G. and Zaccarini, F. (2005) Postmagmatic Origin of Platinum Deposits in the Ural-Alaskan Type Ultramafites: T-fO2 Conditions and Role of Fluids. In: Törmänen, T.O. and Alapieti, T.T., Eds., Transactions of the 10th International Platinum Symposium, 223-226.
145. Nixon, G.T., Cabrill, J. and Gilles Laflamme, J.H. (1990) Platinum Group Element Mineralization in Lode and Placer Deposits Associated with the Tulameen Alaskan Type Complex, British Columbia. Canadian Mineralogist, 28, 503-535.
146. Benham, A.J., Kovac, P., Petterson, M.G., Rojkovic, I., Styles, M.T., Gunn, A.G., McKervey, J.A. and Wasy, A. (2009) Chromite and PGE in the Logar ophiolite Complex, Afghanistan. Trans. of the Inst. of Mining and Metallurgy Section B Applied Earth Science, 118, 45-58.
147. Ismail, S.A., Mirza, T.M. and Carr, P.F. (2010) Platinum-Group Elements Geochemistry in Podiform Chromitites and Associated Peridotites of the Mawat ophiolite, Northeastern Iraq. Journal of Asian Earth Sciences, 37, 3-41.
148. Orovetskii, Yu.P. (1999) Mantle Plumes. A. A. Belkema, Rotterdam, The Netherlands.
149. Marshall, J.S., Pounder, E.R. and Stewart, R.W. (1967) Physics. 2nd Edition, Macmillan of Canada, Toronto.
150. Dingwell, D.B. and Mysen, B.O. (1985) Effects of Water and Fluorine on the Viscosity of Albite Melt at High Pressure: A Preliminary Investigation. Earth and Planetary Science Letters, 74, 266-274.
151. Dingwell, D. B. (1987) Melt viscosities in the System NaAlSi3O8-H2O-F2O-1. In: Mysen, B.O., Ed., Magmatic Processes: Physicochemical Principles. The Geochemical Society, Special Publication No. 1, 423-431.
152. Toplis, M.J. Dingwell, D.B., Hess, K.-U. and Lenchi, T. (1997) Viscosity, Fragility, and Configurational Entropy of Melts along the Join SiO2-NaAlSiO4. American Mineralogist, 82, 979-990.
153. Scarfe, Ch.M. and Cronin, D.J. (1986) Viscosity-Temperature Relationships of Melts at 1 atm in the System Diopside-Albite. American Mineralogist, 71, 767-771.
154. Hummel, W. and Arndt, J. (1985) Variation of Viscosity with Temperature and Composition in the Plagioclase System. Contrib. to Mineralogy and Petrology, 90, 83-92.
155. Deer, W.A., Zussman, J. and Howie, R.A. (2001) Rock-Forming Minerals: Framework Silicates: Feldspars. Vol. 4a, 2nd Edition, The Geological Society of London, London.
156. Ganguly, J. (2008) Thermodynamics in Earth and Planetary Sciences. Springer-Verlag, Berlin-Heidelberg.
157. Ashwal, L. (1998) Anorthosites: Classification, Mythology, Trivia, and a Simple Unified Theory. LPI Technical Report #88-06. Houston, Texas, Lunar and Planetary Inst., 30-32.
158. Shaw, H.R. (1972) Viscosities of Magmatic Silicate Liquids: An Empirical Method of Prediction. American Journal of Science, 272, 870-893.
159. &OUML;hman, T. and Kring, D.A. (2012) Photogeologic Analysis of Impact Melt-Rich Lithologies in Kepler Crater That Could Be Sampled by Future Missions. Journal of Geophysical Research, 117, E00H08.
160. Hiemstra, S.A. (1979) The Role of Collectors in the Formation of the Platinum Deposits in the Bushveld Cornplex. Canadian Mineralogist, 17, 469-482.
161. Pilchin, A. (2005) The Role of Serpentinization in Exhumation of High- to Ultra-High-Pressure Metamorphic Rocks. Earth and Planetary Science Letters, 237, 815-828.
162. Matveev, S. and Ballhaus, Ch. (2002) Role of Water in the Origin of Podiform Chromitite Deposits. Earth and Planetary Science Letters, 203, 235-243.
163. Prichard, H.M., Barnes, S.-J., Maier, W.D. and Fisher, P.C. (2004) Variations in the Nature of the Platinum-Group Minerals in a Cross-Section through the Merensky Reef at Impala Platinum: Implications for the Mode of Formation of the Reef. Canad. Mineralogist, 42, 423-437.

164. Page, P., Barnes, S.-J., Zientek, M.L., Prichard, H.M. and Fisher, P.C. (2009) IPGE (Os, Ir, Ru) Are Not in Chromite. Proceedings of 24th IAGS (Fredericton), 197-200.
165. Pagé, P., Barnes, S.-J., Zientek, M.L., Prichard, H.M. and Fisher, P.C. (2010) Chromite Does Not Control IPGE (Os, Ir, Ru). Proceedings of the 11th International Platinum Symposium Ontario Geological Survey.
166. Ballhaus, Ch. and Sylvester, P. (2000) Noble Metal Enrichment Processes in the Merensky Reef, Bushveld Complex. Journal of Petrology, 41, 545-561.
167. Economou-Eliopoulos, M., Eliopoulos, D.G. and Chryssoulis, S. (2008) A Comparison of High-Au Massive Sulfide Ores Hosted in Ophiolite Complexes of the Balkan Peninsula with Modern Analogues: Genetic Significance. Ore Geology Reviews, 33, 81-100.
168. Yudovskaya, M., Kinnaird, J., Naldrett, A.J., Mokhov, A., Kuznetsova, M. and McDonald, I. (2010) Facies Variability of PGE Mineralization in Platreef Chromitites. Transactions of the 11th INternational Platinum Symposium, Ontario Geological Survey.
169. Huppert, H.E. and Sparks, R.S.J. (1985) Komatiites I: Eruption and Flow. Journal of Petrology, 26, 694-725.
170. Dingwell, D. B., Courtial, P., Giordano, D. and Nichols, A.R.L. (2004) Viscosity of Peridotite Liquid. Earth and Planetary Science Letters, 226, 127-138.
171. Liebske, Ch., Schmickler, B., Terasaki, H., Poe, B.T. Suzuki, A., Funakoshi, K.-I., Ando, R. and Rubie, D.C. (2005) Viscosity of Peridotite Liquid up to 13 GPa: Implications for Magma Ocean Viscosities. Earth and Planetary Science Letters, 240, 589-604.
172. Adjaoud, O., Steinle-Neumann, G. and Jahn, S. (2011) Transport Properties of Mg2SiO4 Liquid at High Pressure: Physical State of a Magma Ocean. Earth and Planetary Science Letters, 312, 463-470.
173. Fedortchouk, Y., LeBarge, W., Barkov, A.Y., Fedele, L. and Bodnar, R.J. (2010) Major- and Trace-Element Composition of Platinum Group Minerals and Their Inclusions from Several Yukon Placers. In: MacFarlane, K.E., Weston, L.H. and Blackburn, L.R., Eds., Yukon Exploration and Geology 2009. Yukon Geological Survey, 185-196.
174. Maier, W.D., de Klerk, L., Blaine, J., Manyeruke, T., Barnes, S.-J., Stevens, M.V.A. and Mavrogenes, J.A. (2008) Petrogenesis of Contact-Style PGE Mineralization in the Northern Lobe of the Bushveld Complex: Comparison of Data from the Farms Rooipoort, Townlands, Drenthe and Nonnenwerth. Mineralium Deposita, 43, 255-280.
175. Cameron, E.N. (1977) Chromite in the Central Sector of the Eastern Bushveld Complex, South Africa. American Mineralogist, 62, 1082-1096.
176. McDonald, I., Holwell, D.A. and Armitage, P.E.B. (2005) Stratigraphy and Platinum-Group Element Mineralization in the Northern Lobe of the Bushveld Complex, South Africa. In: Törmänen, T.O. and Alapieti, T.T., Eds., Transactions of the 10th International Platinum Symposium, Extended Abstracts, 193-196.
177. Boudreau, A. (1999) Fluid Fluxing of Cumulates: The J-M Reef and Associated Rocks of the Stillwater Complex, Montana. Journal of Petrology, 40, 755-772.
178. Boudreau, A.E. and McCallum, I.S. (1986) Investigations of the Stillwater Complex; III, The Picket Pin Pt/Pd Deposit. Economic Geology, 8, 1953-1975.
179. Saini-Eidukat, B., Weiblen, P.W., Bitsianes, G. and Glascock, D. (1990) Contrasts between Platinum Group Element Contents and Biotite Compositions of Duluth Complex Troctolitic and Anorthositic Series Rocks. Mineralogy and Petrology, 42, 121-140.
180. Talkington, R.W. and Watkinson, D.H. (1984) Trends in the Distribution of the Precious Metals in the Lac-Des-Iles Complex, Northwestern Ontario. Canadian Mineralogist, 22, 125-136.
181. Pettigrew, N.T. and Hattori, K.H. (2001) Geology of the Palladium-Rich Legris Lake Mafic-Ultramafic Complex, Western Wabigoon Subprovince, Northwestern Ontario. Exploration and Mining Geology, 10, 35-49.
182. Bayanova, T., Ludden, J. and Mitrofanov, F. (2009) Timing and Duration of Palaeoproterozoic Events Producing Ore-Bearing Layered Intrusions of the Baltic

Shield: Metallogenic, Petrological and Geodynamic Implications. Geological Society, London, Special Publications, 323, 165-198.
183. Orsoev, D.A., Kislov, E.V. and Konnokov, E.G. (1995) PGE-Bearing Anorthosites of the Ioko-Dovyren Layered Massif, Northern Transbaikalia, Russia. In: Pasava, J., Kríbek, B. and Zák, K., Eds., Mineral Deposits: From Their Origin to Their Environmental Impacts, A. A. Balkema, Rotterdam, The Netherlands, 181-182.
184. Oktyabrsky, R.A., Solyanik, A.N., Lennikov, A.M., Gvozdev, V.I. and Ivanov, V.V. (2010) Compositional Variations in Platinum-Group Minerals and Gold in Lenses of Massive Pyrrhotite in the Dzhugdzhur Anorthosite, Far-Eastern Russia. Canad. Mineralogist, 48, 991-1004.
185. Latypov, R.M., Chistyakova, S.Yu. and Alapieti, T.T. (2008) PGE Reefs as an In Situ Crystallization Phenomenon: The Nadezhda Gabbronorite Body, Lukkulaisvaara Layered Intrusion, Fennoscandian Shield, Russia. Mineralogy and Petrology, 92, 211-242.
186. Dora, M.L., Nair, K.K.K. and Shasidharan, K. (2011) Occurrence of Platinum Group Minerals in the Western Bastar Craton, Chandrapur District, Maharashtra. Current Science, 100, 399-405.
187. Karazinov, N.V. (2005) Decriptometric Signs of presence Platinoid-Bearing Autonomous Anorthosites on Example of Geransky Massif (Far East). Proceedings of 9th International Scientific Symposium, Tomsk Polytechncal University, Tomsk, Russia. (In Russian)
188. Collins, P.G. and Wilton, D.H.C. (2005) Evaluation and Characterization of the Ni-Cu Sulfide and PGE Potential in Three Mafic-Ultramafic Intrusive Complexes: Preliminary Results. Current Research, Newfoundland and Labrador Department of Natural Resources, Geological Survey, Report 05-1, 131-149.
189. Pilchin, A.N. (2014) The Involvement of Forsterite and Dunite Layers of the Oceanic Lithosphere in Obduction during Formation of the Appalachians. Geological Association of Canada and Mineralogical Association of Canada Joint Annual Meeting, Fredericton, 37, 223-224.
190. Cawthorn, R.G. and Ashwal, L.D. (2009) Origin of Anorthosite and Magnetitite Layers in the Bushveld Complex, Constrained by Major Element Compositions of Plagioclase. Journal of Petrology, 50, 1607-1637.
191. Severson, M.J. and Hauck, S.A. (2003) Platinum Group Elements (PGEs) and Platinum Group Minerals (PGMs) in the Duluth Complex. Minerals Coordinating Committee and the Natural Resources Research Inst. Technical Report NRRI/TR-2003/37.
192. Marma, J.C. (2002) Magmatic and Hydrothermal PGE Mineralization of the Birch Lake Cu-Ni-PGE Deposit in the South Kawishiwi Iutrusion, Duluth Complex, Northeast Minnesota. M.Sci. Thesis, University of Wisconsin, USA.
193. Ghisler, M. (1976) The Geology, Mineralogy and Geochemistry of the Pre-Orogenic Archean Stratiform Chromite Deposits at Fiskenæsset, West Greenland. Monog. Series on Mineral Deposits, 14.
194. Polat, A., Fryer, B.J., Appel, P.W.U., Kalvig, P., Kerrich, R., Dilek, Y. and Yang, Z. (2011) Geochemistry of Anorthositic Differentiated Sills in the Archean (~2970 Ma) Fiskenæsset Complex, SW Greenland: Implications for Parental Magma Compositions, Geodynamic Setting, and Secular Heat Flow in Arcs. Lithos, 123, 50-72.
195. Mukherjee, R., Mondal, S.K., Frei, R., Rosing, M.T., Waight, T.E., Zhong, H. and Kumar, G.R.R. (2012) The 3.1 Ga Nuggihalli Chromite Deposits, Western Dharwar craton (India): Geochemical and Isotopic Constraints on Mantle Sources, Crustal Evolution and Implications for Supercontinent Formation and Ore Mineralization. Lithos, 155, 392-409.
196. Puchtel, I.S. and Humayun, M. (2001) Platinum Group Element Fractionation in a Komatiitic Basalt Lava Lake. Geochimica et Cosmochimica Acta, 65, 2979-2993.

197. Stephen, R., Sparks, J. and Huppert, H.E. (1984) Density Changes during the Fractional Crystallization of Basaltic Magmas: Fluid Dynamic Implications. Contributions to Mineralogy and Petrology, 85, 300-309.
198. Eales, H.V., Field, M., De Klerk, W.J. and Scoon, R. N. (1988) Regional Trends of Chemical Variation and Thermal Erosion in the Upper Critical Zone, Western Bushveld Complex. Mineralogical Magazine, 52, 63-79.
199. Latypov, R.M., Chistyakova, S.Yu. and Alapieti, T.T. (2008) Fine-Grained Mafic Bodies as Preserved Portions of Magma Replenishing Layered Intrusions: The Nadezhda Gabbronorite body, Lukkulaisvaara Intrusion, Fennoscandian Shield, Russia. Mineralogy and Petrology, 92, 165-209.
200. Papike, J., Ryder, G. and Shearer, Ch. (1998) Lunar Samples. Reviews in Mineralogy and Geochemistry, 36, 5.1-5.234.
201. Czamanske, G.K. and Bohlen, S.R. (1990) The Stillwater Complex and Its Anorthosites: An Accident of Magmatic Underplating. American Mineralogist, 75, 37-45.
202. Blatt, H. and Tracy, R.J. (1996) Petrology: Igneous, Sedimentary and Metamorphic. 2nd Edition, Macmillan, USA.
203. Mottana, A., Crespi, R. and Liborio, G. (1977) Guide to Rocks and Minerals. Simon & Schuster Inc., New York.
204. Boudreau, A.E. and McCallum, I.S. (1984) The Picket Pin Pt/Pd Zone, Stillwater Complex, Montana. Lunar and Planetary Science, XV, 78-79.
205. Mitchell, A.A. and Manthree, R. (2002) The Giant Mottled Anorthosite: A Transitional Sequence at the Top of the Upper Critical Zone of the Bushveld Complex. South African Journal of Geology, 105, 15-24.
206. Kruger, F.J. (2010) The Merensky and Bastard Cyclic Units and the Platreef of the Bushveld Complex: Consequences of Main Zone Magma Influxes and Dynamics. Transactions of the 4th International Platinum Conference, Platinum in Transition 'Boom or Bust', The South African Institute of Mining and Metallurgy, 43-46.
207. Weiblen, P.W. and Morey, G.B. (1980) A Summary of the Stratigraphy, Petrology, and Structure of the Duluth Complex. American Journal of Science, 280-A, 88-133.
208. Scoon, R.N. and Mitchell, A.A. (2004) The Platiniferous Dunite Pipes in the Eastern Limb of the Bushveld Complex: Review and Comparison with Unmineralized Discordant Ultramafic Bodies. South African Journal of Geology, 107, 505-520.
209. Scoon, R.N. and Mitchell, A.A. (2011) The Principal Geological Features of the Mooihoek Platiniferrous Dunite Pipe, Eastern Limb of the Bushveld Complex, and Similarities with Replaced Merensky Reef at the Amandelbult Mine, South Africa. South African Journal of Geology, 114, 15-40.
210. Scoon, R.N. and Mitchell, A.A. (2010) The Principal Geological Features of the Onverwacht Platiniferrous Dunite Pipe, Eastern Limb of the Bushveld Complex. South African Journal of Geology, 113, 155-168.
211. Barnes, S.J. (1998) Chromite in Komatiites, 1. Magmatic Controls on Crystallization and Composition. Journal of Petrology, 39, 1689-1720.
212. Cotterill, P. (1969) The Chromite Deposits of Selukwe, Rhodesia. Transactions of Symposium on Magmatic Ore Deposits. Society of Economic Geologists. Monograph, 4, 154-186.
213. Cotterill, P. (1979) The Selukwe Schist BELT and Its Chromitite Deposits. Geological Society of South Africa Special Publication, 5, 229-245.
214. Hudgins, J.A., Walton, E.L. and Spray, J.G. (2007) Mineralogy, Petrology, and Shock History of Lunar Meteorite Sayh al Uhaymir 300: A Crystalline Impact-Melt breccia. Meteoritics & Planetary Science, 42, 1763-1779.
215. Scarfe, Ch.M., Cronin, D.J., Wenzel, J.T. and Kauffman, D.A. (1983) Viscosity-Temperature Relationships at I atm in the System Diopside-Anorthite. American Mineralogist, 6E, 1083-1088.
216. Williams-Jones, A.E. and Heinrich, Ch.A. (2005) Vapor Transport of Metals and the Formation of Magmatic-Hydrothermal Ore Deposits. Economic Geology, 100, 1287-1312.

217. Cawthorn, R.G. (2005) Stratiform Platinum-Group Element Deposits in Layered Intrusions. In: Mungall, J.E., Ed., Exploration for Platinum-Group Elements Deposits, Mineralogical Association of Canada, Short Course Series, 35, 57-73.
218. Iljina, M.J. and Lee, C.A. (2005) Chapter 4: PGE Deposits in the Marginal Series of Layered Intrusions. In: Mungall, J.E., Ed., Exploration for Platinum-Group Element Deposits, Ottawa, Mineralogical Association of Canada Short Course Series, 35, 75-96.
219. Starostin, V.I. and Sorokhtin, O.G. (2011) A New Interpretation for the Origin of the Norilsk Type PGE-Cu-Ni Sulfide Deposits. Geoscience Frontiers, 2, 583-591.
220. Taylor Jr., H.P. (1967) The Zoned Ultramaphic Complexes of Southern Alaska. In: Wyllie, P.J., Ed., Ultramafic and Related Rocks, John Wiley & Sons, New York, 97-121.
221. Tolstykh, N.D., Sidorov, E.G. and Krivenko, A.P. (2005) Platinum-Group Element Placers Associated with Ural-Alaska Type Complexes. In: Mungall, J.E., Ed., Exploration for Platinum-Group Elements Deposits, Mineralogical Association of Canada Short Course Series, 35, 113-143.
222. Wilde, A. (2005) Descriptive Ore Deposit Models: Hydrothermal and Supergene Pt & Pd Deposits. In: Mungall, J.E., Ed., Exploration for Platinum-Group Elements Deposits, Mineralogical Association of Canada, Short Course Series, 35, 145-161.
223. Melcher, F., Lodziak, J. and Oberthur, T. (2005) Modifications of Detrital PGM from the Eastern Bushveld Complex, South Africa. In: Törmänen, T.O. and Alapieti, T.T., Eds., Transactions of the 10th International Platinum Symposium, 201-204.
224. Wilson, A.H. (1996) The Great Dyke of Zimbabwe. In: Cawthorn, R.G., Ed., Layered Intrusions, Elsevier, 365-402.
225. Li, C. and Ripley, E.M. (2006) Formations of Pt-Fe Alloy by Desulfurization of Pt-Pd Sulfide in the J-M Reef of the Stillwater Complex, Montana. Canadian Mineralogist, 44, 895-903.
226. Weiblen, P.W. and Morey, G.B. (1976) Textural and Compositional Characteristics of Sulfide Ores from the Basal Contact Zone of the South Kawiskiwi Intrusion, Duluth Complex, Northeastern Minnesota. Transactions of the 37th Annual Mining Symposium, Minnesota Geological Survey, Minnesota University, Reprint No. 32.
227. Watkinson, D.H. and Dunning, G. (1979) Geology and Platinum-Group Mineralization, Lac-Des-Iles Complex, Northwestern Ontario. Canadian Mineralogist, 17, 453-462.
228. Sobolev, P.O. (1990) Orientation of Acicular Iron-Ore Mineral Inclusions in Plagioclase. Intern. Geology Review, 32, 616-628.
229. McKelson, J.F., Thalhammer, O.A.R. and Paliulionyte, V. (2005) The Dunite Complex of the Guli Massif, Northern Siberia, Russia: A Multidisciplinary Study. In: Törmänen, T.O. and Alapieti, T.T., Eds., Transactions of the 10th International Platinum Symposium, Extended Abstracts, 197-200.
230. Varlakov, A.S. (1995) Riftogenic Ophiolites, Metamorphism of Hyperbasites and Structure of Vishnevogorsk-Il'menogorsky Complex. Institute of Mineralogy, The Urals Branch of the Russian Academy of Sciences. (In Russian)
231. Varlakov, A.S. (1996) Riftogenic Ophiolites in Geological Evolution of the Earth. Bulletin of the Moscow Society of Nature Investigators, Series: Geology, 71, 19-30.
232. Pilchin, A.N. (1986) On the Role of Pressures in Tectonic Processes. VINITI Press, Russian Academy of Sciences, No. 3723-86, 1-23.
233. Pilchin, A.N. and Eppelbaum, L.V. (2002) Some Peculiarities of Thermodynamic Conditions of the Earth Crust and Upper Mantle. Scientific Israel, 4, 117-142.
234. Peck, D.C. and Theyer, P. (1998) PGE-Copper-Nickel Potential of Mafic-Ultramafic Intrusions in the Bird River Greenstone Belt (Parts of NTS 52L). Manitoba Energy and Mines, Geological Services, Report of Activities, 151-160.
235. Cameron, E.N. (1980) Evolution of the Lower Critical Zone, Central Sector, Eastern Bushveld Complex, and Its Chromite Deposits. Economic Geology, 75, 845-871.
236. Hatton, C.J. and von Gruenewaldt, G. (1987) The Geological Setting and Petrogenesis of the Bushveld Chromitite Layers. In: Stowe, C.W., Ed., Evolution of Chromium Ore Fields. N. Y. Van Nostrand Reinhold, 109-143.

237. Mondal, S.K. and Mathez, E.A. (2007) Origin of the UG2 Chromitite Layer, Bushveld Complex. Journal of Petrology, 48, 495-510.
238. Pilchin, A. and Pilchin, M. (2014) Repeating Obduction in the Process of Forming the Canadian Appalachians. Transactions of Geological Association of Canada and Mineralogical Association of Canada Joint Annual Meeting, Frederiction, 37, 222-223.
239. Pilchin, A.N. (1996) Tectonic and Petrologic Peculiarities of the Precambrian Evolution of the Baltic Shield. Transactions of the 58th Conference of European Association of Geophysicists and Engineers, Amsterdam, The Netherlands, 513.
240. Pilchin, M., and Pilchin, A. (1998) Some Peculiarities of Ocean Crust Activity in the Appalachian Region. Transactions of Geological Society of America Annual Meeting, Toronto, A-352.
241. Hughes, S. and Luetgert, J. H. (1992) Crustal Structure of the Southeastern Grenville Province Northern New York State and Eastern Ontario. Journal of Geophysical Research, 97, 17455-17479.
242. Ebel, J.E. (2010) Resolving Lateral Variations in Upper Mantle Structure of Eastern North America: An Opportunity for USARRAY. Transactions of the 82th Annual Meeting of the Eastern Section of the Seismological Society of America, Boston, 18-19 October 2010, 18.
243. Németh, B., Clowes, R.M. and Hajnal, Z. (2005) Lithospheric Structure of the Trans-Hudson Orogen from Seismic Refraction-Wide-Angle Reflection Studies. Canadian Journal of Earth Sciences, 42, 435-456.
244. Prodehl, C. and Mooney, W.D. (2012) Exploring the Earth's Crust: History and Results of Controlled-Source Seismology. The Geological Society of America, Memoir 208.
245. Keen, M.J., Loncarevic, B.D. and Ewing, G.N. (2005) Continental Margin of Eastern Canada: Georges Bank to Kane Basin. In: Maxwell, A.E., Bullard, E. and Worzel, J.L., Eds., New Concepts of Sea Floor Evolution: Regional Observations Concepts, Harvard University Press, Ch. 7, 251-292.
246. Dainty, A.M., Keen, C.E., Keen, M.J. and Blanchard, J.E. (1966) Review of Geophysical Evidence on Crust and Upper-Mantle Structure on the Eastern Seaboard of Canada. In: Steinhart, J.S. and Smith, T.J., Eds., The Earth Beneath the Continents, A Volume of Geophysical Studies in Honor of Merle A. Tuve, AGU, Geophys. Monogr. Ser., Washington, D. C., 10, 349-369.
247. Rankin, D.S., Ravindra, R. and Zwicker, D. (1969) Preliminary Interpretation of the First Arrival Refraction Arrivals in Gaspé from Shots in Labrador and Quebec. Canadian Journal of Earth Sciences, 6, 771-774.
248. https://doi.org/10.1139/e69-071
249. Hughes, S., Hall, J. and Luetgert, J.H. (1994) The Seismic Velocity Structure of the Newfoundland Appalachian Orogen. Journal of Geophys. Research: Solid Earth, 99, 13633-13653.
250. Chian, D., Marillier, F., Hall, J. and Quinlan, G. (1998) An Improved Velocity Model for the Crust and Upper Mantle along the Central Mobile Belt of the Newfoundland Appalachian Orogen and Its Offshore Extension. Canadian Journal of Earth Sciences, 35, 1238-1251.
251. Max, M.D. and Ohta, Y. (1988) Did Major Fractures in Continental Crust Control Orientation of the Knipovich Ridge—Lena Trough Segment of the Plate Margin? Polar Research, 6, 85-93.
252. Mjelde, R., Goncharov, A. and Dietmar Müller, R. (2013) The Moho: Boundary above Upper Mantle Peridotites or Lower Crustal Eclogites? A Global Review and New Interpretations for Passive Margins. Tectonophysics, 609, 636-650.
253. Van Avendonk, H.J.A., Holbrook, W.S., Okaya, D., Austin, J.K., Davey, F. and Stern, T. (2004) Continental Crust under Compression: A Seismic Refraction Study of South Island Geophysical Transect I, South Island, New Zealand. Journal of Geophys. Research, 109, B06302.

254. Kostyuchenko, S.L., Fedorow, D.L., Solodilov, L.N., Egorkin, A.V. and Zolotov, E.E. (1999) Deep Structure and Evolution for the Pre-Caspian Basin. Transactions of the EUG, 10, Strasbourg, France, 4, G01, 3A/14, G3.

255. Piip, V.B., Rodnikov, A.G. and Buvaev, N.A. (2012) The Deep Structure of the Lithosphere along the Caucasus—South Caspian Basin—Absheron Threshold—Middle Caspian Basin—Turan Plate Seismic Profile. Moscow University Geology Bulletin, 67, 125-132.

256. De Souza, S., Tremblay, A., Ruffet, G. and Pinet, N. (2012) Ophiolite Obduction in the Quebec Appalachians, Canada—40Ar/39Ar Age Constraints and Evidence for Syn-Tectonic Erosion and Sedimentation. Canadian Journal of Earth Sciences, 49, 91-110.

257. Hannan, B.B. and Sinha, A.K. (1989) Petrology and Tectonic Affinity of the Baltimore Mafic Complex, Maryland. In: Mittwede, S.K. and Stoddard, E.F., Eds., Ultramafic Rocks of the Appalachian Piedmont. Geological Society of America Special Paper 231, 1-18.

258. Laurent, R. and Hébert, Y. (1979) Paragenesis of Serpentine Assemblages in Harzburgite Tectonite and Dunite Cumulate from the Quebec Appalachians. Canadian Mineralogist, 17, 857-869.

259. Condie, K.C. and Madison, J.A. (1969) Compositional and Volume Changes Accompanying Progressive Serpentinization of Dunites from the Webster-Addie Ultramafic Body, North Carolina. American Mineralogist, 51, 1173-1179.

260. Minarik, W.G., Gale, A. and Booker, C. (2003) Mont Albert to Buck Mountain: Provenance of Appalachian Ophiolite Chromites Using Osmium Isotopes. Transactions of the American Geophysical Union Fall Meeting, Abstract #V11E-0534.

261. Salisbury, M.H. and Christensen, N.I. (1985) Olivine Fabrics in the Bay of Islands Ophiolite: Implications for Oceanic Mantle Structure and Anisotropy. Canadian Journal of Earth Sciences, 22, 1757-1766.

262. Paiement, J.-Ph., Gagne, J., Duplessis, C., Rousseau, G., Gagnon, G. and Dagbert, M. (2013) Pre-Feasibility Study Mine Arnaud Inc. Sept-Iles Deposit, Québec. Final Report. SGS Canada Inc.

263. Page, P., Bedard, J.H., Tremblay, A. and Schroetter, J. (2003) The Thetford Mines Ophiolite Complex: Focus on the Petrology, Mineralogy and Geochemistry (REE, PGE) of a Supra-Subduction Mantle Section. Transactions of the American Geophysical Union, Fall Meeting, Abstract #V22H-01.

264. Scott, P.W., Jackson, T.A. and Dunham, A.C. (2000) Ore Mineral Associations and Industrial Minerals in the Ultramafic Rocks of Jamaica and Tobago. Caribbean Journal of Earth Science, 34, 5-16.

265. Sole, J.P.J. and Melgarejo, J.C. (1999) Uvarovite in Podiform Chromitite: The Moa-Baracoa Ophiolitic Massif, Cuba. Canadian Mineralogist, 37, 679-690.

266. Brem, A.G., Lin, S. and van Staal, C.R. (2002) Humber Zone—Dunnage Zone Relationships and the Long Range Fault, South of Grand Lake, Western Newfoundland: Preliminary Results. In: Current Research. Newfoundland and Labrador Department of Mines and Energy, Geological Survey Branch, Report: 02-1, 135-144.

267. Pettingill, H.S., Sinha, A. K. and Tatsumoto, M. (1984) Age and Origin of anorthosites, Charnockites, and Granulites in the Central Virginia Blue Ridge: Nd and Sr Isotopic Evidence. Contributions to Mineralogy and Petrology, 85, 279-291.

268. Murthy, G.S. and Rao, K.V. (1976) Paleomagnetism of Steel Mountain and Indian Head Anorthosites from Western Newfoundland. Canadian Journal of Earth Sciences, 13, 75-83.

269. Haworth, R.T. (1978) Interpretation of Geophysical Data in the Northern Gulf of St. Lawrence and Its Relevance to Lower Paleozoic Geology. Geological Society of America Bulletin, 89, 1091-1110.

270. https://doi.org/10.1130/0016-7606(1978)89<1091:IOGDIT>2.0.CO;2

271. Rao, K.V. and Van der Voo, R. (1980) Paleomagnetism of a Paleozoic Anorthosite from the Appalachian Piedmont, Northern Delaware: Possible Tectonic Implications. Earth and Planetary Science Letters, 47, 113-120.
272. McLelland, J.M., Selleck, B.W., Hamilton, M.A. and Bickford, M.E. (2010) Late-to Post-Tectonic Setting of Some Major Proterozoic Anorthosite—Mangerite—Charnockite—Granite (AMCG) Suites. Canadian Mineralogist, 48, 1025-1046.
273. Aleinikoff, J.N., Horton Jr., J.W. and Walter, M. (1996) Middle Proterozoic Age for the Montpelier Anorthosite, Goochland Terrane, Eastern Piedmont, Virginia. Geological Society of America Bulletin, 108, 1481-1491.
274. Crawford, W.A., Robelen, P.G. and Kalmbach, J.H. (1971) The Honey Brook Anorthosite. American Journal of Science, 271, 333-349.
275. Owens, B.E., Iriarte, K.E. and Dymek, R.F. (2004) The Smallest, but Most Potassic Massif Anorthosite on Earth: Geochemistry of the Montpalier Pluton, Goochland Terrane, Virginia. Geological Society of America, Denver Annual Meet. Abstracts with Programs, 36, 220.
276. John, J.St. (2012) Rocks from Ophiolites.
277. http://www.newark.osu.edu/facultystaff/personal/jstjohn/Documents/Cool-Rocks/Ophiolite-rocks.htm
278. Barnes, S. and Jones, S. (2013) Deformed Chromitite Layers in the Coobina Intrusion, Pilbara Craton, Western Australia. Economic Geology, 108, 337-354.
279. Hutcinson, Ch.S. (1972) Alpine-Type Chromite in North Borneo, with Special Reference to Darvel Bay. American Mineralogist, 57, 835-856.
280. Page, P., Barnes, S.J. and Cox, R. (2007) Laser Ablation Study of Trace Elements in Chromite: Thetford Mines Ophiolite Chromitite Ores. Geochimica et Cosmochimica Acta, 71, A748-A748.
281. Tsoupas, G. and Economou-Eliopulos, M. (2008) High PGE Contents and Extremely Abundant PGE-Minerals Hosted in Chromitites from the Veria Ophiolite Complex, Northern Greece. Ore Geology Reviews, 33, 3-19.
282. Ahmed, A.H. and Arai, S. (2003) Platinum-Group Elements Mineralogy of Late Proterozoic Podiform Chromitites from the Eastern Desert of Egypt: A Preliminary Result. Proceed. of 7th Biennial SGA Meeting, Athens, Greece. In: Eliopoulos, et al., Eds., Mineral Exploration and Sustainable Development, Millpress, Rotterdam, 555-558.
283. Kapsiotis, A., Grammatikopoulos, T.A., Tsikouras, B. and Hatzipanagiotou, K. (2009b) Platinum-Group Mineral Characterization in Concentrates from High- Grade PGE Al-Rich Chromitites of Korydallos Area in the Pindos Ophiolite Complex (NW Greece) Resource Geology, 60, 178-191.
284. Zaccarini, F., Garuti, G. and Pushkarev, E.V. (2011) Unusually PGE-Rich Chromitite in the Butyrin Vein of the Kytlym Uralian–Alaskan Complex, Northern Urals, Russia. Canadian Mineralogist, 49, 1413-1431.
285. Zaccarini, F., Pushkarev, E.V., Fershtater, G.B. and Garuti, G. (2004) Composition and Mineralogy of PGE-Rich Chromitites in the Nurali Lherzolite-Gabbro Complex, Southern Urals, Russia. Canadian Mineralogist, 42, 545-562.
286. González-Jiménez, J.M., Augé, T., Gervilla, F., Bailly, L., A. Proenza, J. and Griffin, W.L. (2011) Mineralogy and Geochemistry of Platinum-Rich Chromitites from the Mantle-Crust Transition Zone at Quin Island, New Caledonia Ophiolite. Canad. Mineralogist, 49, 1549-1569.
287. Prichard, H.M., Neary, C.R., Fisher, P.C. and O'Hara, M.J. (2008) PGE-Rich Podiform Chromitites in the Al'Ays Ophiolite Complex, Saudi Arabia: An Example of Critical Mantle Melting to Extract and Concentrate PGE. Economic Geology, 103, 1507-1529.
288. Çina, A., Neziraj, A., Karaj, N., Johan, Z. and Ohnenstetter, M. (2002) PGE Mineralization Related to Albanian Ophiolitic Complex. Geologica Carpathica, 53, 1-7.
289. Routledge, R.E., Cox, J.J., Scott, K.C. and Hwozdyk, L.R. (2010) Technical Report on the Lac Des Iles Mine Property. Thunder Bay, Ontario, Canada. NI 43-101 Report.

290. Prichard, H.M., Fisher, P.C., McDonald, I., Zhou, M.-F. and Wang, C.Y. (2005) Platinum-Group Minerals in the Jinchuan Complex, China. In: Törmänen, T.O. and Alapieti, T.T., Eds., Transactions of the 10th International Platinum Symposium, 219-222.
291. Smith II, R.C. and Barnes, J.H. (2011) Geochemistry and Mineralogy of Platinum Group Elements in Some Chromite Occurrences in the State Line District, Chester and Lancaster Counties, Pennsylvania. Pennsylvania Geology, 41, 3-13.
292. Gervilla, F., Frei, R., González-Jiménez, J.M., Garrido, C.J., Proenza, J., Melgarejo, J.C., Meibom, A., Batista, J.A. and Ruiz, R. (2005) Platinum-Group Elements and Os Isotope Systematics in Chromite Ores from Mayarí-Baracoa Ophiolitic Belt (Eastern Cuba). Transactions of the 10th International Platinum Symposium, Oulu, Finland, 86-89.
293. Foley, J.Y. (1991) Metallogeny of Ophiolitic and Other Mafic-Ultramafic Terranes in Alaska. US Department of the Interior, Geological Survey. US Bureau of Mines Unpublished Report.
294. Kapsiotis, A., Grammatikopoulos, T.A., Tsikouras, B., Hatzipanagiotou, K., Zaccarini, F. and Garuti, G. (2009a) Chromian Spinel Composition and Platinum-Group Element Mineralogy of Chromitites from the Milia Area, Pindos Ophiolite Complex, Greece. Canadian Mineralogist, 47, 1037-1056.
295. Gervilla, F., Frei, R., González-Jiménez, J.M., Kerestedjian, T. and Glavev, B. (2008) Geochemistry of platinum Group Elements in Chromitites from the Rhodope Massif (Bulgaria). Revista de la Sociedad Española de Mineralogía, No. 9, 115-116.
296. Volchenko, Yu.A., Koroteev, V.A. and Neustroeva, I.I. (2009) Platinum-Group Elements in Alpine-Type Ultramafic Rocks and Related Chromite Ores of the Main Ophiolite Belt of the Urals. Geology of Ore Deposits, 51, 162-178.
297. Gurskaya, L.I., Smelova, L.V. and Shahova, S.N. (2005) PGE Mineralization Associated with Ultramafic Intrusions of the Polar Urals: New Data to Beneficiation. In: Törmänen, T.O. and Alapeiti, T.T., Eds., Trans. of the 10th Intern. Plat. Symp. "Platinum-Group Elements—From Genesis to Beneficiation and Environmental Impact", Finland, 359-361.
298. Uysal, I., Sadiklar, M.B., Zaccarini, F., Garuti, G., Tarkian, M. and Meisel, T. (2010) Cr-PGE Mineralization in the Turkish Ophiolites. Transactions of the State of Art, 11th International Platinum Symposium, 1-3.
299. Uysal, I., Zaccarini, F., Sadiklar, M.B., Tarkian, M., Thalhammer, O.A.R. and Garuti, G. (2009) The Podiform Chromitites in the Dagküplü and Kavak Mines, Eskisehir Ophiolite (NW-Turkey): Genetic Implications of Mineralogical and Geochemical data. Geologica Acta, 7, 351-362.
300. Proenza, J.A., Zaccarini, F., Lewis, J.F., Longo, F. and Garuti, G. (2007) Chromian Spinel Composition and the Platinum-Group Minerals of the PGE-Rich Loma Peguera Chromitites, Loma Caribe Peridotite, Dominican Republic. Canadian Mineralogist, 45, 631-648.
301. Jowitt, S.M., Keays, R.R. and Peck, D.C. (2010) New Insights into the Geology and Mineral Potential of the West Raglan Ni-Cu-PGE Project, Cape Smith Fold Belt, Ontario Geological Survey, Miscellaneous Release-Data.
302. Paktunc, A.D. (1990) Comparative Geochemistry of platinum-Group Elements of Nickel-Copper Sulfide Occurrences Associated with Mafic-Ultramafic Intrusions in the Appalachian Orogen. Journal of Geochemical Exploration, 37, 101-111.
303. Tanguay, S. and Hebert, R. (1990) Distribution of PGE in Pyroxene-Bearing Ultramafic Cumulates in the Thetford Mines ophiolitic Complex, Quebec. Canadian Mineralogist, 28, 597-605.
304. Papunen, H., Distler, V. and Sokolov, A. (1992) PGE in the Upper Proterozoic Dovirensky Layered Complex, North Baikal Area, Siberia. Australian Journal of Earth Sciences, 39, 327-334.
305. Lipin, B.R. (1984) Chromite from the Blue Ridge Province of North Carolina. Amer. Journal of Science, 284, 507-529.

306. Lasley, S. (2012) From Placer Streams to Historical Mines, Hints of Elusive Metal span Alaska. Petroleum News, Newspaper, Anchorage, Alaska.
307. Sahoo, R.K., Mohanty, J.K., Das, S.K. and Paul, A.K. (2009) Chromites of India Their Textural and Mineralogical Characteristics. ISMO-2009. Transactions of the International Symposium on Magmatic Ore Deposits. Bhubaneswar, Orissa, India.
308. Mondal, S.K. and Baidya, T.K. (1997) Platinum-Group Minerals from the Nuasahi Ultramafic-Mafic Complex, Orissa, India. Mineralogical Magazine, 61, 902-906.
309. Endress, M. and Bischoff, A. (1993) Mineralogy, Degree of Brecciation, and Aqueous Alteration of CI Chondrites Orgueil, Ivuna, and Alais. Meteoritics, 28, 345-346.
310. Brearley, A.J. (1992) Mineralogy of Fine-Grained Matrix in the Ivuna CI Carbonaceous Chondrite. Lunar and Planetary Science, 23, 153-154.
311. Johnson, C.A. and Prinz, M. (1991) Chromite and Olivine in Type II Chondrules in Carbonaceous and Ordinary Chondrites: Implications for Thermal Histories and Group Differences. Geochimica et Cosmochimica Acta, 55, 893-904.
312. Buseck, P.R. and Hua, X. (1993) Matrices of Carbonaceous chondrite Meteorites. Annual Review of Earth and Planetary Sciences, 21, 255-305.
313. https://doi.org/10.1146/annurev.ea.21.050193.001351
314. Shukolyukov, A., Lugmair, G.W. and Bogdanovski, O. (2003) Manganese-Chro- mium Isotope Systematics of Ivuna, Kainsaz and Other Carbonaceous Chondrites. Lunar and Planetary Science, XXXIV, Abstract 1279.
315. Cloutisa, E.A., Hiroi, T., Gaffey, M.J. and Alexander, C.M. O'D. and Mann, P. (2011) Spectral Reflectance Properties of Carbonaceous Chondrites: 1. CI Chondrites. Icarus, 212, 180-209.
316. Visa, R.D. and Heymann, D. (1999) On the Q-Phase of Carbonaceous Chondrites. Nuclear Instruments and Methods in Physics Research B, 158, 538-543.
317. Ivanova, M.A., Taylor, L.A., Clayton, R.N., Mayeda, T.K., Nazarov, M.A., Brandstätter, F. and Kurat, G. (2002) Dho 225 vs. the CM Clan: Metamorphosed or New Type of Carbonaceous Chondrite. 33rd Lunar and Planetary Science Conferences, Abstract #1437. CD-ROM.
318. Riebe, M. (2009) Spinel Group Minerals in Carbonaceous and Ordinary Chondrites. BA Thesis, Lunds University, No. 250.
319. Ramdohr, P. (1973) The Opaque Minerals in Stony Meteorites. Elsevier, Amsterdam.
320. Bridges, J.C., Schmitz, B., Huchison, R., Greenwood, R.C., Tassinari, M. and Franchi, I.A. (2007) Petrographic Classification of Middle Ordovician Fossil Meteorites from Sweden. Meteoritics and Planetary Science, 42, 1781-1789.
321. Göpel, C., Birck, J.-L. and Zanda, B. (2011) Mn/Cr Systematics in Carbonaceous chondrites: Mineral Isochrons versus Stepwise Dissolution. Mineralogical Magazine, 7, A32.
322. Rubin, A.E. (1997) Mineralogy of Meteorite Groups. Meteoritics & Planetary Science, 32, 231-247.
323. Wlotzka, F. (2005) Cr Spinel and Chromite as Petrogenetic Indicators in Ordinary Chondrites: Equilibrium Temperatures of Petrologic Types 3.7 to 6. Meteoritics and Planetary Science, 40, 1673-1702.
324. Khain, V.Ye. (1984) Regional Geotectonics—Alpine Mediterranean Belt. Nedra, Moscow. (In Russian)
325. Huang, Q.S., Shi, R.D., Liu, D.L., Zhang, X.R., Fan, S.Q. and Ding, L. (2013) Os Isotopic Evidence for a Carbonaceous Chondritic Mantle Source for the Nagqu Ophiolite from Tibet and Its Implications. Chinese Science Bulletin, 58, 92-98.
326. Yang, W., Teng, F.-Zh. and Zhang, H.-F. (2009) Chondritic Magnesium Isotopic Composition of the Terrestrial Mantle: A Case Study of Peridotite Xenoliths from the North China Craton. Earth and Planetary Science Letters, 288, 475-482.
327. Büchl, A., Brügmann, G.E., Batanova, V.G. and Hofmann, A.W. (2003) Os Mobilization during Melt Percolation: The Evolution of Os Isotope Heterogeneities in the Mantle Sequence of the Troodos Ophiolite, Cyprus. Geochimica et Cosmochimica Acta, 68, 3397-3408.

328. Batanova, V.G. and Sobolev, A.V. (2000) Compositional Heterogeneity in Subduction-Related Mantle Peridotites, Troodos massif, Cyprus. Geology, 28, 55-58.

CHAPTER 12

Microchemical and Structural Evidence for Space Weathering in Soils from Asteroid Itokawa

Michelle S Thompson[1*], Roy Christoffersen[2], Thomas J Zega[1] and Lindsay P Keller[3]

[1]*Lunar and Planetary Laboratory, University of Arizona, 1629 E University Blvd, Tucson, AZ 85721, USA.*

[2]*Jacobs, NASA Johnson Space Center, Mail Code KR, Houston, TX 77058, USA.*

[3]*NASA JSC Mail Code KR, Houston, TX 77058, USA.*

ABSTRACT

Here we report microchemical and microstructural features indicative of space weathering in a particle returned from the surface of asteroid Itokawa by the Hayabusa mission. Space weathering features include partially and completely amorphous rims, chemically and structurally heterogeneous multilayer rims, amorphous surface islands, vesiculated rim textures, and nanophase iron particles. Solar-wind irradiation is likely responsible for the amorphization as well as the associated vesiculation of grain rims. The multilayer rims contain a nanocrystalline outer layer that is underlain by an amorphous inner layer, and both have compositions that are distinct from the underlying, crystalline orthopyroxene grain. The multilayer rim features could be derived from either radiation-induced sputter deposition or vapor deposition from micrometeorite impact events. The amorphous islands on grain surfaces have a distinctive morphology and composition suggesting that they represent surface deposits of melt derived from micrometeorite impact events. These observations indicate that both irradiation damage and micrometeorite impacts play a role in surface modification and space weathering on asteroid Itokawa.

Keywords: Hayabusa Itokawa Space weathering Irradiation Micrometeorite impact Airless body Transmission electron microscopy Sample return analysis

INTRODUCTION

Due to their exposure to interplanetary space, the surfaces of airless bodies are continually being modified by solar ions, predominantly from the solar wind, and through micrometeorite impacts. These modifications are collectively known as space weathering, and they cause changes in the chemistry, mineralogy, microstructure, and morphology of soils exposed on airless body surfaces. The culmination of these processes alters the optical properties of surface materials by reddening and darkening soil reflectance spectra and attenuating characteristic absorption bands (Hapke 2001; Clark et al. 2002; Chapman 2004). Understanding these spectral changes is important as they are theorized to contribute to the discrepancies between reflectance spectra acquired from meteorites and asteroids that prevent us from matching meteorite samples to their parent-body sources, e.g., ordinary chondrites and S-type asteroids (Binzel et al. 1996). Investigation of the nature of the space weathering process and quantification of its effects can provide information fundamental to understanding the evolution of airless body surfaces from remote sensing data.

Space weathering processes are occurring on the surfaces of airless bodies across the solar system, including the Moon and both near-Earth and main-belt asteroids. Until the recent return of the Hayabusa mission from S-type asteroid Itokawa (Nakamura et al. 2011), sample-based studies of space weathering focused exclusively on lunar soils returned from the Apollo missions and select asteroidal and lunar regolith breccia meteorites (Bibring et al. 1972; Hapke et al. 1975; Keller and McKay 1993, 1997; Bernatowicz et al. 1994; Pieters et al. 2000; Taylor et al. 2001; Noble et al. 2005, 2011). These investigations indicated that space weathering features form as a product of two primary processes: (i) solar wind ion irradiation, which can cause partial or complete amorphization of material and possible chemical changes in grain rims, as well as erosion, transport, and redeposition of material on grain surfaces induced by ion sputtering, and (ii) micrometeorite impact events which promote melting, vaporization, and the recondensation of the target and impactor material. Each of these processes has the potential to produce reduced iron nanoparticles ($npFe^0$), which can be formed through vapor or sputter deposition onto the rims of surface grains and also as interior particles in lunar agglutinate glass grains. This $npFe^0$ contributes to the changes in optical properties we see in space-weathered materials (Noble et al. 2007).

The Hayabusa mission successfully returned thousands of dust particles from the surface of S-type asteroid Itokawa (Nakamura et al. 2011). Mineralogical analysis of several particles revealed compositions consistent with LL5-6 chondrites (Nakamura et al. 2011). The link between these particles and LL chondrites was further confirmed by oxygen isotope analyses (Yurimoto et al. 2011; Nakashima et al. 2013). The return of these samples has provided the strongest evidence yet for the long-held idea that the ordinary chondrites, the most common meteorites in our collections, are sourced from the populous S-type asteroids. The dissimilarity, however, between the reflectance spectra of Itokawa's surface and LL-type chondrites (mainly reddening in the asteroid spectrum) suggests that space weathering processes have occurred on the surface of Itokawa (Hiroi et al. 2006).

Previous investigations of Hayabusa particles for detection of space weathering features were initially completed on ten particles, five of which exhibited characteristics consistent with space weathering processes (Noguchi et al. 2011). A further detailed analysis of 12 particles was performed, and these results have shown that several grains exhibit space weathering features including radiation-damaged rims (partially and completely amorphous), Fe nanoparticles, vesiculated textures, and chemically distinct, surface-deposited rims (Noguchi et al. 2014). Thus far, Noguchi et al. (2014) suggest that the primary mechanism for surface alteration on asteroid Itokawa is solar wind irradiation including sputtering and ion implantation. However, given the small number of particles examined, further detailed investigation is required for a comprehensive understanding of space weathering processes on Itokawa as well as other airless bodies. Here we report on previously unrecognized space weathering characteristics in a particle returned from the surface of Itokawa.

METHODS

We investigated three samples (RA-QD02-0042-01, RA-QD02-0042-02, and RA-QD02-0042-03) of one particle, which was collected from Room A of the sample catcher on the Hayabusa spacecraft. These samples were prepared for analysis using ultramicrotomy by the Hayabusa mission curation team. The grain, originally measuring 96 μm in diameter, was sliced into many ultrathin (<100 nm) sections. Each sample contains multiple slices that were placed on C-coated Cu-support grids for analysis using transmission electron microscopy (TEM). The original particle was exposed to the atmosphere during its preparation for analysis in the TEM, but its total exposure time was less than 6 h (Noguchi et al. 2014). These samples, and a subset of the space weathering features they exhibited, were described in previous investigations (Noguchi et al. 2011, 2014).

We investigated the samples using several analytical scanning transmission electron microscopes (STEM instruments) including (1) a 200-keV FEI Osiris ChemiSTEM equipped with a high brightness field-emission gun (X-FEG), quad-annular silicon drift detector (SDD) for energy-dispersive X-ray spectroscopy (EDS), and both dark-field (DF) and bright-field (BF) STEM detectors previously located at FEI headquarters (Portland, OR, USA); (2) a 200-keV JEOL 2500SE analytical S/TEM (JEOL Ltd., Tokyo, Japan) at NASA Johnson Space Center equipped with a Thermo thin-window Li-drifted Si [Si(Li)] EDS detector and BF and DF STEM detectors (Thermo Fisher Scientific, Waltham, MA, USA); and (3) a 200-keV JEOL 2010F S/TEM at Arizona State University equipped with a DF STEM detector and EDAX thin-window Si(Li) EDS system. Analysis of the assemblages within the slices included imaging via conventional TEM techniques, STEM BF and DF, and high-angle annular dark field (HAADF). Phase identification was determined using selected-area electron-diffraction (SAED) patterns and high-resolution TEM (HRTEM) imaging. All SAED patterns were measured both manually with Adobe Photoshop and with the Crystallographic Processing Software Package (CRISP) based on calibrated camera constants. Indexing of SAED patterns was based on appropriate lattice constants and verified through simulated patterns calculated with the CrystalMaker software package. Grain compositions were measured with EDS on the JEOL 2500SE and the JEOL 2010 F using spectral imaging with scanned probe sizes between 1 and 3 nm, count rates varying between 5,200 and 20,000 cps, and a dwell time of 50 μs for both instruments. EDS spectra were quantified using the Cliff-Lorimer method, with K-factors determined from in-house standards.

RESULTS

General petrography

The particle studied here is primarily composed of an orthopyroxene grain that exhibits high strain contrast with smaller, subsidiary grains of olivine (<5 μm) and plagioclase (<15 μm) attached to the orthopyroxene. These minerals are traceable through adjacent slices in the microtome sections. There are also Fe- and Fe-Ni sulfide minerals (<1 μm in size) both embedded within the pyroxene grain and located on its periphery, in multiple adjacent slices on the grid. Individual grains of Fe-Ni metal, identified as kamacite, were also found on the exterior rim of this assemblage. The presence of each phase was confirmed with EDS spectral imaging and SAED (Figure 1). The lath-like nature of this assemblage (shards separated by void space) is characteristic of chattering, an artifact produced during microtome sample preparation and, in this respect, is similar to samples observed from the STARDUST mission (Zolensky et al. 2006).

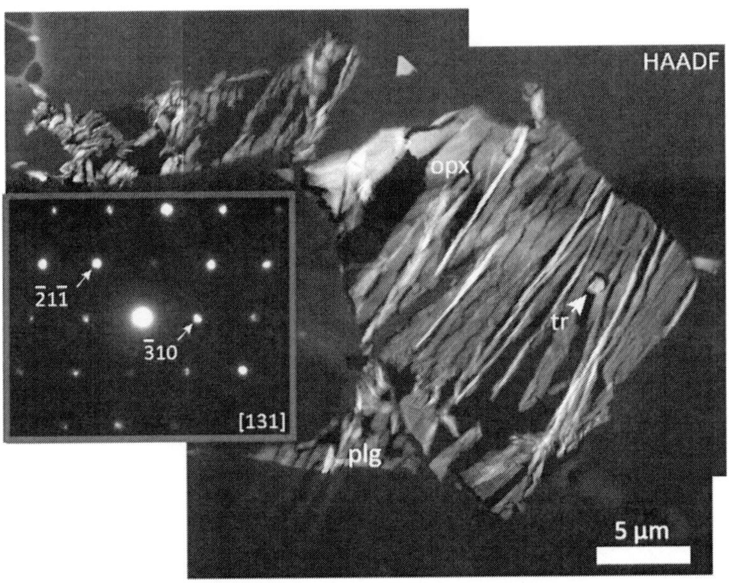

Figure 1. TEM data and EDS maps of particle RA-QD02-0042-02. HAADF image of one slice of the assemblage. The particle is primarily composed of orthopyroxene (opx) with minor plagioclase (plg), and troilite (tr). The lathic morphology is a result of chattering, a common texture observed in ultramicrotomed samples of small particles. The SAED pattern (inset) was collected from the region outlined by the red circle on the HAADF image and is indexed to orthopyroxene.

Space weathering features

We analyzed over 30 individual slices for microchemical and microstructural evidence of space weathering. We observed several features indicative of space weathering that occur in very close spatial proximity to one another along the outer rims of the grains that collectively comprise the particle within these slices. Nearly all the individual slices that we examined exhibit some degree of space weathering. While space weathering features occur in several different mineral types, the extent and degree of their manifestation differ significantly. We discuss the space weathering features in detail below.

Partial amorphization

There are zones in the particle rim that have characteristics consistent with partial-to-complete solid-state amorphization of the underlying crystalline grains. The width and nature of these zones vary significantly depending on the mineralogy of the underlying grain. The partially amorphous regions are not continuous over the entire rim of the particle but instead occur in localized regions that extend laterally along the rims of individual grains in this assemblage for hundreds of nanometers (Figure 2). In the low-Ca orthopyroxene grain, there is a partially amorphous zone that forms an irregular boundary with the underlying crystalline grain and extends between 10 to 60 nm into the interior (Figure 2b). Within these zones, there are localized nanocrystalline regions which range from 5 to 15 nm in size. Amorphous material locally surrounds and intermixes with the crystalline domains. EDS mapping shows that the partially amorphous rims on the orthopyroxene have compositions identical to the underlying crystalline material, indicating that their only difference from the underlying material is their lack of long-range structural order. In addition, these zones of partial amorphization contain isolated Fe nanoparticles, the details of which we describe below.

Complete amorphization

An outer rim of complete amorphization occurs on the plagioclase grain which contrasts to the zones of partial amorphization found on the orthopyroxene. This rim is also thinner than that on the orthopyroxene, extending only approximately 20 nm from the outer edge into the crystalline material, and is of more uniform thickness (Figure 2c,d). The composition of the plagioclase rim is identical to the underlying crystal.

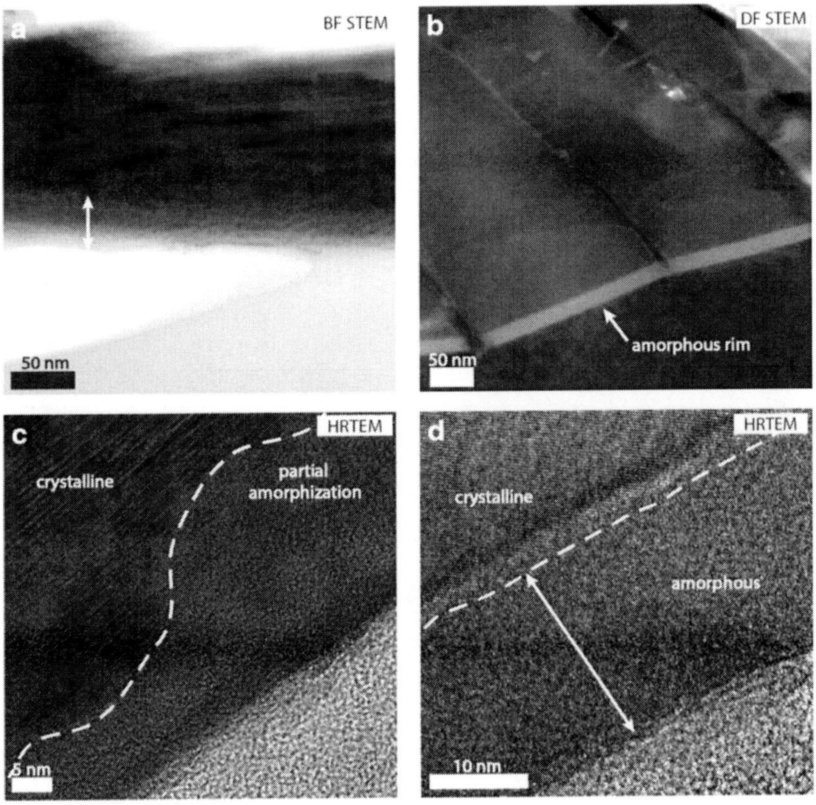

Figure 2. TEM data on the orthopyroxene and plagioclase grains. (a) BF STEM image of a region of the rim of orthopyroxene with arrow indicating the zone of partial amorphization. **(b)** DF STEM image of the amorphous rim on the plagioclase grain, indicated by the arrow. **(c)** HRTEM image of the highly variable depth and degree of amorphization within the orthopyroxene. **(d)** HRTEM image of the rim of complete amorphization showing uniform the thickness on the plagioclase.

We note that we looked for solar flare tracks in both the orthopyroxene and plagioclase grains, but none were observed.

Multilayer rims

Local regions along the outer margin of the orthopyroxene grain are comprised of altered rim material with both underlying and overlying layers, each 2 to 5 nm thick (Figure 3a,b). Laterally, the layers extend over tens to hundreds of nanometers of the grain rim. The multilayer feature is traceable through multiple adjacent slices and occurs in the same region of the grain rim in each slice. EDS spectral imaging shows that the outer and inner layers have compositions that differ from one another and from the underlying grain interior (Figure 3). Quantified compositional profiles were extracted from EDS spectral images along a line through the rims and into the grain interior (*cf.* Figure 3c,d,e,f,g). The outermost layer in the multilayer rim displays an approximately 7 at% depletion in Si relative to the bulk crystal interior and approximately 4 at% enrichments in both Fe and Mg. In contrast, the inner layer has an approximately 10 at% enrichment in Si, an approximately 3 at% depletion in Fe, and an approximately 10 at% depletion in Mg relative to the grain interior. The estimated errors are <1.7 at% (1σ) without an absorption correction.

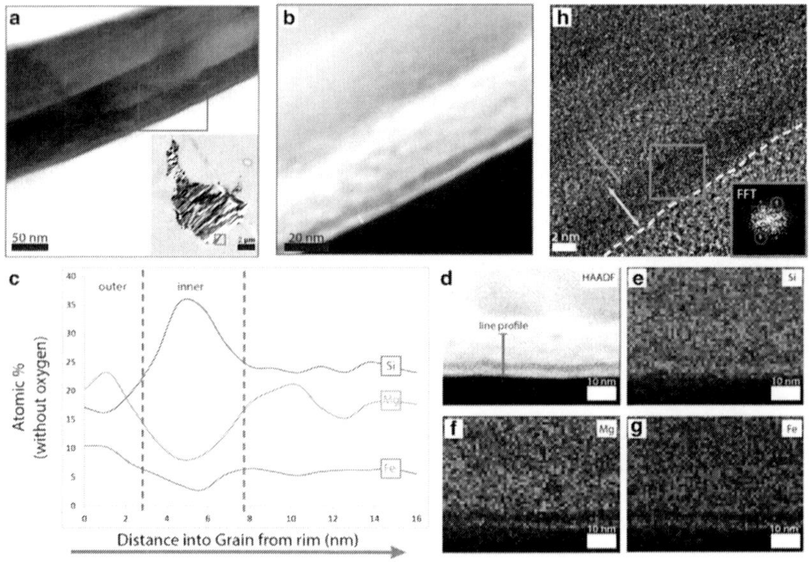

Figure 3. TEM and EDS data of the multilayer rims on orthopyroxene. (a) BF image of the multilayer rim on the edge of the pyroxene grain. **(b)** HAADF image of the rim with outer and inner layers indicated by arrows. **(c)** Line profile acquired from the edge of the grain, through the multilayer rim to the interior. Red dashed lines indicate boundaries between layers on the grain exterior and the underlying grain interior. **(d)** HAADF image of the multilayer rim and location of the line profile. **(e-g)** EDS spectral images of the multilayer rim. **(h)** HRTEM image of the multilayer rim with yellow arrows indicating the zone of nanocrystallinity and blue arrows indicating amorphous region. The white dashed line indicates the edge of the grain, and the inset FFT shows discrete reflections measuring 0.223 nm, indicating nanocrystalline nature of the outermost layer.

The multilayer rim exhibits microstructural heterogeneity that correlates with the chemical differences. The outer, Si-depleted, Fe- and Mg-enriched layer is composed of nanocrystalline domains, as shown in the HRTEM image in Figure 3h. Fourier transform diffractograms of several HRTEM images of this nanocrystalline region give reflections with *d*-spacings of 0.223 nm, which is close (within error) to several orthopyroxene spacings, e.g., (122); olivine spacings, e.g., (140) or (122); Fe metal spacings, e.g., (101); or MgO spacings, e.g., (111) or (200). HRTEM images show that the underlying Si-enriched layer is completely amorphous and abruptly transitions into the crystalline grain interior.

Amorphous islands

In localized areas of the grain rim, there are isolated regions of material that protrude tens of nanometers from the grain surface. These island-like regions are superimposed on top of the Si-enriched layer of the multilayer rims, suggesting that they were deposited on top of the grain edge. The chemical composition of these islands is different from the underlying bulk crystal (Figure 4b,c,d,e). The islands contain significant S (approximately 4 at%) that is not present in the grain interior. In addition, EDS spectral imaging reveals that the islands are enriched in Fe and Mg and depleted in Si relative to the underlying bulk crystal (Figure 4b,c,d,e). HRTEM images of these islands show that they are completely amorphous.

Vesiculated textures

There are both well- and poorly developed vesicles in the partially amorphous rim of the orthopyroxene grain (Figure 4f,g). These vesicles occur 10 to 30 nm below the grain rim surface and vary in size between 5 and 20 nm. The well-developed vesicles are emplaced in a partially amorphous matrix and cause bulging of the grain rim around them. These features are not the product of sample preparation or beam damage, having been identified in pyroxene, a relatively beam-stable mineral in the TEM, in both ultramicrotomed and focused-ion beam sections. These vesicles are present in very low concentrations, occurring in localized regions of only a few slices.

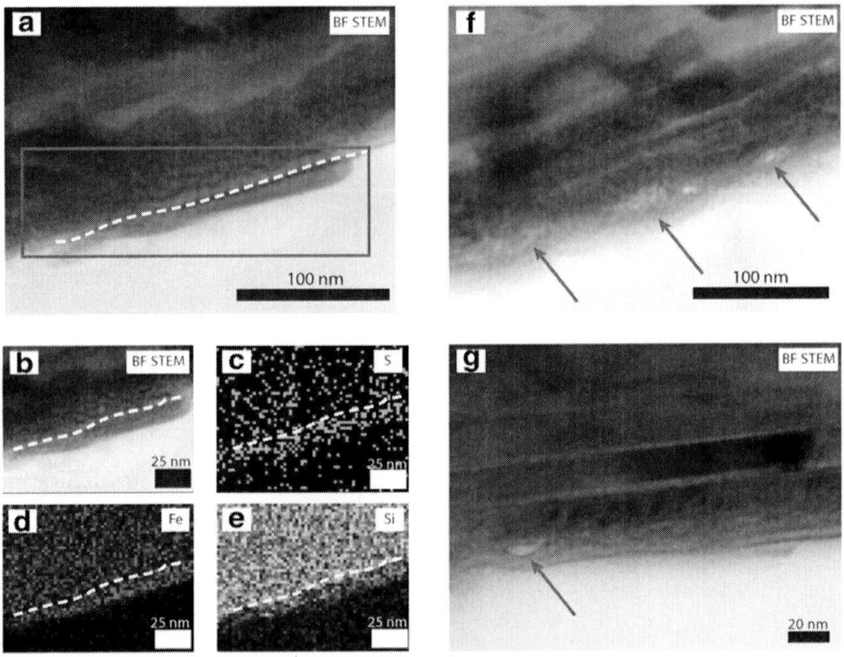

Figure 4. TEM and EDS data of other space weathering features in these grains. (a) BF STEM image of the amorphous island deposited on the grain rim. The white dashed line indicates the original grain surface. **(b-e)** BF STEM and EDS spectral images of the island feature showing enrichments in Fe and S relative to the underlying grain, as located by the red box in **(a)**. **(f)** BF STEM image of poorly developed vesicles, as indicated by arrows in the partially amorphous matrix of the orthopyroxene rim. **(g)** BF STEM image of a well-developed vesicle, indicated by red arrow in the partially amorphous orthopyroxene rim.

Nanophase iron particles

In three of the individual slices analyzed, Fe nanoparticles were identified in high-resolution EDS spectral mapping in regions of the orthopyroxene rim, but were not identified in any other mineral type sections. These findings indicate that the overall abundance of Fe nanoparticles in the analyzed sample as a whole is low (e.g., Figure 5). Where they occur, the nanophase Fe grains are uniformly distributed in their confined regions, occurring up to 60 nm from the exterior grain rim, and are emplaced in a partially amorphous matrix. The individual particles are uniform in size, each measuring between 2 and 3 nm. EDS mapping on an area containing a particularly high concentration of nanophase Fe particles indicates that the grains are composed primarily of Fe, with no associated Mg or S signatures (Figure 5b,c,d,e,f). These nanoparticles may be Fe oxide or metallic Fe, though the relative depletion in O (Figure 5f) in the profile of the grain may suggest metallic Fe.

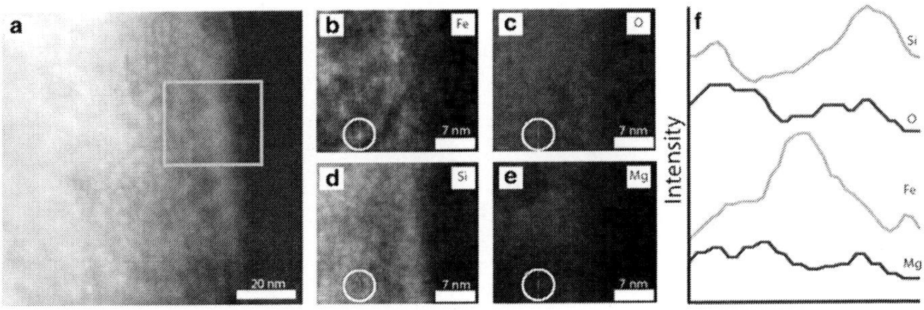

Figure 5. TEM and EDS data of nanophase Fe in the rim of orthopyroxene. (a) HAADF image of the orthopyroxene grain rim showing a high concentration of Fe nanoparticles. **(b-e)** EDS maps of the region indicated by the yellow box in **(a)**. An individual particle is indicated by a white circle in **(b-e)**. **(f)** Intensity profiles for Si, O, Fe, and Mg taken across the identified particle, indicated by a dashed yellow line. Note the obvious Fe peak, the relatively consistent Mg and O concentrations, and the Si enrichment bordering the particle.

DISCUSSION

The variety of individual space weathering features we observed in these samples indicates that these grains have experienced a complex processing history on the surface of Itokawa. The nature of the observed features suggests that they developed with contributions from both ion radiation processing, almost certainly associated with solar wind ions, and micrometeorite impact events. Below we discuss the specific implications of each of the observed features for surface processing on Itokawa.

Amorphization features

The amorphous zones (both partial and complete, which have identical compositions to the underlying grain) are attributed to solar wind ion irradiation effects. Solar ions, mostly 1 to 4 keV H^+ and He^+, have been estimated by Noguchi et al. (2014) using the Stopping and Range of Ions in Matter (SRIM) Monte Carlo atomic collisions code (Ziegler et al. 2008) to penetrate between 27 and 51 nm (H and He, respectively) into the grain interior from the rim. These simulations were performed for forsteritic (Fo_{70}) olivine (Noguchi et al. 2014). The simulations and the observed thickness of the orthopyroxene rims on our samples (thickness <60 nm) are also consistent with the width of ion-amorphized layers in experimental samples, the width of rims on lunar regolith grains, and previously described Itokawa samples (Keller and McKay 1997; Carrez et al. 2002; Loeffler et al. 2009; Noguchi et al. 2011, 2014). In comparison, the plagioclase grain rim exhibits a more consistent degree of damage (i.e., complete amorphization) and is uniformly more narrow (<25 nm) than the average width of the orthopyroxene rim.

Plagioclase is, however, more susceptible to electron beam damage that may have artificially widened the rim or altered its chemistry and microstructure. In order to account for this sensitivity, we imaged the rim for structure before completing any prolonged analytical work that would have otherwise damaged the sample before we had a chance to assess long-range order. This experimental approach ensured that the thickness and amorphous nature of the grain rim was indigenous to the samples and not an artifact of TEM analysis. After completing spectral imaging, we did observe the effects of beam damage on the plagioclase as the interface between the rim and underlying crystalline material retreated further into the grain interior, revealing that the amorphous layer had increased in thickness.

The discrepancy between the nature and thickness of the orthopyroxene and plagioclase rims suggests three possibilities. (1) The grains had relatively similar solar wind exposure times, and the different characteristics of the rims primarily reflect variations in the solid-state response of the two phases to solar wind ion irradiation. Most notably, this could include differences in the critical ion fluence needed to amorphize the two phases. (2) The solid-state response of both phases to irradiation is relatively similar, and the grains were simply exposed directly to solar wind ions for different lengths of time. Finally, (3) both factors contributed to the development of these features.

We are currently not aware of any direct, quantitative experimental comparison of the ion processing behavior or relative amorphization fluences of orthopyroxene compared to plagioclase under conditions applicable to the solar wind. The ion processing behavior of pyroxene relative to a wide range of other silicates has, however, been investigated under irradiation similar to the regime expected for the solar wind (Wang and Ewing 1992; Wang et al. 1998a,[b]; Christoffersen and Keller 2011). These studies suggest that structures in which the Si-O tetrahedral sites are more polymerized, as in plagioclase, with a greater variety of interstitial sites, are generally much more susceptible to radiation-induced amorphization (Wang and Ewing 1992). This prediction is consistent with the well-known greater susceptibility of feldspars to electron-beam damage relative to pyroxene. As such, the differences in the degree of partial or complete amorphization between the plagioclase and orthopyroxene may reflect this difference.

With regard to the differences in width between the feldspar and pyroxene grain rims, irrespective of their degree of structural amorphization, several factors may play a role. The intrinsic penetration range of ions into the two minerals is determined by differences in the nuclear-elastic and electronic stopping power of their collective constituent atoms and by the structure's atomic density (Ziegler et al. 2008). Based on SRIM calculations, the maximum in deposited collision energy by H^+ and He^+ at solar wind energies occurs at respective depths of approximately 11 and 23 nm for plagioclase and 8 and 20 nm for orthopyroxene. This maximum is likely to correspond to the rim width, because the deposited energy falls off rapidly beyond the maximum. Based on these relations, rims in feldspar can be expected to be wider than those in pyroxene, all other factors being equal. These simulations assume ion bombardment at fixed energies with one ion type, unlike the energy range and multi-ion composition of the solar wind, and only account for normal incidence of irradiation. Nonetheless, these are important first-order estimates for understanding the development of irradiated rims on airless body materials.

As noted above, there is the possibility that differences in exposure time may also play a role that is superposed on differences in the ion processing behavior of the two minerals. The narrow width of the rim on the plagioclase grain analyzed here (<25 nm) is significantly less than those observed on plagioclase grains in mature lunar soils, which range from 10 to 100 nm in thickness, with an average thickness of approximately 50 nm (Zhang and Keller 2012). The smaller rim thickness on the plagioclase grain from Itokawa would seem to suggest that its exposure time is therefore less than that for a mature lunar soil. However, although the orthopyroxene and plagioclase grains are part of one coherent assemblage now, it is possible that they did not exist cohesively during their entire surface exposure lifetime. The observed spatial relationship could be a product of sample collection methods or the amalgamation of grains during a recent surface event, e.g., an impact. Such an event could group together grains with vastly different exposure times resulting in variable radiation damage features.

The lack of distinct solar flare tracks may speak to the exposure history of this individual particle which our microtome slices are sourced from. It is possible that the lack of tracks identified here is a result of the inherent difficulty in identifying solar flare tracks in pyroxene minerals that exhibit a high concentration of strain features, including this grain. Another possibility, however, is that these grains were exposed for a very short period of time on the surface of Itokawa, preventing the accumulation of tracks in high enough density to measure. However, we note the difficulty in constraining relevant surface residence times of material on Itokawa from the analysis of one single particle.

Multilayer rim features

The chemically heterogeneous layers we report here are distinct from the composite redeposition rims previously reported (Noguchi et al. 2011, 2014). The rims previously described are comprised of an outermost totally amorphous region, whereas the rims we report on here contain a multilayer, mixed nanocrystalline, and amorphous morphology. We hypothesize two possible formation mechanisms for these features, or a combination of these processes, described below. The significantly different composition of the multilayer rims relative to the underlying grain (Figure 3) suggests that the rims could be the product of vapor deposition from a micrometeorite impact of an adjacent grain with a different chemical composition. However, similar features in lunar soils attributed to micrometeorite impact events are typically amorphous because the vapor-deposited material is quenched at low temperatures, preventing the development of crystalline domains. As a result, the nanocrystalline nature of the outer layer in the multilayer rims may argue against this formation mechanism as it suggests that there was sufficient time for at least small grains to crystallize. A nearby heating event, e.g., a subsequent micrometeorite impact in the local surface region, could have provided enough thermal energy to slow quenching and allow for the outer layer to recrystallize.

Alternative processes for forming the chemically heterogeneous multilayer rims could also involve preferential solar ion sputtering of the grain surface itself, sputter deposition of material derived from surrounding grains, or a combination of the two processes (Hapke 2001). For silicate minerals, the compositional relations that can develop in the outer rims of grains involved in either of these processes, especially involving relative changes in the constituent cation concentrations like those observed here in the orthopyroxene rims, are still not well understood. Therefore, it remains possible that either of these sputtering processes, or a sequence of the two processes, may have contributed to forming the compositional relationships in the inner and outer layers in the orthopyroxene rims.

Amorphous islands

The amorphous islands are similar to lunar glass features that have been attributed to micrometeorite impact events (Keller and McKay 1997). Their relief from the underlying grain, completely amorphous crystal structure, and different chemistry when compared to the bulk grain beneath are similar to impact glasses in lunar soils that are 'splashed' onto the surface from adjacent material. We therefore infer that micrometeorite impacts produced the amorphous islands that we observe in these Itokawa samples.

Fe nanoparticles

$NpFe^0$ particles have been produced experimentally through both irradiation experiments and simulated micrometeorite impacts using laser irradiation (Sasaki et al. 2001). Previous studies have attributed individual formation mechanisms to nanoparticles with specific chemistries. $NpFe^0$ particles associated with Mg and S are attributed to micrometeorite impact events and subsequent vapor deposition (Noguchi et al. 2011). These particles often occur close to the surface of the grain, in a matrix of completely amorphous material. In contrast, particles emplaced deeper into the grain that are not associated with Mg or S but are instead purely Fe are suggested to have formed through radiation processing events (Noguchi et al. 2011). The data on the Fe nanoparticles observed in our Hayabusa particles suggest that they were produced through ion irradiation events because they occur in a matrix that is isochemical with the underlying grain and are also located deeper in the grain in a matrix of partially amorphous material. The relatively low concentration of these particles when compared to lunar samples may suggest a short exposure time on the surface of Itokawa, preventing the accumulation of high volumes of Fe nanoparticles. If this low concentration were characteristic of the population of space-weathered grains on Itokawa as a whole, it may indicate that the spectral reddening characteristic of S-type asteroids can be produced with only a small amount of $npFe^0$.

Vesiculated textures

The vesiculated textures observed in our assemblages have been reproduced in experimental samples subjected to irradiation (Hishmeh et al. 1994). Such high concentrations of vesicles are rare in grain rims from mature lunar soils (Keller and McKay 1997), although vesiculated grain rims are commonplace in friable lunar breccias (Noble et al. 2005). Noble et al. (2005) proposed that the vesiculated rims result from thermal processing of the solar wind implanted grains. A similar thermal event may have produced the vesiculated rims we observe on the Itokawa grains.

The space weathering features observed in samples from Itokawa suggest that while radiation from solar energetic ions does play a significant role in surface alteration of this body, observed impact glasses suggest that micrometeorite impacts also contribute to its significant and complex processing history. The surface environments of the Moon and Itokawa are different, e.g., potentially higher resurfacing rates on Itokawa resulting in shorter grain exposure times and lower surface gravity resulting in lower retention of impact products. However, the data presented here suggest that the *processes* altering their surface characteristics are similar, although the *features* these processes generate are variable.

CONCLUSIONS

We examined the detailed microstructure and chemistry of particle RA-QD02-0042 returned by the Hayabusa mission. The particle contains an assemblage with a diverse mineralogy including orthopyroxene, plagioclase, olivine, Fe and Fe-Ni sulfides, and Fe-Ni metal. The data suggest that the grains on the surface of Itokawa have a complex processing history with contributions from both micrometeorite impacts and radiation processing events. Several characteristics, including the differences we observe in concentration and distribution of $npFe^0$ particles in these samples, vary from previous observations of lunar soils. The discrepancies in these features are likely a product of different surface modification environments between the Moon and Itokawa and indicate that we should revisit our lunar soil collection for analysis in a new context.

ACKNOWLEDGEMENTS

We thank JAXA for the allocation of Hayabusa samples for this study. We gratefully acknowledge the use of facilities within the Leroy Eyring Center for Solid State Science at Arizona State University. We acknowledge the use of facilities and assistance from Huikai Cheng at FEI. The TEM work at the Johnson Space Center was supported in part by a grant to LPK from the NASA Laboratory Analysis of Returned Samples (LARS) program. Funding for MS Thompson is provided by the Natural Sciences and Engineering Research Council of Canada (NSERC). This research was supported in part by NASA. Three anonymous reviewers are acknowledged for their constructive comments.

REFERENCES

1. Bernatowicz TJ, Nichols RH, Hohenberg CM, Maurette M: Vapor deposits in the lunar regolith. Science 1994, 264(5166):1779–1780. doi:10.1126/science.264.5166.1779-a
2. Bibring JP, Duraud JP, Durrieu L, Jouret C, Maurette M, Meunier R: Ultrathin amorphous coatings on lunar dust grains. Science 1972, 175(4023):753–755. 10.1126/science.175.4023.753
3. Binzel RP, Bus SJ, Burbine TH, Sunshine JM: Spectral properties of near-Earth asteroids: evidence for sources of ordinary chondrite meteorites. Science 1996, 373: 946–948.
4. Carrez P, Demyk K, Cordier P, Gengembre L, Grimblot J, D'Hendecourt L, Jones AP, Leroux H: Low–energy helium ion irradiation–induced amorphization and chemical changes in olivine: insights for silicate dust evolution in the interstellar medium. Meteorit Planet Sci 2002, 37(11):1599–1614.
5. Chapman CR: Space weathering of asteroid surfaces. Ann Rev Earth Planet Sci 2004, 32: 539–567.
6. Christoffersen R, Keller LP: Space radiation processing of sulfides and silicates in primitive solar systems materials: comparative insights from in situ TEM ion irradiation experiments. Meteorit Planet Sci 2011, 46(7):950–969. 10.1111/j.1945-5100.2011.01203.x
7. Clark BE, Hapke B, Pieters C, Britt D: Asteroid space weathering and regolith evolution. In Asteroids III. Edited by: Bottke W, Cellino A, Paolicchi P, Binzel RP. Tucson, Arizona: University of Arizona Press; 2002.
8. Hapke B: Space weathering from Mercury to the asteroid belt. J Geophys Res Planet 2001, 106(E5):10039–10073.
9. Hapke B, Cassidy W, Wells E: Effects of vapor-phase deposition processes on the optical, chemical, and magnetic properties of the lunar regolith. Moon 1975, 13(1–3):339–353.
10. Hiroi T, Abe M, Kitazato K, Abe S, Clark BE, Sasaki S, Ishiguro M, Barnouin-Jha OS: Developing space weathering on the asteroid 25143 Itokawa. Nature 2006, 443(7107):56–58.
11. Hishmeh G, Cartz L, Desage F, Templier C, Desoyer J, Birtcher R: Rare gas bubbles in muscovite mica implanted with xenon and krypton. J Mater Res 1994, 9(12):3095–3107.
12. Keller LP, McKay DS: Discovery of vapor deposits in the lunar regolith. Science 1993, 261(5126):1305–1307. 10.1126/science.261.5126.1305 10.1126/science.261.5126.1305
13. Keller LP, McKay DS: The nature and origin of rims on lunar soil grains. Geochim Cosmochim Acta 1997, 61(11):2331–2341.
14. Loeffler M, Dukes C, Baragiola R: Irradiation of olivine by 4 keV He+: simulation of space weathering by the solar wind. J Geophys Res Planet 2009, 114: E03003.
15. Nakamura T, Noguchi T, Tanaka M, Zolensky ME, Kimura M, Tsuchiyama A, Nakato A, Ogami T, Ishida H, Uesugi M, Yada T, Shirai K, Fujimura A, Okazaki R, Sandford SA, Ishibashi Y, Abe M, Okada T, Ueno M, Mukai T, Yoshikawa M, Kawaguchi J: Itokawa dust particles: a direct link between S-type asteroids and ordinary chondrites. Science 2011, 333(6046):1113–1116.
16. Nakashima D, Kita NT, Ushikubo T, Noguchi T, Nakamura T, Valley JW: Oxygen three-isotope ratios of silicate particles returned from asteroid Itokawa by the Hayabusa spacecraft: a strong link with equilibrated LL chondrites. Earth Planet Sci Lett 2013, 379: 127–136.
17. Noble SK, Keller LP, Pieters CM: Evidence of space weathering in regolith breccias I: lunar regolith breccias. Meteorit Planet Sci 2005, 40(3):397–408.

18. Noble SK, Pieters CM, Keller LP: An experimental approach to understanding the optical effects of space weathering. Icarus 2007, 192(2):629–642.
19. Noble SK, Keller LP, Pieters CM: Evidence of space weathering in regolith breccias II: asteroidal regolith breccias. Meteorit Planet Sci 2011, 45(12):2007–2015.
20. Noguchi T, Nakamura T, Kimura M, Zolensky ME, Tanaka M, Hashimoto T, Konno M, Nakato A, Ogami T, Fujimura A, Abe M, Yada T, Mukai T, Ueno M, Okada T, Shirai K, Ishibashi Y, Okazaki R: Incipient space weathering observed on the surface of Itokawa dust particles. Science 2011, 333(6046):1121–1125.
21. Noguchi T, Kimura M, Hashimoto T, Konno M, Nakamura T, Zolensky ME, Okazaki R, Tanaka M, Tsuchiyama A, Nakato A, Ogami T, Ishida H, Sagae R, Tsujimoto S, Matsumoto T, Matsuno J, Fujimura A, Abe M, Yada T, Mukai T, Ueno M, Okada T, Shirai K, Ishibashi Y: Space weathered rims found on the surfaces of the Itokawa dust particles. Meteorit Planet Sci 2014, 49(2):188–214.
22. Pieters CM, Taylor LA, Noble SK, Keller LP, Hapke B, Morris RV, Allen CC, McKay DS, Wentworth S: Space weathering on airless bodies: resolving a mystery with lunar samples. Meteorit Planet Sci 2000, 35(5):1101–1107.
23. Sasaki S, Nakamura K, Hamabe Y, Kurahashi E, Hiroi T: Production of iron nanoparticles by laser irradiation in a simulation of lunar-like space weathering. Nature 2001, 410(6828):555–557.
24. Taylor LA, Pieters CM, Keller LP, Morris RV, McKay DS: Lunar mare soils: space weathering and the major effects of surface–correlated nanophase Fe. J Geophys Res Planet 2001, 106: 27985–27999.
25. Wang LM, Ewing RC: Ion-beam-induced amorphization of complex ceramic materials—minerals. MRS Bull 1992, 17(05):38–44.
26. Wang SX, Wang LM, Ewing RC, Doremus RH: Ion beam-induced amorphization in MgO–Al2O3–SiO2. I. Experimental and theoretical basis. J Non-Cryst Solids 1998, 238(3):198–213.
27. Wang SX, Wang LM, Ewing RC, Doremus RH: Ion beam-induced amorphization in MgO–Al2O3–SiO2. II. Empirical model. J Non-Cryst Solids 1998, 238(3):214–224.
28. Yurimoto H, K-i A, Abe M, Ebihara M, Fujimura A, Hashiguchi M, Hashizume K, Ireland TR, Itoh S, Katayama J: Oxygen isotopic compositions of asteroidal materials returned from Itokawa by the Hayabusa mission. Science 2011, 333(6046):1116–1119.
29. Zhang S, Keller LP: Rates of space weathering in lunar regolith grains. In Abstracts of the Meteoritical Society Meeting. Cairns, Australia; 2012.
30. Ziegler JF, Biersack JP, Ziegler MD: SRIM, the Stopping and Range of Ions in Matter. Lulu Press, Raleigh; 2008.
31. Zolensky ME, Zega TJ, Yano H, Wirick S, Westphal AJ, Weisberg MK, Weber I, Warren JL, Velbel MA, Tsuchiyama A, Tsou P, Toppani A, Toka N, Tomeoka K, Teslich N, Taheri M, Susini J, Stroud R, Stephan T, Stadermann FJ, Snead CJ, Simon SB, Simionovici A, See TH, Robert F, Rietmeijer FJM, Rao W, Perronnet MC, Papanastassiou DA, Okudaira K: Mineralogy and petrology of comet 81P/wild 2 nucleus samples. Science 2006, 314(5806):1735–1739.

CHAPTER 13

Titan as the Abode of Life

Christopher P. McKay

Space Science Division, NASA Ames Research Center, Moffett Field, CA 94035, USA;

ABSTRACT

Titan is the only world we know, other than Earth, that has a liquid on its surface. It also has a thick atmosphere composed of nitrogen and methane with a thick organic haze. There are lakes, rain, and clouds of methane and ethane. Here, we address the question of carbon-based life living in Titan liquids. Photochemically produced organics, particularly acetylene, in Titan's atmosphere could be a source of biological energy when reacted with atmospheric hydrogen. Light levels on the surface of Titan are more than adequate for photosynthesis, but the biochemical limitations due to the few elements available in the environment may lead only to simple ecosystems that only consume atmospheric nutrients. Life on Titan may make use of the trace metals and other inorganic elements produced by meteorites as they ablate in its atmosphere. It is conceivable that H_2O molecules on Titan could be used in a biochemistry that is rooted in hydrogen bonds in a way that metals are used in enzymes by life on Earth. Previous theoretical work has shown possible membrane structures, azotosomes, in Titan liquids, azotosomes, composed of small organic nitrogen compounds, such as acrylonitrile. The search for a plausible information molecule for life in Titan liquids remains an open research topic—polyethers have been considered and shown to be insoluble at Titan temperatures. Possible search strategies for life on Titan include looking for unusual concentrations of certain molecules reflecting biological selection. Homochirality is a special and powerful example of such biology selection. Environmentally, a depletion of hydrogen in the lower atmosphere may be a sign of metabolism. A discovery of life in liquid methane and ethane would be our first compelling indication that the universe is full of diverse and wondrous life forms.

Keywords: Titan; life; methane; second genesis

1. INTRODUCTION

The search for life on other worlds is often characterized by the strategy of "follow the water". Water provides a convenient shorthand for the range of attributes that support life on Earth, including environmental processes associated with cycling of water, biochemistry of carbon in the medium of water, and the ecological systems that water-based life creates in water-cycling environments. Many of these attributes of the Earth, as an abode for life, are taken as given once the presence of liquid water is established. Thus, we can motivate a search for life on early Mars simply by the clear indication of liquid water on its surface, secure in the background knowledge of the environmental, biochemical, and ecological correlates of the widespread presence of liquid water.

It is important to keep in mind that one cannot really separate chemistry from environment. The role of water as a solvent for life on Earth depends on its chemical reactivity and on its physical properties, and the two are closely related. Thus, considering other liquids requires addressing the needs life has for chemical functionality, both in terms of physicochemical and reaction-chemistry-related requirements.

If we turn to the possibility of life based on another liquid, each of these background assumptions must be reexamined. The only other world we know that has a liquid on its surface is Titan, where we find liquids of methane and ethane. Titan is the largest moon of the planet Saturn but is small compared to Earth, with a surface gravity 1/7 that of Earth. Its surface atmospheric pressure is about 1.5 times that of Earth, and the surface temperature is 95 K. The atmosphere is dominated by N_2 (95%) with CH_4 (5%) and H_2 (0.1%) and various trace organic compounds. The lower atmosphere has an active hydrological cycle of liquid methane, including convective clouds and rain, both of which vary with the seasons. Titan is locked in a synchronous orbit around Saturn with a period of 16 days (the cycle of light and dark). The tilt of Saturn's spin axis to its orbit plane is ~27°, resulting in seasonal changes in the position of the sun in Titan's sky over the ~30-year period of Saturn's orbit. The lower atmosphere of Titan is too thick to respond to the 16-

day light-dark cycle but does respond on the 30-year seasonal timescale. Thick convective clouds congregate in the summer polar region.

Photochemical reactions beginning with the dissociation of the atmospheric N_2 and CH_4 create an array of organic molecules in Titan's atmosphere and produce a solid organic haze in the upper atmosphere that obscures the lower atmosphere and surface. The organic haze particles eventually settle on the surface. A major product of the photochemistry is ethane, which accumulates on the surface and mixes with liquid methane. There are large lakes of methane and ethane on Titan, ranging in size up to ~1000 km across. Observations of the Huygens lander indicated moisture (methane and ethane) in the ground at the landing site. It appears that liquid methane and ethane are widespread and actively cycled on Titan. Is a "follow the methane" strategy plausible?

In a seminal paper that considered the range of possible chemistries for life, Benner *et al.* [1] (see also Bains [2]) speculated that hydrocarbons that are naturally liquid on Titan could be a solvent for life. Benner *et al.* [1] noted that the organic reactivity in hydrocarbon solvents is no less versatile than in water, and indeed the ability to exclude water is an important aspect of many catalytic sites. Perhaps there could be life in Titan's liquids.

The purpose of this paper is to develop an approach to characterizing Titan as a possible abode of life by examining the possibilities of physical environments created by the presence and cycling of Titan's liquid, the biochemistry of carbon in that liquid, and the ecological systems that life might develop in that liquid. The case of water and life on Earth is the only guide we have for Titan as an abode of life, and so we must try to separate those features of habitability on Earth that are specific to Earth from those features that might be generalized to other liquids on other worlds.

2. THE GLOBAL ENVIRONMENT

On Earth, life is connected to the global environment. Habitability is determined by the physico-chemical interactions of the fluids (water on Earth) in a specific environment, and by the solids present in that environment. Through physical processes such as sunlight and volcanism and their interaction with cycles of water, important physical conditions for life are established on the Earth. These include: (1) sources of chemical or light energy suitable for life; (2) nutrients; (3) liquid habitats; and (4) transport cycles of liquid moving nutrients and wastes. Do these also exist on Titan?

2.1. Sources of Chemical or Light Energy Suitable for Life

The area related to the habitability of Titan that has received the most detailed attention is the possibility of chemical energy sources. Schulze-Makuch and Grinspoon [3] and McKay and Smith [4] noted that the photochemically produced organics in Titan's atmosphere would produce energy if reacted with atmospheric H_2, and that this could be a source of biological energy. McKay and Smith [4] quantified the energy released from such reactions (Table 1). Hydrogenation of C_2H_2 provided a particularly energetic reaction, with 334 kJ per mole of C_2H_2 consumed. This can be compared to the minimum energy required to power methanogen growth on Earth of ~40 kJ mole^{-1}, determined by Kral *et al.* [5], or the energy from the reaction of O_2 with CH_4, which produces ~900 kJ mole^{-1}. The reactions listed in Table 1 are exothermic but kinetically inhibited at Titan temperatures. This is ideal for biology. For example, the reaction of O_2 with CH_4 is kinetically inhibited at Earth temperatures, but methane-oxidizing microorganisms present in the environment catalyze the reaction and thus derive energy from it.

Table 1. Free energy released by hydrogenation of hydrocarbons on Titan, per mole of hydrocarbon consumed [4].

Reaction	Energy Released (kJ mole^{-1})
$C_2H_2 + 3H_2 = 2CH_4$	334
$C_2H_6 + H_2 = 2CH_4$	57
$R{-}CH_2 + H_2 = R + CH_4$	54

Given that redox couples (e.g., C_2H_2, $3H_2$) are produced in the atmosphere, and that these small molecules will be widespread on Titan and are readily soluble in liquid ethane and methane [6], there may not be any need for photosynthesis on Titan. However, nonetheless, it is interesting to consider the availability of sunlight on Titan's surface. Models of light penetration in Titan's atmosphere [7], and direct measurements by the Huygens Probe [8] provide a good understanding of the availability of sunlight in the atmosphere and at the surface. Because of Titan's distance from the sun (10 AU) and the haze in the atmosphere, the maximum level of sunlight on the surface of Titan is about 0.1% that of the overhead sun on Earth's surface [9]. Despite the haze, the distribution of solar flux with wavelength roughly follows the solar spectrum, with the peak flux occurring at about 0.6 um (Figure 11 of [8]). This low-light level is more than adequate for photosynthesis on Earth—which process can utilize light levels as low as 10^{-6} of the Earth noonday solar flux [10,11]. Thus, photosynthesis should be possible on Titan with pigments that would not be dissimilar to those used on Earth. Even with attenuation by the atmospheric haze and the large distance from the sun, the energy in sunlight reaching the surface of Titan is orders of magnitude larger than the chemical energy in hydrocarbons descending from the upper atmosphere. As on Earth, the biosphere on Titan would be most productive if it were to use solar energy directly. On Earth, photosynthesis is used to produce organic material primarily from CO_2 and H_2O. On Titan, presumably, photosynthesis should produce organic material from CH_4, and H_2 would be a byproduct—the reverse of the reactions listed in Table 1. This is similar to the suggestion for photosynthesis in hydrogen-dominated atmospheres [12].

2.2. Nutrients

Organic compounds are common on Titan, including N containing organics. Thus, C, H, and N are available in multiple compounds. The surface contains H_2O as ice but no other compounds with O atoms are common at the surface. Thus, life on Titan may have a restricted set of elements as nutrients, compared to the wide variety available and used by life on Earth. Bains and Seager [13] have pointed out that a small set of elements may have implications for the complexity of possible biochemistry. Small hydrocarbons and N_2 are readily soluble in liquid methane [6], and some N and O containing molecules may be present in the liquid as well, as listed in Table 2.

Table 2. Molecules with O or N in liquid CH_4 and C_2H_6 [6].

Compound	Solubility, Mole Fraction	
	Methane	Ethane
H_2O	10^{-22}–10^{-18}	10^{-18}–10^{-15}
CO_2	10^{-5}–10^{-3}	10^{-5}–10^{-3}
HCN	10^{-12}–10^{-10}	10^{-10}–10^{-8}
C_2N_2	10^{-8}–10^{-7}	10^{-7}–10^{-6}
CH_3 CN: acetonitrile	10^{-12}–10^{-10}	10^{-10}–10^{-8}
C_2H_3CN: acrylonitrile	10^{-12}–10^{-10}	10^{-10}–10^{-8}

Schulze-Makuch *et al.* [14] suggested that a critical problem for life on Titan would be access to inorganic elements, such as Fe, Cu, Mn, Zn, Ni, S, Ca, Na, and K, *etc.*, which life on Earth can access through their solubility in water. These inorganic elements are used, often in trace levels, in various life functions. Of particular interest is the use of metals in the active sites of enzymes.

We can consider two approaches for nutrients to support life on Titan: (1) the conservative use of elements that are hard to access; and (2) the use of H_2O to fill some of the roles of the inorganic elements.

The biology and environmental cycling of phosphorous on Earth provides an example of the conservative use of an element that is not readily available in the environment and is not naturally cycled by the physical transport systems in the environment. A similar conservative use may be needed on Titan for inorganic elements. A possible source of inorganic elements is the influx of meteorites and comets in the upper atmosphere. This source is known to be responsible for CO, CO_2, and H_2O in the atmosphere [15]. Thus, a small but possibly biologically useful flux of inorganic elements is descending from the upper atmosphere with the haze and may be adequate for life forms that carefully recycle and reuse these elements.

The alternative to a conservative use of trace inorganic elements for biology on Titan is to eliminate the need for these elements entirely. If life on Titan does not employ photosynthesis, and does not need to fix nitrogen from N_2 (N is available in organic compounds produced photochemically), then two of the major needs for metal-based catalysis used by Earth life are removed. Furthermore, it is conceivable that H_2O molecules on Titan could be used in a way that metals are used in enzymes by life on Earth. Because of the low temperatures on Titan, hydrogen bonds provide enough binding strength to form useful structures. Benner *et al.* [1] suggested that hydrogen bonding is difficult to use in the assembly of supramolecular structures in water, and they speculated that life in Titan liquids would be able to use hydrogen bonding more, and that these bonds would have the strength appropriate for the low temperature. The hydrogen bond (5 to 30 kJ mole^{-1}) is stronger than a van der Waals interaction, but weaker than covalent (~300 kJ mole^{-1}) or ionic bonds (20–30 kJ mole^{-1}). Thermal energy, RT, at the Titan temperature of 95 K equals about 1 kJ mole^{-1}. As a polar molecule (one of the few in the Titan environment), H_2O is particularly good at affecting hydrogen bonds. Individual H_2O molecules or small clusters held in hydrocarbon cages could play the catalytic role for hydrogen-bonded structures that metals do for redox reactions in Earth biochemistry.

2.3. Liquid Habitats

On Earth, life is widespread because habitable liquid water is widespread. Even in the driest desert on Earth, the Atacama Desert of Chile, there is occasional liquid water. Liquid on Titan is also widespread. There are large lakes in the northern hemisphere (Figure 1) and at least one large lake in the southern hemisphere [16]. Observations from the Huygens Probe indicated that the soil at the equatorial landing site was moist with methane and ethane [17,18] and, by extension with orbital spectral data, most of the low latitude soils are moist [19]. There may be low latitude lakes [20,21].

The nature of the lakes, the processes that fill them, and the factors that set their distribution are still not fully understood. With the exception of Ontario Lacus, the lakes on Titan occur in the northern hemisphere (these are shown in Figure 1) and 97% of the lakes on Titan are clustered in a region of 900 by 1800 km—about 2% of Titan's surface area. The large lakes and small lacustrine depressions seen in Figure 1 can be roughly grouped into two principle modes based on size. The large lakes (several hundred kilometers in width) are thought to be deep (up to hundreds of meters in depth) [22,23,24]. They possess fractal shorelines and are connected to fluvial channels (e.g., Ligeia Mare, Figure 2) [25,26,27]. In contrast, the small lacustrine depressions appear to be more shallow depressions with rounded shorelines and show a gradient with elevation [28,29]. Empty depressions, which appear similar to the small lakes are found at elevations about 250 m above the similarly shaped small lakes (as discussed by Cornet *et al.* [6]). This could be indicative of the presence of aquifers and subsurface connectivity establishing defined liquid levels. [6,28,30].

The geological processes that form the lakes, both large and small, on Titan are still unclear. Recently Cornet *et al.* [6] reviewed the current state of knowledge of the lakes and proposed karst-like formation processes resulting from the dissolution of solid organics by the liquid methane and ethane.

Orbital spectral data have lead to the characterization of Titan's surface into five terrain types: (1) bright terrain; (2) dark equatorial dune fields, or dark brown units; (3) blue units; (4) 5 μm bright units; and the (5) dark lakes [31,32]. The spectral data in the regions inside and around some polar (and equatorial) lacustrine depressions indicate the presence of various hydrocarbons and nitriles [33,34] and are not compatible with

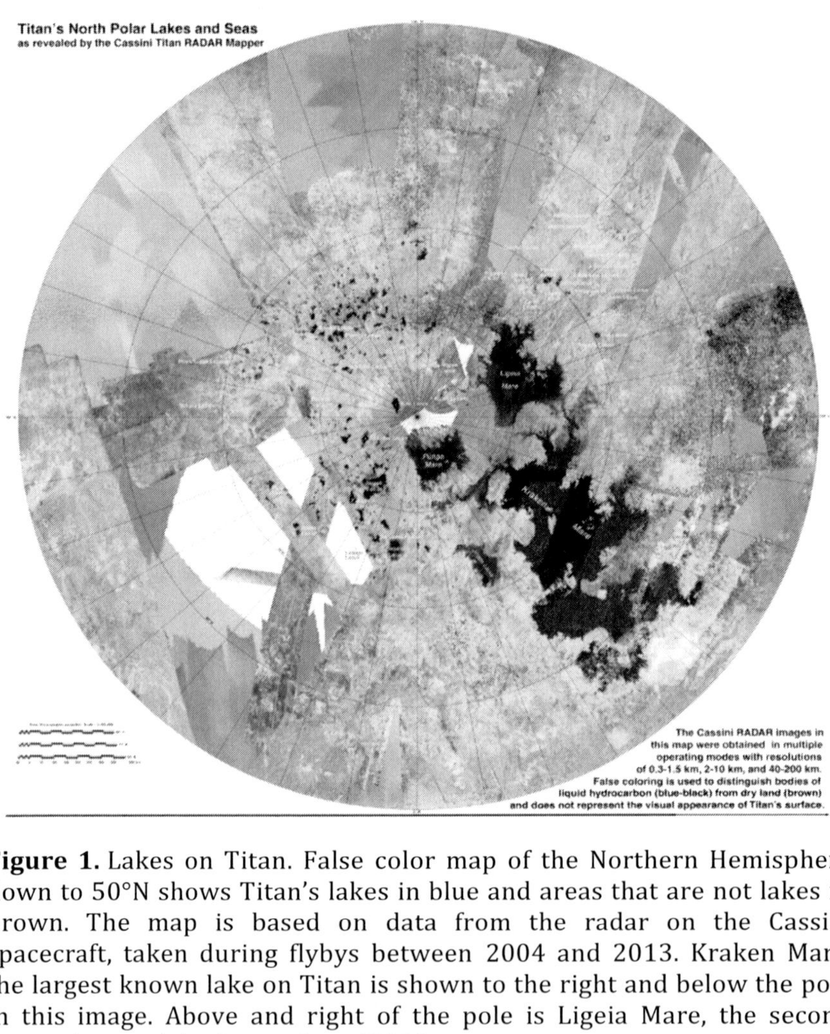

Figure 1. Lakes on Titan. False color map of the Northern Hemisphere down to 50°N shows Titan's lakes in blue and areas that are not lakes in brown. The map is based on data from the radar on the Cassini spacecraft, taken during flybys between 2004 and 2013. Kraken Mare, the largest known lake on Titan is shown to the right and below the pole in this image. Above and right of the pole is Ligeia Mare, the second largest known lake. Image from NASA.

the presence of water ice exposed on the surface [35]. Rannou *et al.* [19] find that the surface spectrum at the Huygens landing site indicates a layer of water ice grains overlaid by a moist layer of weakly compacted haze particles. The spectrum is inconsistent with dry haze particles. Rannou *et al.* [19] could not locate the Huygens landing site, shown in Figure 3, on a specific type of terrain, because it is on the border between light and dark surface units. Extrapolating their spectral analysis of the Huygens landing to other equatorial regions, Rannou *et al.* [19] conclude that the ground at low latitudes is persistently wet, resulting either from a subsurface rich in liquid or from frequent enough rains to maintain moist surface conditions against evaporation. Williams *et al.* [36] suggested that moist surface conditions could persist for 5 to 50 days after a rain event, depending on moisture level. Rain in equatorial latitudes has been observed [37]. The CH_4 and C_2H_6 liquids are widespread on Titan in lakes and in the moist surface. If life can exist in such liquids, life should be widespread on Titan.

Figure 2. Close-up radar image of Ligeia Mare, showing coastline features and rivers. The lake is ~400 km across. Image from ESA based on Cassini Data.

Figure 3. Titan's surface at the Huygens landing site, 10.2°S, 192.4°W. There are at least eight rocks visible in the image—numbered in red with size indicated for two of them. Distance from the lander is shown in blue and the horizon at 88.5° is labeled in green. The small rocks are thought to be H_2O ice mostly coated by organic solid material. The rounded nature of these rocks suggests past fluvial activity. The right panel shows, approximately, the true color of the scene. Direct measurements by the Huygens probe indicated that the ground at the landing was moist with methane and ethane at the time of the landing, 14 January 2005. Image from ESA.

2.4. Transport Cycles of Liquid Moving Nutrients and Wastes

On Earth, natural waters can be treated as a mixture of water and a non-volatile solid (salt). The amount of dissolved air in the water is negligible.

Evaporation separates the volatile water from the salt-creating fresh water, defining in the process two somewhat different water habitats on Earth: freshwater—rivers and lakes, and salt water—oceans. The extensive surface area of water on Earth and the resulting evaporation ensures that there is an active hydrological cycle continually producing rain and snow.

As discussed above, rain has been observed in a few low-latitude regions [37]. In addition, there is extensive convective cloud activity in the summer polar regions [38,39] and at mid latitudes [40]. Global Circulation Models provide rough indications of the latitudinal and seasonal variations in rain and surface evaporation [41,42,43,44]. Typically, rain much in excess of evaporation is predicted at latitudes >60° higher than 60° during summer, in agreement with the presence of lakes, whereas middle to low latitudes experience net evaporation, in agreement with the absence of lakes—but still consistent with moist-soil conditions persisting following frequent rain.

In contrast, on Titan liquids are composed of three major components: methane, ethane, and dissolved atmospheric nitrogen. The solubility of N_2 in the methane and ethane liquids is significant, up to 20% [45]. Ethane is non-volatile compared to methane and nitrogen and is left behind as the liquid evaporates. Nitrogen and methane (and perhaps ethane to a smaller extent) are present in the atmosphere and can equilibrate with surface liquids. Rain on Titan is expected to be a mixture of these gases [45] and the fraction of methane and ethane depending on the mixing ratio of ethane in the lower atmosphere, which is not known [36].

Tan *et al.* [46,47] and Luspay-Kuti *et al.* [48] have shown that the mixture of three components of differing volatility creates unusual behavior during evaporation and condensation, compared to a single component fluid. In particular, they find that the density of the fluid increases with temperature, and thus liquids in the polar regions are less dense than those near the equator, due to the compositional effects on density, exceeding those of thermal expansion. They also show that the liquid density decreases with pressure, very much unlike water on Earth. These properties affect the global circulation of fluids on Titan. Thus, the cycling of liquid on Titan, and its various compositions, is complex when compared to the cycling of water on Earth.

These variations in composition appear to be reflected on Titan. The composition of the northern lakes appears to be dominated by methane [48], whereas ethane has been identified as a major component in the southern-hemisphere Ontario Lacus [16,48], which may indicate that it is a remnant evaporitic basin [6] analogous to the Dead Sea on Earth. In contrast with Earth however, it is not clear that the liquid becomes less habitable as it is enriched in the lower volatile component. In fact, ethane is a better solvent than methane or nitrogen, and thus ethane-rich lakes may provide a richer more diverse solution for growth. For example, ethane is ~20× better than methane regarding solubility of small organic molecules [49].

Cornet *et al.* [6] suggest that if the lakes are karst features, then Titan's lakes could be considered to be similar in age with some of the youngest features of the moon (less than ~100,000 year). In addition, karst formation rates in the northern mid-latitude (60°) regions should be more than three times faster than in the southern mid-latitude regions because in the present climate, GCM models predict that the southern latitudes receive less total precipitation, albeit with more intense rainstorms than northern latitudes. Cornet *et al.* [6] suggest this could explain why Titan's south polar regions are devoid of well-developed lacustrine depressions, compared to the north.

The timescale of seasonal cycles is 30 times longer on Titan than on Earth just due to the orbital period. The diurnal period on Titan is 16 Earth days but nothing in the environment changes on the diurnal schedule. The CH_4 and C_2H_6liquids on Titan participate in an active hydrological cycle including rain, evaporation, lakes, and surface moisture. The cycles produce variations in composition of the liquids with latitude and seasons. Such cycles seem adequate for the transport of nutrients and wastes needed for life.

3. CARBON BIOCHEMISTRY ON TITAN

Life on Earth is based on carbon biochemistry in the medium of liquid water. The postulated life on the surface of Titan is carbon biochemistry in a medium of liquid methane and ethane. Many of the key biochemical structures that life on Earth uses are specific to solvents with properties like liquid water. These include, for example, lipid bilayers as membrane components, amino acids and their collection into folded proteins, and the shape of DNA. Carbon biochemistry in liquid water on Earth is able to create the following: compartmentalization for autonomy and reproduction, information storage molecules and a way to duplicate them, structural molecules and a way to build them. For life on Titan similar structures must be created in a solution of methane and ethane.

3.1. Compartmentalization for Autonomy and Reproduction

It is generally thought that the ability to create a boundary layer between the interior of a living system and the external environment is a requirement for life [50,51]. On Earth, this is accomplished using a lipid bilayer, where the nature of the membrane is the result of the interaction of the bi-polar lipids with liquid water. Based on computational chemical analysis, Stevenson *et al.* [52] proposed a type of membrane which they termed "azotosome" capable of forming and functioning in liquid methane at cryogenic temperatures. The membrane is composed of small organic nitrogen compounds, such as acrylonitrile. The structural integrity of the membrane results from the attraction between polar heads of short-chain molecules that are rich in nitrogen and the interlocking nitrogen and hydrogen atoms that reinforce the structure. Although the azotosome structures may be difficult to construct in the laboratory, the structures provide a persuasive example of a possible membrane system for life on Titan.

3.2. Information Storage Molecules and a Way to Duplicate Them

Benner *et al.* [1] have observed that polymeric molecules suitable for information storage, such as DNA, must have the property that the shape of the molecule does not depend on the information stored in it. The shape of the biopolymer DNA is insensitive to the sequence encoded by that DNA. In contrast, the shape of a protein, another biopolymer, can be radically altered by the change of even a single amino acid in the sequence. Thus DNA is a suitable information-storage molecule and proteins are not. Following this approach, Benner's research group [53] experimentally investigated the solubility of polyethers in non-polar hydrocarbon solvents at low temperatures. Such polyethers are a possible model for information-storage molecules in non-polar hydrocarbon solvents. While they find solubility in high-temperature hydrocarbons, they find that the solubility becomes immeasurably low below 170 K. McLendon *et al.* [53] note that overall, their results support a perspective elaborated by Bains [2] that water is a good solvent, primarily because it is liquid at high temperature, and secondarily because of its chemical properties (polar, high dielectric constant). In this view, the main difficulties with solution chemistry in Titan liquids are primarily the low temperature and secondarily the non-polar nature of the solvent. The search for a plausible information molecule for life in Titan liquids remains an open research topic. For the reasons discussed above, if such a Titan information molecule is discovered, it is likely that the binding of the information bits is achieved via hydrogen bonds. A possible approach for the molecular structure of the information system is a two-letter code involving hydrogen bonding with polar molecules containing O and N—both of which form hydrogen bonds and are available on Titan (O in H_2O, and N in HCN, as well as in other nitriles). Another possible class of polymers that could store information is conducting polymers, such as polypyrrole and polyaniline, which are currently of much interest in the nanomaterials field [54]. These polymers are composed of C, N and H only and can transition between stable redox states—possibly a basis for information encoding.

3.3. Structural Molecules and a Way to Build Them

Life on Earth uses proteins as the primary structural molecule. Using ~20 amino acids to form proteins, life can construct a combinatorially large number of possible proteins. Due both to their internal interactions and especially to their interaction with water in both hydrophilic and hydrophobic ways, individual proteins fold into specific shapes. These shapes form the basis of structural molecules used by Earth life.

In Titan liquids, the structural substitutes for proteins might include hydrocarbon chains, aromatic ring structures, carbon nanostructures, including graphene, and various types of fullerenes. The addition of N to the carbon structures can potentially add diversity.

4. ECOLOGY

On Earth, organisms, even microorganisms, live in communities. Within such communities organisms exchange matter and genes. Arguably, communities of microorganisms can survive and thrive in conditions that would be difficult for any of the individuals in the community to grow in in isolation. Communities can more efficiently cycle and recycle nutrients and can enhance genetic resilience. The liquid-water environment on Earth provides the medium for an organism to come into physical contact; as well, it mobilizes material released from cells. It worth noting here that viruses play a role in exchanging genetic information.

If life on Titan utilizes a set of biomolecules compatible with low-temperature methane and ethane liquids, then it is plausible that the same sort of ecological communication and exchange that occurs on Earth can occur in communities of Titan life forms. Signaling molecules could be low-molecular weight hydrocarbons that would be mobile in the Titan liquid. The role of NO and H_2S in cell signaling in the human body shows that such signaling molecules need not be complex. If genetic material on Titan is based on soluble polymers as suggested by Stevenson *et al.* [52], then these also could be mobile in Titan liquid. Furthermore a Titan analog to viruses could be possible with hydrocarbon shells encasing raw Titan genetic polymers that attach and then insert the genetic polymers into host organisms. As discussed above, close recycling of elements may be essential for life on Titan, especially for metal atoms and H_2O molecules, if these are used in catalysts.

5. SEARCH

Given how different carbon-based life in liquid methane and ethane must be from carbon-based life in liquid water, it will be a challenge to implement a search strategy for life on Titan. However, some general principles that have emerged in the search for life may be applicable.

It has been suggested that a key feature of life is that it is selective in the basic molecules it uses [55,56]. Molecules with similar chemical properties can be present in biology in vastly different concentrations. Life on Titan would also have to select specific molecules from the possible variations. Thus, the prediction of McKay [56] should hold on Titan that a biological distribution of molecules would be a series of relatively sharp spikes while an abiotic distribution would be smooth. Surface material from Titan, for example the sediments at the shore of a lake, could provide a promising sample. An analysis of the nature of the molecules and their relative abundance could suggest a biological signal. This would be of great interest both as an indication of the presence of a biological system and as an inventory of the molecules used in the biochemistry of that system. One would know a great deal about life on Earth even if all that one knew were that life utilized the following: a certain twenty amino acids; a specific five nucleotide bases; and polysaccharides made up mostly from a few specific simple sugars.

The premier example of biochemical selectivity is chirality. This property is most well known in the amino acids used in proteins. Life on Earth uses only the L version of amino acids in proteins, not the mirror image, the D version. If life on Titan also uses molecules with chiral centers then detection of homochirality is a powerful indication of life. The simplest and most common chirality center is an atom that has four different groups bonded to it in such a manner that it has a non-superimposable mirror image. With N atoms added to hydrocarbons, chiral centers can be expected to form. Chirality can be determined by chromographic and optical methods, and thus may provide a robust and easily determinable method to test whether organic material in sediments on Titan are of biological origin.

Ecological Change and H_2

Life that is widespread will effect the global environment. The majority of the O_2, CO_2, CH_4, and even N_2 in the Earth's atmosphere are products of biology. The CO_2 is of particular interest because its change with seasons reflects active growth. Local sources of CH_4 also reflect biological processes. This approach may provide a method to search for life on Titan by analysis of the lower—possibly an easier task than collecting and analyzing lake-shore sediments. McKay and Smith (2005) suggested this approach and pointed out the H_2 is the most promising atmospheric constituent on Titan to show a biological effect. If life is consuming atmospheric hydrogen, it will have a measurable effect on the hydrogen mixing ratio in the troposphere, if the biological consumption is greater than 10^9 $cm^{-2}\cdot s^{-1}$ [4]. Cornet *et al.* [6] review photochemical models for Titan and list production rates for C_2H_2, from 0.32 to 1.2 × 10^9 $cm^{-2}\cdot s^{-1}$ (e.g., [57,58]), and C_2H_6, from 1.2 to 15 × 10^9 $cm^{-2}\cdot s^{-1}$; thus if ~20% of the available C_2H_2 and C_2H_6 is consumed by methanogens, the corresponding H_2consumption should significantly deplete the near-surface H_2 profile. This is shown schematically in Figure 4.

In addition to H_2, McKay and Smith [4] suggested that methane-based life on Titan could consume acetylene and ethane. There seems to be evidence for depletion of acetylene and ethane on Titan. The data that suggest that there is less ethane on Titan than expected is well established [22]. Photochemical models have predicted that Titan should have a layer of ethane sufficient to cover the entire surface to a thickness of many meters but Cassini has found no such layer. Clark *et al.* [33] find a lack of acetylene on the surface despite its expected production in the atmosphere and subsequent deposition on the ground. There was also no evidence of acetylene in the gases released from the surface after the Huygens Probe landing [17,18]. Thus, the evidence for less ethane and less acetylene than expected seems clear.

The depletion of ethane and acetylene is more significant in the astrobiological sense, because of the report of a hydrogen flux into the surface. This is the key that suggests that these depletions may not be just due to a lack of production but are due to some kind of chemical reaction at the surface.

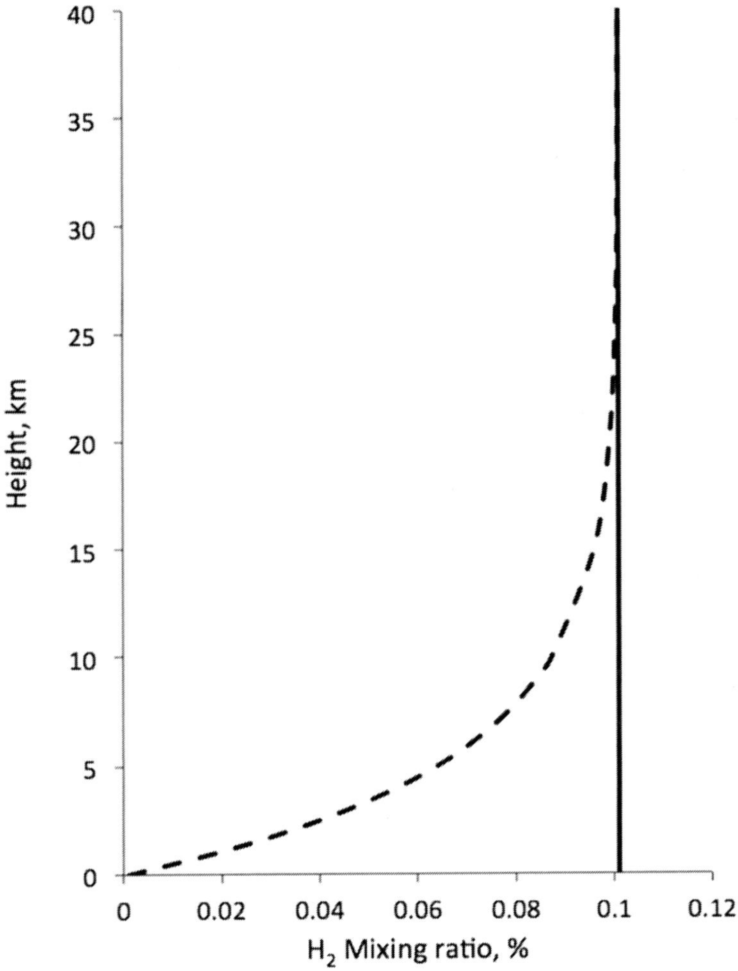

Figure 4. Schematic of expected H_2 concentration in the lower atmosphere without a biological sink at the surface (solid line) and with a biological sink (dotted line).

Strobel [59] predicts a strong flux of hydrogen into the surface. This determination is not the result of a direct observation but rather it is the result of a computer simulation designed to fit measurements of the hydrogen concentration in the lower and upper atmosphere in a self-consistent way. It is not presently clear from Strobel's results how his conclusion of a hydrogen flux into the surface is dependent on the way the computer simulation is constructed or how accurately it simulates the Titan chemistry.

Courtin *et al.* [60] found that the latitudinal distribution of H_2 appears to be non-uniform, with a mole fraction above 50°N larger than the globally-averaged value. This is unexpected, and Courtin *et al.* [60] speculate on an atmospheric process redistributing H_2. One could also speculate that the ethane-enhanced southern hemisphere is biologically consuming more H_2 than the methane-dominated northern hemisphere due to the superior properties of ethane as a solvent.

On Titan, the consumption of H_2 and production of CH_4 (Table 1) is obviously not a net source of CH_4: it merely recycles CH_4, thereby undoing the photolysis of CH_4, and there is no *a priori* reason to expect the resulting CH_4 to exhibit an isotopic shift from these reactions. The C-C bond in acetylene is strong, but this by itself does not imply a strong isotopic selectivity. For example, life on Earth breaks the strong bond between the N atoms in N_2 without leaving a clear isotopic effect. Thus, the isotopic state of C on Titan is not definitive to the question of the presence of Titan methanogens. If there is an isotopic effect in the uptake of H_2 this could be large, given the mass difference between D and H.

In overview, there are four possibilities for the recently reported findings regarding a flux of hydrogen into the surface of Titan. Here they are listed in order of their likely reality.

1. The determination that there is a strong flux of hydrogen into the surface may be mistaken. It will be interesting to see if other researchers, in trying to duplicate Strobel's results, reach the same conclusion.

2. There could exist a physical process that is transporting H_2 from the upper atmosphere into the lower atmosphere. One possibility is adsorption onto the solid organic atmospheric haze particles, which eventually fall to the ground. However, this would be a flux of H_2, and not a net loss of H_2.

3. If the loss of hydrogen at the surface is correct, the non-biological explanation requires that there should be some sort of surface catalyst, presently unknown, that can mediate the hydrogenation reaction at 95 K, the temperature of the Titan surface. That would be quite interesting and a startling find, although not as startling as the presence of life.

4. The depletion of hydrogen, acetylene, and ethane, could be due to a new type of liquid-methane-based life form, as predicted in [1,3,4].

Further measurements of hydrogen, as well as of acetylene and ethane abundances in the lower atmosphere of Titan, may be the most promising immediate strategy for a life search on that world.

Photosynthetic life might produce H_2 and offset the biological sources. But the results would be seasonal changes in $H_2$by analogy with CO_2 which is both produced and consumed on Earth but the relative rates vary with season (on Titan seasons are only changes in light levels).

In the context of role of H_2 and life on Titan, it is interesting to note that proposals that energy derived from H_2 reactions may have played a key role in the origin of life on Earth [61] may be modified and applied to the origin of life on Titan.

6. SUMMARY AND CONCLUSIONS

In this paper we have considered Titan as an abode of life based, of necessity, on aspects and the phenomenon of Earth and of life on Earth. The presence of a widespread liquid on its surface, the abundance of light and chemical energy, and the continual production of organic gases and solids all encourage the concept of indigenous life on Titan. Theoretical simulations have indicated a possible membrane system. However, laboratory investigations have failed to suggest a plausible information containing molecule. In sum, the question of Titan as an abode of life remains unresolved but indications so far warrant further investigation.

Based on what we do know, we can list the following fundamental challenges to life on Titan, in approximate order of severity.

1. The small diversity of elements available in the environment.
2. The low temperature of solution and the resulting negligible solubility.
3. The non-polar nature of the methane and ethane solvent, further lowering solubility.
4. The limited diversity of hydrocarbon structural molecules (compared to proteins).

Given these limitations, it may be that if there is life in the liquids on Titan's surface it may be simple, heterotrophic, slow to metabolize, and slow to adapt with limited genetic and metabolic complexity. The simple molecules needed for metabolism may be widespread in the environment and in the methane/ethane liquids, but the complex organics needed for structural or genetic systems may be hard to obtain or synthesize. The communities formed may be ecologically simple—perhaps analogous to the microbial ecosystems found in extreme cold and dry environments on Earth.

The advantages of the Titan environment for life include: the free food from the sky in the form of organics (principally C_2H_2), the chemically benign nature of the non-polar solvent in terms of attacking biomolecules (in contrast to water), the lack of ultraviolet or ionizing radiation at the surface, and low rates of thermal decomposition due to the low temperature. Titan may have only a simple trophic system, probably without primary producers and without predators. Photosynthesis may be beyond the complexity that can be achieved with the limited element and hence genetic diversity—and food should be free. The simple low temperature life forms and communities envisaged would have very low energy demands and would grow slowly. Life on Titan may be not much more than auto-catalytic reactions encased in azotomes. However, if it had genetics, and were thus Darwinian, then what a wonderful life it would be: a second genesis different enough from Earth life to suggest that our Universe is full of diverse and wondrous life forms.

In terms of ethical considerations, the search for a second genesis of life on Titan contrasts with the similar search on Mars. On Mars, issues of contamination both with respect to the Martian environment and back contamination of the Earth are important concerns. Furthermore, it has been suggested that if there were a second genesis of life on Mars the global conditions are not favorable for it and humans may choose to intervene to improve them [62]. In contrast, possibilities of contamination of Titan with water-based life from Earth are essentially zero, as are the possibilities for back contamination of the Earth with life from Titan. In addition, if there is methane or ethane based life on Titan, the global environment is probably quite suitable for it, and it will not need human help for many billions of years.

ACKNOWLEDGMENTS

Funding for this research and its publication has come from the NASA Astrobiology Program.

CONFLICTS OF INTEREST

The author declares no conflict of interest.

REFERENCES

1. Benner, S.A.; Ricardo, A.; Carrigan, M.A. Is there a common chemical model for life in the universe? Curr. Opin. Chem. Biol. 2004, 8, 672–689.
2. Bains, W. Many chemistries could be used to build living systems. Astrobiology 2004, 4, 137–167.
3. Schulze-Makuch, D.; Grinspoon, D.H. Biologically enhanced energy and carbon cycling on Titan? Astrobiology 2005, 5, 560–567.
4. McKay, C.P.; Smith, H.D. Possibilities for methanogenic life in liquid methane on the surface of Titan. Icarus 2005, 178, 274–276.
5. Kral, T.A.; Brink, K.M.; Miller, S.L.; McKay, C.P. Hydrogen consumption by methanogens on the early Earth. Orig. Life Evol. Biosph. 1998, 28, 311–319.
6. Cornet, T.; Cordier, D.; Le Bahers, T.; Bourgeois, O.; Fleurant, C.; Le Mouélic, S.; Altobelli, N. Dissolution on Titan and on Earth: Towards the age of Titan's karstic landscapes. J. Geophys. Res.: Planets 2015, 120, 1044–1074.
7. McKay, C.P.; Pollack, J.B.; Courtin, R. The thermal structure of Titan's atmosphere. Icarus 1989, 80, 23–53.
8. Tomasko, M.G.; Bézard, B.; Doose, L.; Engel, S.; Karkoschka, E.; Vinatier, S. Heat balance in Titan's atmosphere. Planet. Space Sci. 2008, 56, 648–659.
9. McKay, C.P.; Pollack, J.B.; Courtin, R. The greenhouse and antigreenhouse effects on Titan. Science 1991, 253, 1118–1121.
10. Raven, J.A.; Kübler, J.E.; Beardall, J. Put out the light, and then put out the light. J. Mar. Biolog. Assoc. UK 2000, 80, 1–25.
11. McKay, C.P. Requirements and limits for life in the context of exoplanets. Proc. Natl. Acad. Sci. USA 2014, 111, 12628–12633.
12. Bains, W.; Seager, S.; Zsom, A. Photosynthesis in Hydrogen-Dominated Atmospheres. Life 2014, 4, 716–744.
13. Bains, W.; Seager, S. A combinatorial approach to biochemical space: Description and application to the redox distribution of metabolism. Astrobiology 2012, 12, 271–281.
14. Schulze-Makuch, D.; Fairén, A.G.; Davila, A. Locally targeted ecosynthesis: A proactive in situ search for extant life on other worlds. Astrobiology 2013, 13, 674–678.
15. De Kok, R.; Irwin, P.G.J.; Teanby, N.A.; Lellouch, E.; Bézard, B.; Vinatier, S.; Nixon, C.A.; Fletcher, L.; Howett, C.; Calcutt, S.B.; et al. Oxygen compounds in Titan's stratosphere as observed by Cassini CIRS. Icarus 2007, 186, 354–363.
16. Brown, R.H.; Soderblom, L.A.; Soderblom, J.M.; Clark, R.N.; Jaumann, R.; Barnes, J.W.; Sotin, C.; Buratti, B.; Baines, K.H.; Nicholson, P.D. The identification of liquid ethane in Titan's Ontario Lacus. Nature 2008, 454, 607–610.
17. Niemann, H.B.; Atreya, S.K.; Bauer, S.J.; Carignan, G.R.; Demick, J.E.; Frost, R.L.; Gautier, D.; Haberman, J.A.; Harpold, D.N.; Hunten, D.M.; et al. The abundances of constituents of Titan's atmosphere from the GCMS instrument on the Huygens probe. Nature 2005, 438, 779–784.
18. Lorenz, L.D.; Niemann, H.B.; Harpold, D.N.; Way, S.H.; Zarnecki, J.C. Titan's damp ground: Constraints on Titan surface thermal properties from the temperature evolution of the Huygens GCMS inlet. Meteorit. Planet. Sci. 2006, 41, 1705–1714.
19. Rannou, P.; Toledo, D.; Lavvas, P.; D'Aversa, E.; Moriconi, M.L.; Adriani, A.; Le Mouélic, S.; Sotin, C.; Brown, R. Titan's surface spectra at the Huygens landing site and Shangri-La. Icarus 2016.
20. Griffith, C.A.; Lora, J.M.; Turner, J.; Penteado, P.F.; Brown, R.H.; Tomasko, M.G.; Doose, L.; See, C. Possible tropical lakes on Titan from observations of dark terrain. Nature 2012, 486, 237–239.
21. Vixie, G.; Barnes, J.W.; Jackson, B.; Rodriguez, S.; Le Mouélic, S.; Sotin, C.; MacKenzie, S.; Wilson, P. Possible Temperate Lakes on Titan. Icarus 2015, 257, 313–323.

22. Lorenz, R.D.; Mitchell, K.L.; Kirk, R.L.; Hayes, A.G.; Aharonson, O.; Zebker, H.A.; Paillou, P.; Radebaugh, J.; Lunine, J.I.; Janssen, M.A.; et al. Titan's inventory of organic surface materials. Geophys. Res. Lett. 2008, 35.
23. Lorenz, R.D.; Kirk, R.L.; Hayes, A.G.; Anderson, Y.Z.; Lunine, J.I.; Tokano, T.; Turtle, E.P.; Malaska, M.J.; Soderblom, J.M.; Lucas, A.; et al. A Radar map of Titan seas: Tidal dissipation and ocean mixing through the throat of Kraken. Icarus 2014, 237, 9–15.
24. Marco, M.; Valerio, P.; Alexander, H.; Ralph, L.; Jonathan, L.; Giovanni, P.; Roberto, S.; Enrico, F.; Giuseppe, M.; Claudia, N.; et al. The bathymetry of a Titan sea. Geophys. Res. Lett. 2014, 41, 1432–1437.
25. Stofan, E.R.; Elachi, C.; Lunine, J.I.; Lorenz, R.D.; Stiles, B.; Mitchell, K.L.; Ostro, S.; Soderblom, L.; Wood, C.; Zebker, H.; et al. The lakes of Titan. Nature 2007, 445, 61–64.
26. Sotin, C.; Lawrence, K.J.; Reinhardt, B.; Barnes, J.W.; Brown, R.H.; Hayes, A.G.; le Mouélic, S.; Rodriguez, S.; Soderblom, J.M.; Soderblom, L.A.; et al. Observations of Titan's northern lakes at 5 microns: Implications for the organic cycle and geology. Icarus 2012, 221, 768–786.
27. Wasiak, F.C.; Androes, D.; Blackburn, D.G.; Tullis, J.A.; Dixon, J.; Chevrier, V.F. A geological characterization of Ligeia Mare in the northern polar region of Titan. Planet. Space Sci. 2013, 84, 141–147.
28. Hayes, A.; Aharonson, O.; Callahan, P.; Elachi, C.; Gim, Y.; Kirk, R.; Lewis, K.; Lopes, R.; Lorenz, R.; Lunine, J.; et al. Hydrocarbon lakes on Titan: Distribution and interaction with a porous regolith. Geophys. Res. Lett. 2008, 35.
29. Lorenz, R.D.; Stiles, B.W.; Aharonson, O.; Lucas, A.; Kirk, R.L.; Zebker, H.A.; Turtle, E.P.; Neish, C.D.; Stofan, E.R.; et al. A global topographic map of Titan. Icarus 2013, 225, 367–377.
30. Cornet, T.; Bourgeois, O.; le Mouélic, S.; Rodriguez, S.; Gonzalez, T.L.; Sotin, C.; Tobie, G.; Fleurant, C.; Barnesf, J.W.; Brown, R.H.; et al. Geomorphological significance of Ontario Lacus on Titan: Integrated interpretation of Cassini VIMS, ISS and RADAR data and comparison with the Etosha Pan (Namibia). Icarus 2012, 218, 788–806.
31. Barnes, J.W.; Brown, R.H.; Soderblom, L.; Buratti, B.J.; Sotin, C.; Rodriguez, S.; le Mouèlic, S.; Baines, K.H.; Clarke, R.; Nicholson, P. Global-scale surface spectral variations on Titan seen from Cassini/VIMS. Icarus 2007, 186, 242–258.
32. Stephan, K.; Jaumann, R.; Karkoschka, E.; Kirk, R.L.; Barnes, J.W.; Tomasko, M.G.; Turtle, E.P.; Le Corre, L.; Langhans, M.; Le Mouélic, S.; et al. Mapping products of Titan's surface. In Titan from Cassini-Huygens; Brown, R.H., Lebreton, J.-P., Waite, J.H., Eds.; Springer: Dordrecht, The Netherlands, 2009; pp. 489–510.
33. Clark, R.N.; Curchin, J.M.; Barnes, J.W.; Jaumann, R.; Soderblom, L.; Cruikshank, D.P.; Brown, R.H.; Rodriguez, S.; Lunine, J.; Stephan, K.; et al. Detection and Mapping of Hydrocarbon Deposits on Titan. J. Geophys. Res. 2010, 115.
34. Moriconi, M.L.; Lunine, J.I.; Adriani, A.; D'Aversa, E.; Negrao, A.; Filacchione, G.; Coradini, A. Characterization of Titan's Ontario Lacus region from Cassini/VIMS observations. Icarus 2010, 210, 823–831.
35. Barnes, J.W.; Soderblom, J.M.; Brown, R.H.; Buratti, B.J.; Sotin, C.; Baines, K.H.; Clark, R.N.; Jaumann, R.; McCord, T.B.; Nelson, R.; et al. VIMS spectral mapping observations of Titan during the Cassini prime mission. Planet. Space Sci. 2009, 57, 1950–1962.
36. Williams, K.E.; McKay, C.P.; Persson, F. The surface energy balance at the Huygens landing site and the moist surface conditions on Titan. Planet. Space Sci. 2012, 60, 376–385.
37. Turtle, E.P.; Perry, J.E.; Hayes, A.G.; Lorenz, R.D.; Barnes, J.W.; McEwen, A.S.; West, R.A.; del Genio, A.D.; Barbara, J.M.; Lunine, J.I.; et al. Rapid and extensive surface changes near Titan's equator: Evidence of April showers. Science2011, 331, 1414–1417.
38. Porco, C.C.; Baker, E.; Barbara, J.; Beurle, K.; Brahic, A.; Burns, J.A.; Charnoz, S.; Cooper, N.; Dawson, D.D.; del Genio, A.D.; et al. Imaging of Titan from the Cassini spacecraft. Nature 2005, 434, 159–168.

39. Schaller, E.L.; Brown, M.E.; Roe, H.G.; Bouchez, A.H. A large cloud outburst at Titan's South Pole. Icarus 2006, 182, 224–229.
40. Roe, H.G.; Bouchez, A.H.; Trujillo, C.A.; Schaller, E.L.; Brown, M.E. Discovery of temperate latitude clouds on Titan. Astrophys. J. Lett. 2005, 618.
41. Rannou, P.; Montmessin, F.; Hourdin, F.; Lebonnois, S. The latitudinal distribution of clouds on Titan. Science 2006, 311, 201–205.
42. Mitchell, J.L. The drying of Titan's dunes: Titan's methane hydrology and its impact on atmospheric circulation. J. Geophys. Res.: Planets 2008, 113.
43. Tokano, T. Impact of seas/lakes on polar meteorology of Titan: Simulation by a coupled GCM-Sea model. Icarus 2009, 204, 619–636.
44. Schneider, T.; Graves, S.D.B.; Schaller, E.L.; Brown, M.E. Polar methane accumulation and rainstorms on Titan from simulations of the methane cycle. Nature 2012, 481, 58–61.
45. Graves, S.D.B.; McKay, C.P.; Griffith, C.A.; Ferri, F.; Fulchignoni, M. Rain and hail can reach the surface of Titan. Planet. Space Sci. 2008, 56, 346–357.
46. Tan, S.P.; Kargel, J.S.; Jennings, D.E.; Mastrogiuseppe, M.; Adidharma, H.; Marion, G.M. Titan's liquids: Exotic behavior and its implications on global fluid circulation. Icarus 2015, 250, 64–75.
47. Tan, S.P.; Kargel, J.S.; Marion, G.M. Titan's atmosphere and surface liquid: New calculation using Statistical Associating Fluid Theory. Icarus 2013, 222, 53–72.
48. Luspay-Kuti, A.; Chevrier, V.F.; Cordier, D.; Rivera-Valentin, E.G.; Singh, S.; Wagner, A.; Wasiak, F.C. Experimental constraints on the composition and dynamics of Titan's polar lakes. Earth Planet. Sci. Lett. 2015, 410, 75–83.
49. Tiffin, D.L.; Kohn, J.P.; Luks, K.D. Solid hydrocarbon solubility in liquid methane-ethane mixtures along three-phase solid-liquid-vapor loci. J. Chem. Eng. Data 1979, 24, 306–310.
50. Davies, P.C.W. The Fifth Miracle: The Search for the Origin and Meaning of Life; Simon and Schuster: New York, NY, USA, 2000.
51. Koshland, D.E. The seven pillars of life. Science 2002, 295, 2215–2216.
52. Stevenson, J.; Lunine, J.; Clancy, P. Membrane alternatives in worlds without oxygen: Creation of an azotosome. Sci. Adv. 2015, 1, e1400067.
53. McLendon, C.; Opalko, F.J.; Illangkoon, H.I.; Benner, S.A. Solubility of Polyethers in Hydrocarbons at Low Temperatures. A Model for Potential Genetic Backbones on Warm Titans. Astrobiology 2015, 15, 200–206.
54. Abdulla, H.S.; Abbo, A.I. Optical and electrical properties of thin films of polyaniline and polypyrrole. Int J. Electrochem. Sci. 2012, 7, 10666–10678.
55. Lovelock, J.E. A physical basis for life detection experiments. Nature 1965, 207, 568–570.
56. McKay, C.P. What is life-and how do we search for it in other worlds? PLoS Biol. 2004, 2, 1260–1262.
57. Toublanc, D.; Parisot, J.P.; Brillet, J.; Gautier, D.; Raulin, F.; McKay, C.P. Photochemical modeling of Titan's atmosphere. Icarus 1995, 113, 2–26.
58. Krasnopolsky, V.A. A photochemical model of Titan's atmosphere and ionosphere. Icarus 2009, 201, 226–256.
59. Strobel, D.F. Molecular hydrogen in Titan's atmosphere: Implications of the measured tropospheric and thermospheric mole fractions. Icarus 2010, 208, 878–886.
60. Courtin, R.; Sim, C.K.; Kim, S.J.; Gautier, D. The abundance of H 2 in Titan's troposphere from the Cassini CIRS investigation. Planet. Space Sci. 2012, 69, 89–99.
61. Russell, M.J.; Hall, A.J.; Martin, W. Serpentinization as a source of energy at the origin of life. Geobiology 2010, 8, 355–371.
62. McKay, C.P. Biologically reversible exploration. Science 2009, 323, 718.

CHAPTER 14

Quantum Meteorites: An Extemporaneous Description of the System of the World

Jean-Paul Auffray

Courant Institute of Mathematical Sciences, New York University, New York, NY, USA

ABSTRACT

We extemporaneously summarize the theoretical scheme we have introduced recently in this note to account for the (hidden) functioning of the System of the World at the quantum level. We further explore the relevance, the pertinence and the potential validity of this scheme. We obtain new fundamental results, bearing notably on the nature and the origin of the "virtual particles" conventionally called upon in quantum field theories to explain the Casimir Effect. We define the concept of quantum meteorites in this context. We disable the Heisenberg Uncertainty Principle, judged to be a (mis) interpretation of otherwise legitimate observations, and we replace it with our new XQP Absolute Certainty Principle.

Keywords: Xonic Quantum Physics, Casimir Effect, Virtual Photons, Uncertainty Principle, Planck Length, Quantum Meteorites

1. INTRODUCTION

Extemporaneous speaking is said to be "a limited-preparation speech event based on research and original analysis", perhaps "previously planned but delivered with the help of few or no notes" [1] .

We endeavor in the present note to extemporaneously review results we obtain in a preceding note concerning the functioning of the System of the World at the quantum level [2] and we derive new results.

To refer to the System of the World, we retrieved the word Scheme, and more precisely the expression theoretical Scheme, from the Greek "σχήμα" (skhēma), shape, or more generally, plan—a less technical definition than the one that Alexander Grothendieck attributed to this word in 1960 in his treatise Eléments de géométrie algébrique [3] . To him, for example, the equations $x = 0$ and $x^2 = 0$ define the same points but different schemes. There is nothing of the sort with us in the present note, which is entirely dedicated to simplicity.

We carry out our quest while taking an imaginary leisurely thoughtful walk in the gardens of the ancient Lyceum in Athens, once a grove of trees named in honor of Apollo Lykeus ("Apollo as a wolf") where Aristotle used to expose his doctrine while walking as he taught—i.e. extemporaneously (Figure 1).

2. CREATING THE WORLD

Just as drawing is the art of projecting Reality on a plane, theoretical physics is the art—the game?—of using words and symbols (and/or equations) to describe how the System of the World (Nature) might be functioning in the shadows of Reality—i.e. in the hidden depths of the famous quantum realm.

Consider the Topological Unit Line Segment (ULS), which is nothing more than a modest straight line segment drawn with a pen on a piece of paper or a blackboard (Figure 2) and which is equipped with two "endpoints" 0 and 1—this is important:

0———1

Hard to believe, but this is our starting point. Speaking of points... The ULS is made up of points glued together so to speak, so as to form a continuum, a concept difficult to define in simple words—there are no "interspaces" between any two contiguous points belonging to the ULS.

Remove (figuratively) all the points the ULS contains, leaving untouched the endpoints 0 and 1. Call what is left behind—i.e. the endpoints 0 and 1 with nothing (no points) left between them—the Void. Fill up the Void with the ULS again—allowing for a likely Imperfect Fit (IF). This IF will become significant later but let us ignore it for now.

We now have Chaos in the Void, so to speak—with one hic as the French would say (a hitch, a snag).

Nature is said to abhor the Void. According to our view, it also dislikes—abhors—continuity (points glued together) in its midst, and here is the hic: generated as we just described, our Chaos (future Universe) is made up of points once again "glued together", thereby forming a continuum.

Figure 1. Aristotle (center to the right in blue) teaching extemporaneously in Athens as depicted by RaffaelloSanzio (Raphael) in his fresco Scuola di Atene.http://www.newbanner.com/AboutPic/Soa3.jpg

Figure 2. How to draw Unit Line Segments. http://en.wikipedia.org/wiki/Line_segment#/media/File:Fotothek_df_tg_0003359_Geometrie_%5E_Konstruktion_%5E_Strecke_%5E_Messinstrument.jpg

To engender out of this continuum an acceptable Chaos, an "active principle", the noX [2] , intervenes randomly repetitiously so as to "liberate" the points the continuum contains from any obligation of being connected by notions of proximity or distance with each other. In performing this task—randomly and repetitiously (iteratively)—the noX generates two kinds or species of (disconnected) points: i-points, which form a substrate (a Dust, a Mist); and e-points, which progressively establish in the Chaos a (hidden) structure in the framework of which they maintain the positions they originally occupied relative to each other in the original continuum.

In the context of his E infinity theory, Alexandria theoretician M. S. El Naschie calls our i-points Substrate the Empty set and the e-points Structure the Zero set.

We now have a somewhat organized Chaos in the Void: it contains a substrate—a Mist, a Dust—and a structure in formation, a most engaging beginning. Our representation of the System of the World painted with words and symbols is shaping up.

3. THE CANTOR MIRACLE

The reason why the description we just presented constitutes the basis for a legitimate representation of the System of the World is fascinating in itself. We briefly recall the story here for the benefit of the inquisitive reader of the present note.

In 1874, Saint-Petersburg born German mathematician Georg Cantor, then twenty-nine years old, published a paper entitled (our translation) "On a Characteristic Property of All Real Algebraic Numbers" [4] thereby founding the new branch of mathematics known since as Set theory. Three years later, in a letter he addressed to his mathematician friend Richard Dedekind, a former pupil of Gauss in Göttingen, Georg Cantor proved that for any positive integer n, there exists a 1-to-1 correspondence between the points on the ULS and all of the points in an n-dimensional space. Stunned by this discovery, Cantor exclaimed to Dedekind: "Je le vois, mais je ne le crois pas!" ("I see it, but I don't believe it!") [5] . This means that our ULS may be used to represent any n-di- mensional point sets (with n integer), for example the space-times currently favored by theoretical physicists. But we are going to have a problem when applying the theorem in the context of our scheme, because our e- points structure—El Naschie's Zero set—has a dimension less than 1—it is not an integer n [2] .

4. WHAT IS A POINT? WHAT IS A PRINCIPLE?

The concept of pointis central in our considerations. It played a more than special role in Newton's treatise, the famous Philosophiæ Naturalis Principia Mathematica. There is in fact much more to say concerning that question than what we were able to report in [2] .

The first two words in the treatise's title translate unambiguously in English to "Of Natural Philosophy"— understand of Philosophy as it concerns Nature, but the last two words, "Principia Mathematica"?

Principia is commonly translated as Principles in English—Principes in French, Grundsätze in German, 原则 in Chinese, принцип in Russian... Very well, but when it is used in conjunction with the qualifier Mathematica the word acquires an altogether different meaning. As we showed in [2] , Mathematica for Newton meant alchemical, i.e. “active”, and Principia something like “ingredients”, e.g. when making bread, yeast is the active ingredient (it causes the wet flour to rise).

The noX is an active ingredient—an active principle—capable of acting on points in our scheme.

5. LITTLE BANGS (ACTION) IN THE I-POINTS SUBSTRATE

As the noX discontinuously and iteratively disrupts the continuum organization which prevails in the Chaos, its counterpart, the Xon, also an active principle (duality), intervenes in the i-points substrate left behind by the noX, generating in its midst fleeting structures which appear as space lengths l and time lapse durationsd, controlled by two very simple relations:

$$lp = h, Ed = h$$

where p measures a momentum, E an energy, and where the symbol h, commonly known as the “Planck constant”, represents the action content of the elementary quantum of dynamical action.

6. CHAOS IN THE CHAOS

Sooner or later, contemporary physicists will have to come to terms with this reality: when he allowed—when he incited—physicists at large to think of his symbol h as representing a constant—Konstant for him, writing in German—the illustrious Founding Father of Quantum Physics, Max Planck, induced his followers to commit a deep error, the effects of which persist with us still today. In brief, he introduced Chaos in the Chaos.

It happened this way.

7. WHEN MAX PLANCK WENT WILD

On May 18th 1899, then in his forty's and a great believer in the fundamental merits of the thermodynamic concepts of Entropy and Energy, Max Planck presented a report to the Berlin Academy self-proclaiming enthusiastically his recently formulated logarithmic law of entropy to be a "Universal law of nature". Following in the footsteps of several illustrious predecessors—James Maxwell and Karl Friedrich Gauss notably—he announced that he had undertaken the strenuous task of constructing a "Natural system of units" assigning units to all the physical entities known in his days. How? This is where he went wild: as functions of four "universal constants" (constants of Nature—Naturkonstante)—of his choosing, which he designated as a, b, c and f.

In embarking on this ambitious project, Planck had a grandiose plan in mind: "[These units], he asserted (our translation, our emphasis), will necessarily maintain their significance for all times and all civilizations, including extra-terrestrial and non-humans and hence can properly be called "Natural units" [Nature units]" [4] .

The four "universal constants" Planck selected to serve as building blocks of his Natural system of units are known today under modernized symbols (suggested by Planck himself, on second thoughts): a has become k and is known as the Boltzmann constant (even though Boltzmann had nothing to do with its invention); b has become h and is known as the Planck constant (see below); c remained c and is known as measuring the speed of light in vacuum; f has become G and is known as the Newton gravitational constant (even though Newton never used it in his calculations).

The use of the qualifier "constant" [Konstant] by Planck in his ambitious undertaking and the abusive use of the word quantum—which signifies nothing more than "quantity" (amount)—are at the roots of the uncontrolled erratic way QP has developed.

Planck's h has the dimension of angular momentum. In calling it a "constant", Planck presumably meant to say (to suggest) that, in all likelihood, it would preserve its significance, its value as a "quantity"—its value as "the quantum"—for all times and all civilizations. At the same time, he was well aware of the true significance of his invention. In a paper he published in 1907 after he had discovered that his h is a "Poincaré relativistic invariant", raising the possibility, wrote Planck, that "each and every change in nature corresponds to a definite number of action elements completely independent of the choice of coordinate system."

Each and every change in Nature… A remarkable statement indeed, which is in full agreement with the basic concepts of our Xonic Quantum Physics project (XQP) and with the results we have obtained so far [2] .

But physicists understood and continue to understand the meaning of Planck's word "Konstant" differently. The "Planck constant" has appeared ever since in thousands and thousands of equations published around the world in every kind of position within the equations, precisely as one would be allowed to do mathematically if, instead of h, the constant involved was the Archimedes' constant π, for example.—or Euler's number e, or Pythagoras' constant $\sqrt{2}$, or Descartes' imaginary unit i, or the golden ratio φ, or, if one wants to be a bit fancy, the Champernowne constant C_{10} = 0.12345678910111213141516 · · · which, like the golden ratio, is an irrational number dotted of extraordinary properties; for example its value is reproduced by the banal ratio 10/81 with an error of the order of 1×10^{-9} only; and if you expand it as a continued fraction keeping the first 18 terms, the error is reduced to approximately 1×10^{-190}. And to concludethis non-exhaustive list of legitimate "constants" with a touch of humor (we are doing theoretical physics, which is a game, isn't?), why not include in our list of "true" constants the number Zero, which came into the English language via the French zéro from the Arabic Sifr (صفر) meaning "empty", translation of the Sanskrit śūnya (शून्य).

One of the early victims of the misconstrued play on words concerning the "Planck constants" and his "constant h" in particular, was a certain Albert Einstein, then working as a modest Technical Expert Third Class at the Swiss Patent Office in Bern.

Einstein published his second academic paper in 1905. It covers seventeen pages and proposes some fifty-two equations. In presenting them, Einstein uses some standard notation—E for energy, S for entropy—but he also uses some symbols of his own—L for the speed of light (?) and β to represent the "ratio" h/k of the "Planck constant" h and the "Boltzmann constant" k. And here is the decisive, the telling detail: even though these two "constants" have different dimensions, Einstein assigned to their ratio the numerical value

$$\beta = 4866 \times 10^{-31}.$$

placing surreptitiously a period after the equation without providing any indication of dimensions—as if β was a pure number [6] .

As a result of the generalization of such malpractices, the Edifice of contemporary Quantum Physics (EQP) has become a Leaning Tower perilously tilted to one side (Figure 3).

In our scheme, h can appear in an equation only if the equation is written in the form "something" = h with "something" having the dimension of dynamical action, and more precisely of angular momentum—spin.

The relations lp = h and Ed = h which prevail in our scheme differentiate this scheme from all other quantum theoretical models of Reality, which allow the Planck "constant" h to enter their equations in any odd way, assuredly one of the most extravagant (mis) uses of a concept in the history of theoretical physics. (The misuse of the word "Energy" is also flagrant).

Speaking of Energy... let us note, passim, that—at least at this stage of our investigations—the two relations lp = h and Ed = h are independent of each other. Each generates Little Bangs on its own.

Let us leave this question aside for now and turn our attention to something else.

What is missing in our Chaos at this stage for it to become the Cosmos—the Universe?

Figure 3. The Huqiu Tower which has been named China's Leaning Tower of Pisa. http://www.china.org.cn/images/38257.jpg

Particles (i.e. matter).

Maybe particles are there, hidden in our scheme and we have not noticed them yet...

That would help.

Before we look for particles in our System of the World, let us investigate the possibility that our scheme contains something significant, but simpler than particles, that we have failed to recognize hitherto.

8. PROSPECTS

A bright new idea sometimes emerges as the result of a chance encounter. While pondering these questions, we recalled an intriguing proposal made some time ago by Californian theoretician Steve Olah under the appellation The Kinetic Universe. We retrieved S. Olah's paper [7] .

Presenting his theory, Steve Olah writes: "For the Kinetic Universe we revive and redefine Aether", asserting: "Aether provides the infrastructure for our existence."

Aether... another word of Greek origin...

Olah's Kinetic model relies on kinetic energy and avoids "the use of ghostly fields and massless mediators" a fact which makes it very attractive from the viewpoint of our scheme. Indeed, the fundamental motors which run the System of the World in our scheme are fueled by dynamical action, not by kinetic energy, but let not this prevent us from exploring the road S. Olah opened.

9. GENERATING MOMENTUM IN THE CHAOTIC SUBSTRATE

The relation $lp = h$ describes how the xon generates in the System of the World space lengths l associated with (linear) momenta p, and/or conversely (Figure 4).

This raises two related fundamental questions that we shall now address.

The first is: in the relation $lp = h$ does the length l refer to a physical space length composed of i-points "glued together" again so as to form true (geometric) line segments—i.e. localized, fleeting (mini) continuums?

Or does it represent—measure—the length of the path that a single i-point—an aethon!—follows in the Chaos while driven by the momentum p?

We used the word "path" deliberately here, in memory of the great Richard Feynman who used this word in his QED theory to designate trajectories in action, not in space or time—precisely as we do in our scheme.

In a subtle way, the two interpretations might be equivalent, just as the strike of light a meteor traces in the sky as it crosses the atmosphere is both, the meteorite and its path.

Figure 4. Generating momentum. http://sr.photos3.fotosearch.com /bthumb/CSP/CSP992/k14094769.jpg

The second related question is: momentum is normally carried by entities—by particles—which possess a mass. As far as his Kinetic Universe is concerned, S. Olah says it clearly: "In the Kinetic Universe force and pressure is [sic] created with collisions of elementary particles that have mass", assuring: "Collisions can't pull, collisions can only push."

Where are the masses in our scheme?

The standard relativistic momentum expression for a particle with rest mass m_0 moving at velocity v is

$$p = m_0 v / \sqrt{\left(1 - v^2/c^2\right)}$$

with c the speed of light. As v approaches c, this expression approaches 0/0 and cannot be used. One uses instead the alternate relativistic energy relation

$$E = mc^2 / \sqrt{\left(p^2 c^2 + m_0^2 c^4\right)}$$

which yields, for $m_0 = 0$

$$p = E/c = hv/c = h/l$$
$$lp = h.$$

This is our fundamental relation!

Thus, by the standard use of well-established relativistic relations, we find that our i-points do not require being equipped with a rest mass to be able to carry momentum. This is in fact precisely the reasoning physicists use to explain and justify the fact that photons, which possess no rest mass, carry a momentum nevertheless.

From these considerations, new aspects of our undertaking emerge that we shall now examine.

10. WIRKUNG

At the end of his life, Max Planck expressed a regret: "All matter originates and exists only by virtue of a force which brings the particles of an atom to vibration and holds this most minute solar system of the atom together [8] ." He had not fully realized, even then, that he had identified the said "force" at the very beginning of his academic career—and that he had given it almost the right name, calling it Das Elementares Wirkungsquantum, even though instead of the word Wirkung—which translates to "Effect" in English—he would have been better inspired to use the word Aktion—Action in English—since he was referring very precisely to the concept forged in 1689-1692 by Wilhelm Gottfried Leibniz under the Latin name Actio—"Dynamical action" for us today in this note.

Speaking of Wirkung... Let us examine the possibility that some "Effect" already discovered might hold the key to a better understanding of the workings of our scheme.

We did not have to look very far. Dr M. S. El Naschie called our attention to an Effect he felt was apt to do the trick [9] .

11. AN EFFECT WITHOUT A CAUSE

The ancient Greeks used the word phenomenon (φαινόμενον, phainomenon, from the verb φαίνειν, phainein, "to show, shine, appear, to be manifest (or manifest itself)") to refer to events occurring in the celestial spheres (solar eclipses for instance). They used the word meteor (μετέωρον, metéōron, "lofty") to designate events occurring in the terrestrial spheres—wind, rain, snow, meteorites... Our contemporary use of the word (and the concept) Meteorology reflects this ancient subtle difference

In 1948, a stunning "meteor"—now called an "Effect"—was conceived by two Dutch theoretical physicists, Hendrik Brugt Gerhard Casimir and Dirk Polder, then working at the Philips Natuurkundig Laboratorium (Philips Physics Laboratory, NatLab) in Eindhoven in the Netherlands. It is known to us today after the name of one of the two dreamers as the Casimir Effect. The "force" said to be involved in the Effect is the Casimir-Polder Force. The basic functioning of the Casimir Effect is easily described. Place two polished plates (two mirrors) facing each other a short distance apart in a container equipped with a vacuum pomp. Empty the container of all the air it contains as best you can so as to realize a vacuum in the container and watch what happens then. The plates start moving toward each other as if mutually attracted to each other by the simple presence of the vacuum.

12. WHAT TAKES PLACE IN THE VACUUM? OFFICIAL EXPLANATION(S)

Quantum field theoreticians tell us that fundamental "fields"—notably the "electromagnetic field"—present in the vacuum are "quantized" at each and every point in space, causing these fields to act locally as "virtual particles". Then, they say, if there are fewer of these virtual particles in the space between the plates than outside that space, the "pressure" the virtual particles generate against the plates will be less on one side of the plates and the plates will be "pushed" toward each other—as if "attracting" each other.

In their original prediction of the Casimir effect, Hendrik Casimir and Dirk Polder reasoned in terms of "energy", not in term of "dynamical action" as we do in our scheme. They (more or less implicitly) assumed:

1) That the virtual particles in their proposed experiments are "virtual photons"—after all photons are supposed to be the carriers of energy in this conception of Quantum Physics; and

2) That only those virtual photons whose wavelengths fit a whole number of times into the gap between the plates should be counted when calculating the "vacuum energy".

Hendrik Casimir and Dirk Polder did not know it then, but they were playing in our courtyard! For what better incarnation of "virtual particles" than our moving i-points? Particularly when, to "explain" their Effect, our learned Dutch physicists also spoke of "vacuum fluctuations"—a well-chosen designation indeed, to describe the deconstruction-construction processes which occur discontinuously in our i-points substrate—in El Naschie's Empty set which in our scheme represents the vacuum.

13. AN I-POINT IN MOTION IS A QUANTUM METEORITE!

In the framework of our scheme, what takes place between the plates follows a simple scenario. The Little Bangs which discontinuously deconstruct and (re)construct space lengths l in the i-points substrate generate "real" virtual particles in the form of i-points which follow paths of lengths l, the specific value being determined randomly. Lengths greater than the distance L separating the plates cannot properly fit in the space between the plates, too narrow for them. As a consequence, the Little Bangs accumulate randomly fewer moving i-points in the space between the plates than outside that space; a (random) selection occurs which generates what M. S. El Naschie calls a "topological pressure gradient" which drives the plates toward each other—once again in accordance with the Olah diktat: "Collisions can't pull, collisions can only push". In the end we are led to make this stunning observation: a ni-point in motion is a quantum meteorite! (Figure 5, Figure 6). As for the mathematics... insofar as they are relevant, they apply unchanged in the frame of our scheme since it is only the nature and the origin of the "virtual particles" involved that we have redefined.

14. WHAT'S NEXT?

The result we just obtained is so delightful that it invites us to conclude this note without further ado. We'll do so... immediately after we indulge in one last fun endeavor.

In [10] , we inaugurated the Physics Museum of Obsolete Concepts, the PMOC, and we placed General Relativity space-times in good stand in its midst. We shall open here today a new gallery in the PMOC, the Gallery of Specious Good Ideas (GSGI) and we shall move in it one of the pillars of the QP Leaning Tower Edifice... the Heisenberg Uncertainty Principle. Here is why.

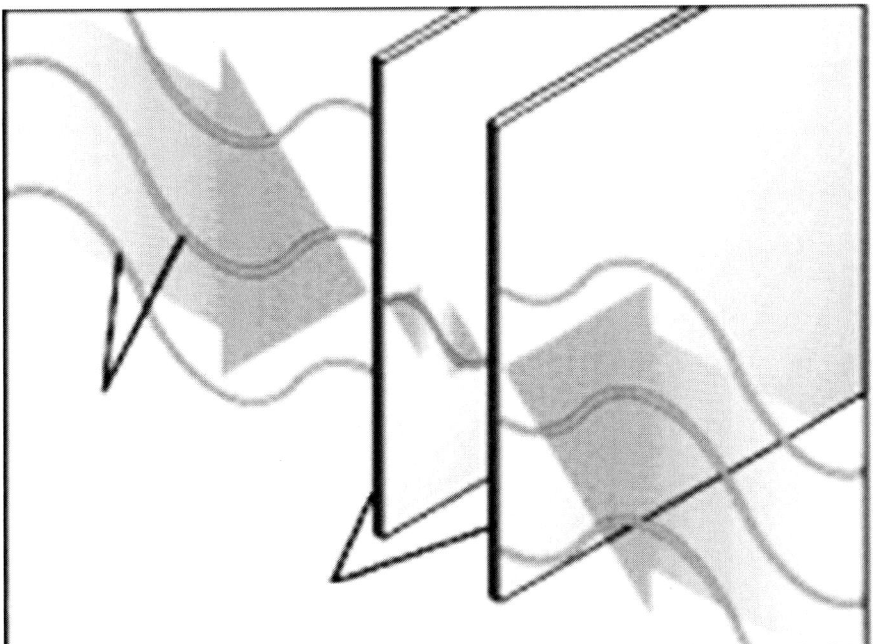

Figure 5. The Casimir effect. http://www.scientificamerican.com/article/what-is-the-casimir-effec/

Figure 6. Meteorites striking their paths into the night. https://sp.yimg.com/ib/th?id=JN.sffs2avmEvpD2vZo2imb8A&pid=15.1&w=184&h=114&p=0

Born in 1901, young Werner Heisenberg conceived his Uncertainty Principle in 1927 on the basis of a deep personal philosophical conviction. He asserted (our translation): "Even in principle, we cannot know the present in all detail", leading him to make this remark which concerns Quantum Physics directly: "The more precisely the position [of an electron] is determined, the less precisely the momentum is known in this instant, and vice versa." [11] .

To sustain his argument, Heisenberg imagined a thought experiment in which a gamma-ray microscope is used to observe an electron and determine its position, He wrote (our translation, his notation): "Let q_1 measure the precision with which the value q is known (q_1 is something like the average error on q), in our case the wavelength of the gamma ray, p the precision with which the value p can be determined, here the discontinuous change in p_1 produced by the Compton effect, then by the elementary Compton effect formulae p_1 and q_1 satisfy the following relation $p_1q_1 \sim h$."

When a Magician wants to execute atour de passe-passe—a conjuring trick in English, Zauberei in German, 魔术伎俩in Chinese, truco de magia in Spanish, Заклинаятрюк in Russian... he calls your attention to something else at the instant he performs the trick. The surreptitious hypothesis

Heisenberg made in developing his attack against the Principle of causality is "hidden" within parentheses in the declaration quoted above. When he writes "(q_1 is something like the average error on q)", he expresses not a fact, but a predetermined desire to find uncertainties in QP measurements.

We don't need to go further. In a slightly different notation, Heisenberg's relation $p_1q_1 \sim h$ is our fundamental relation $lp = h$. The only difference is in the interpretation given.

As per our scheme, the Heisenberg Uncertainty Principle is a specious interpretation of the basic relation $lp = h$ and as such deserves to be placed in in our PMOC-SGI gallery and we shall replace it with...

The Xonic Absolute Certainty Principle (ACP)

An i-point moving with momentum p follows a Feynman-like path of length l such that the composition of l with p satisfies the relation $lp = h$ with an Absolute Certainty (Figure 7).

By the way...Where in the world—where on Earth—should we install our SGI gallery? At the core of an intricate Labyrinth built on the model of the elaborate structure the legendary Daedalus once constructed at Knossos for King Minos of Crete to serve as a prison from which the Minotaur would be unable to escape...?A better idea came to mind. To say it differently, why not place our precious SGI under the protection of the mischievous creature-god "Man in the Maze", the litoi, said to reside in a cave below the peak of the Baboquivari Mountain in the Tohono O'odham Nation in Arizona (Figure 8).

And let us not forget, even though we have not stressed it during the course of this leisurely visit in the inner hidings of the System of the World, that everything in it is spinning—at the quantum level our two active principles, the noX and the Xon, in a way are pure spin and convey this intrinsic characteristic to everything they touch— our quantum meteorites carry spin as they travel!

It is noteworthy that in the universe at large, everything seems to be rotating, as pointed out to us by Canadian philosophically-minded Xiang He (private communication). Spin and (classical) rotation might arise from the same cause—for us the noX and the Xon—at this stage a pure speculation (Figure 9).

Figure 7. Following a path with absolute certainty. https://sp.yimg.com/ib/th?id=JN.j2N2hTuJy2hyfpxkR1oBfA&pid=15.1&P=0

Figure 8. The Baboquivari Peak, home of the Litoi. A good place to hide in security "for all times and all civilizations" the Heisenberg Uncertainty Principle. http://www.sangres.com/arizona/blm/baboquivaripeak.htm#.VRwuGvmsXxU

Figure 9. When everything in the universe is rotating.https://sp.yimg.com/ib/th?id=JN.dIMIWggIN44j0Bv%2fLaPZcQ&pid=15.1&P=0

Figure 10. Spurious little bangs and flashy quantum meteors... https://sp.yimg.com/ib/th?id=JN.%2bsBy8WFsmYKQ0Nv7KUDN7A&pid=15.1&P=0

15. CONCLUSIONS

Following a suggestion made by M. S. El Naschie, we examine in this note how the Casimir Effect can be—is to be—understood in the framework of the theoretical scheme we have developed step by step in recent publications under the generic name Xonic Quantum Physics (XQP) in conjunction with M. S. El Naschie's E infinity theory. This analysis has led us to the unexpected discovery of a pleasant double new treasure: moving i-points which occur in our scheme may properly be said to be real virtual particles and to constitute quantum meteorites, i.e. spinning point-like "objects" moving along Feynman-type paths in the sky of the Quantum World.

A subjacent System of the World animated by spurious Little Bangs and flashy rotating Quantum Meteorites... who can want anything more? We shall conclude this paper on these high notes and reserve the consideration of other perturbing related topics for forthcoming publications, e.g. the relation Ed = h (Figure 10).

ACKNOWLEDGEMENTS

Jean-Paul Auffray We gratefully acknowledge Professor M. S. El Naschie's contribution to the work reported in this note when he suggested to us the

Casimir Effect as a possible source of inspiration related to our study. We also wish to express our gratitude to Ms. Clara Gao and to Ms. Kelly Sang, JMP Editorial Board Assistants, for their valuable help and advice in properly formatting this note for publication.

REFERENCES

1. http://en.wikipedia.org/wiki/Extemporaneous_speaking
2. Auffray, J.-P. (2015) Journal of Modern Physics, 6, 536-545. http://dx.doi.org/10.4236/jmp.2015.65058
3. http://webusers.imj-prg.fr/~leila.schneps/grothendieckcircle/unpubtexts.php
4. Planck, M. (1899) Sitzungsberichte der Königlich Preußischen Akademie der Wissenschaften zu Berlin, 5, 440-480.
5. See for Instance http://www-history.mcs.st-andrews.ac.uk/Quotations/Cantor.html
6. Einstein, A. (1905) Annalen der Physik, 17, 132-148.
7. Olah, S. (2014) The Kinetic Universe. A View From Outside of the Box. http://gsjournal.net/Science-Journals/Research%20Papers-Unification%20Theories/Download/5528
8. Planck. M. Max Planck Quote. http://izquotes.com/quote/146303
9. El Naschie, M.S. (Private Communication). See also, El Naschie, M.S. (2015) World Journal of Nano Science and Engineering, 5, 26-33.
10. Auffray, J.-P. (2014) Journal of Modern Physics, 6, 5.
11. Heisenberg, W. (1927) Zeitschrift fürPhysik, 43, 172-198. English Translation in (Wheeler and Zurek, 1983), 62-84.

CHAPTER 15

Arab Metallurgy owes much to Meteorites iron —A Special Regard to Damask Saber

Ahmad Bilal

Damascus University, Damascus, Syria;

ABSTRACT

Ancient Arabs have recognized metallic and mineral ores, including their fluid inclusions which were still considered as a new scientific fields in the 70's last century. They discovered metallic ores (gold, silver, copper), in inhabited areas and in delta and river crosses, using different techniques for exploration and exploitation. Metallic industry flourished during the Islamic period, silver and gold were used as currency for commercial exchanges. Meteorites were also recognized by Ancient Arabs. They collected them in the deserted areas, and used them for arms manufacturing, as sabers and daggers. The more famous of these arms is the Damask saber steel, known reputedly Jaohar. It has an extraordinary mechanical properties, and exceptional sharp cutting edge. The Jaohar blades were forged directly from fall meteorites, at temperature of 80°C, using a sophistical thermomechanical of forging, annealing to refine the steel, and giving it this exceptional quality and superelastic behavior. Meteorites using by Ancient Arabs come back to 400 years ago, as confirmed by a recent research on two meteorites samples, presented in Aleppo museum, found in Ras Shamra, and Umm El Maraa, in Syria.

Keywords: Arabs Science Heritage; Ores Metallic; Meteorites; Jaohar Saber

1. INTRODUCTION

The scientific literatures learn about the Arab contribution in the sciences domain, especially, minerals, gemstones, and metallurgy [1]. Al-Birouni, the great Arab scientist, was the first person to identify and describe fluid inclusions [2]. It happens that at this time (70's of the last century), these were still considered as a new scientific fields, only studied by a handful of researchers [3].

Therefore an interest research was lanced to know more about the contributions of the ancient Arab civilization to modern science, particularly in the fields of exploitation and manufacturing of metals and mineral ores [4], with a special regard to the meteorites iron and Damask saber.

The research on meteorites and theirs parent with the Damsk saber, using a portable analyzer, Niton XL 3t-900, results of recent meteorite exploration in Syria, done in the framework of a cooperative scientific project between Damascus University and the Pierre and Marie Curie University of Paris-France [5].

In this paper, we will present briefly the heritage of Ancient Arabs in these fields during the pre-Islamic and Islamic periods, through the identification of most famous ores mines by visiting most of the Arab country and their literature treasuries, the proof that Ancient Arabs did recognize and collect meteorites in the desert area, and use them for manufacturing tools and arms, especially the Damask saber, giving a brief information on the characteristics of this saber.

2. ANCIENT ARAB METALLURGIC AND ARCHAEOLOGY CONTRIBUTION

The heritage of ancient Arabs in metallic ores is recognized since the early times, during the pre-Islamic and Islamic periods [4,6]. Remnants of numerous ores mines witness extensive commercial change at this time. Less known is the fact that the famous Damascus blades may have been made of meteorite iron.

2.1. The Pre-Islamic Period Owes

During the pre-Islamic period, metallic ores were discovered mostly in inhabited areas of the Arabian Peninsula (today's Yemen, and Hijaz in eastern Saudi Arabia) [7,8]. At that time, Arabs knew and excavated gold, both as an ore and a metal, in the north of Sabaa, Hijaz, and Najran. They used to exchange it for silver and cooper with neighboring countries, especially Yemen [8,9]. Tools used to explore and exploit gold from the Mahd Al Zahab mine have been found in Saudi Arabia (**Figure 1**). During those old times, lead was used to cement stones with pillars to build citadels [10]. This activity made from Arab region, the most famous archaeological reservoir in the world.

2.2. The Islamic Period Heritage

During the Islamic period, the metallic industry flourished. Ibn Khaldun mentions that the early Muslims used silver and gold as currency for commercial exchange between the states and emirates, in addition toother precious stones [11]. They used to pay gold ransoms for POWs, whether soldiers or commanders (as in the case of the conquest of Carthage). In Egypt they found gold in Ancient Pharaonic temples [12]. They knew then how to excavate gold ore from rivers and desert sands, especially in Wadi Al Aallaki, 600 k. m. from Aswan, as mentioned by the famous Arab geographer Al Idrissi [13]. He described how they use to prospect for gold by locating its glimmer when the moon was dimmed. They then washed sand to separate the gold. Several rivers were famous: Sifri, Tajo, and Hadra in Andalusia, and, Sakiat Wadi Al Zahab in Africa. They used to prospect for gold in deltas and river crosses. For silver, they would locate the ore in underground caves using candles until they went off for lack of oxygen. Maghrebian regions have exploited the ore mines using the rooms and pillars method (**Figure 2**). All Arab regions, have known, exploited and used metallic and mineral ores, using the advanced technology constituted of waterwheel to get upon the water, the support pillared for the mine rooftop, tools for crashing the ore, and for water transporting (**Figure 3**), especially North Africa [14-17]. Egypt was the most famous for precious stones.

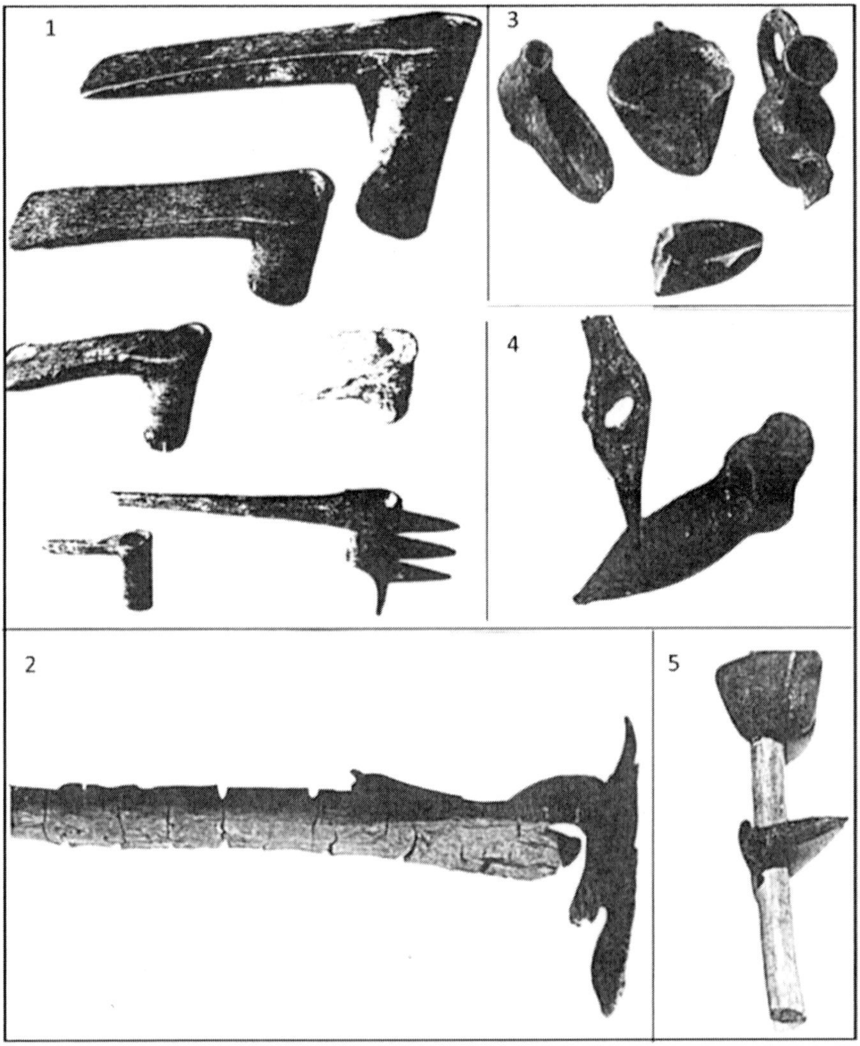

Figure 1. Used tools by ancient Arabs in exploration and exploitation of metallic and mineral ores: 1 —Hatchets, 2—Belle, 3—Candles, 4—Heads of hatchets, 5—Hammer [8,10].

2.3. Ancient Arabs Ores Mines Occurrence

A long research in the literature has permit to me to calculate (107) ancient ores mines in the Arab countries (**Table 1**). They distributed as following:

/37/ gold, /4/ silver, /5/ cooper, /11/ lead, /15/ iron, /35/ mixed (**Table 1**).

3. ARAB HERITAGE IN METEORITES MANUFACTURING-DAMASK SABER

Meteorites are small parts forms stars which fall away and crash the earth. The peoples see them in the space, and they look for them in the erase areas. They collect them, especially in the deserts, and none cultivate areas, where they can are preserved.

Ancient Arabs have recognized meteorites and collected them in the deserted areas, and use them for arms manufacturing.

3.1. Damask Saber Characteristics

The more famous of these arms are sabers, and daggers, very known in Blade Ash Sham countries (Syria, Iraq, Palestine, Lebanon, Jordan).

The ancient Damask saber steel, known as Damask Saber, or more reputedly Jaohar, has features not found in European steels [18]. It has an extraordinary mechanical properties, and exceptionally sharp cutting edge. The Damascus saber blades could be forged directly from fall meteorites at temperature of 800 °C, using a sophisticated thermomechanical treatment of forging and annealing, to refine the steel to its exceptional quality, giving the meteorite its fine-grained, and superplastic behavior. A recent study on the Damask saber steel, has detected a carbon nanotubes, indicating an age of 400-year age [18].

3.2. Syrian Meteorites and Damask Saber Parent

During a recent scientific project between the Pierre and Marie Curie University in France, and the Damascus University in Syria, on the meteorites in Syria, two meteorites samples, in the Aleppo museum, were analyzed in place using a portable analyzer, Niton XL3t-900 [5].

Ras shamra sample: It is a meteorite Hammer d'Ugarite, from the Bronzian age, found in Ras Shamra, in the twentieth years (**Figure 4**). XRF analyses results of the iron blade, at different locations, are reported in **Table 2**. The blade is made of iron with a significant Ni content, in the range from 1.7 wt% to 7.6 wt%, and an average value of 4.33%. This can be compared with the range of 5% -

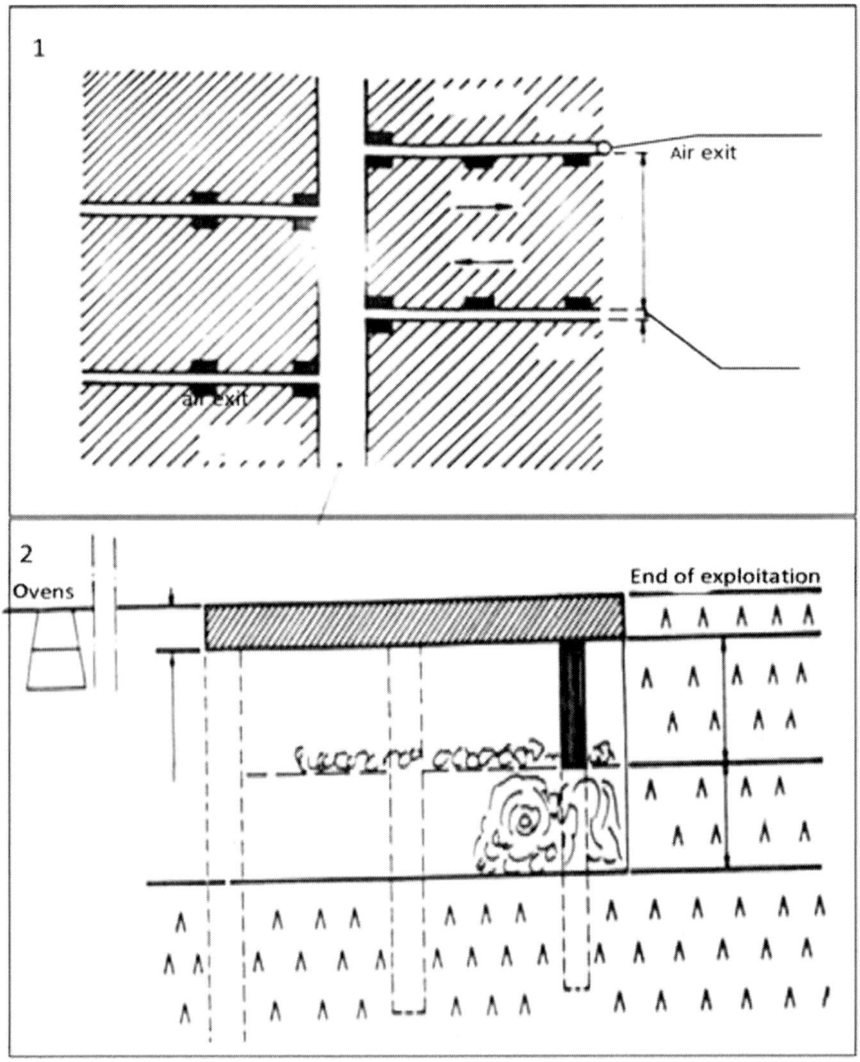

Figure 2. Method of rooms and pillars to exploit the mines in Morocco (1), and its details (2) [4].

Figure 3. Schematic approach to exploit a min of metallic ore and technology of waterwheel to get upon the water (1, 2), support pillared for the min rooftop (3), and some types of used tools for crashing the ore (4), and for transporting water (5, 6) [8,10].

Table 1. Ancient Arabs ores mines occurrence. The digit indicates the number of mines of the relevant ore in each country.

Country	Iron	Zinc	Cooper	Lead	Gold	Silver	Mix
Algeria	1	-	-	3	-	-	1
Egypt	-	1	8	1	20	-	2
Iraq	2	-	-	-	-	-	1
Jordan	1	-	3	-	-	-	-
Morocco	6	-	9	4	-	2	3
Mauritania	-	-	-	-	1	-	1
Saudis Arabia	1	-	-	-	2	2	9
Somalia	-	-	-	1	-	-	-
Sudden	-	-	1	-	13	-	1
Syria	1	-	-	-	-	-	-
Tunisia	2	-	-	2	-	-	-
Yemen	1	-	-	-	1	-	-

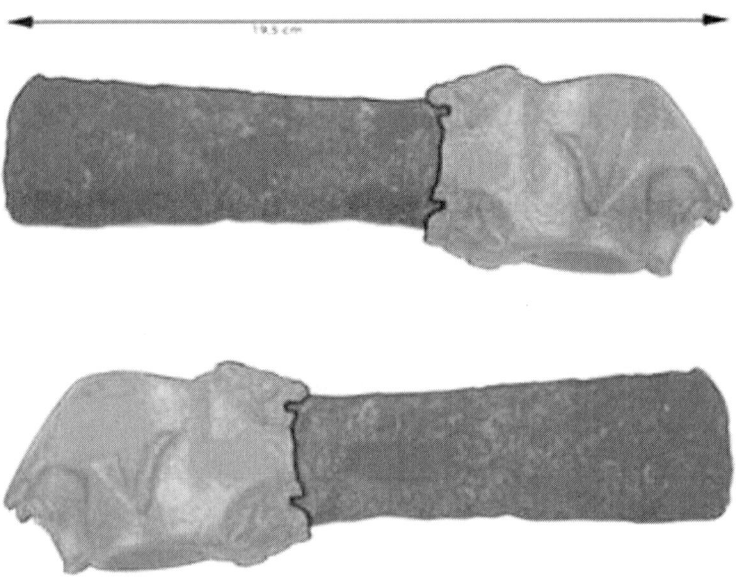

Figure 4. Meteorite Hammer d'Ugarite of Ras Shamra.

Table 2. Comparison of RS Hummer d'Ugarite analyses with data for iron meteorites taken from the literature on fresh samples, and variably weathered samples analyzed by XRF on one hand and terrestrial mantle value on the other.

	Ni	Co	Ni/Fe	Ni/Co
RS	**4.33**	**0.47**	**0.045**	**9.17**
Iron XRF	4.5 - 16	0.41 - 0.78	0.05 - 0.19	9 - 20
Irons literature	6 - 17	0.36 - 0.74	0.07 - 0.21	11 - 24
Mantle	0.20	0.010	0.025	19

20% (average 7%) for iron meteorites. The low values of RS result from oxidization, known to be impoverished in Ni. Thus the Ni signal is partly absorbed by the surface layer of iron oxide leading to an average apparent composition lower in Ni than the fresh metal. This phenolmenon has been checked by analyzing, with the same XRF analyser, iron meteorites with different weathering stages. Consequently it is logical to consider the correlation between the Ni/Fe and Ni/Co of iron meteorites. The range observed for RS overlaps the range observed for iron meteorites. Therefore the blade of Ras Shamra axe is made from meteoritic iron.

Umm El Marra sample: It is a meteorite alter Pendant, rusted iron (a small residual metallic iron of 25 g) (**Figure 5**), from the Bronzian age, found by the team led by Glenn Schwartz and Hans Curvers 2000. The analytical results are reported in **Table 3**. The Ni/Fe ratio is mostly the representative coefficient to checks the meteorites origin. This ratio registers a high value (0.02 to 0.1) whereas iron from ore is in a much lower range (<0.001). The results in **Table 3**, are therefore quite obvious: all analyses present ratios in excess of 0.0, between 0.013 - 0.035. Exempting any contaminations possibilities: addition of Ni from the surrounding to Ni free metal, or selective loss of Ni from a Ni-rich metal. The sample therefore had a high Ni/Fe ratio typical of meteoritic iron.

Damask saber (Jaoher): The chemical composition average of the Damask saber, through several analyses of some sabers from the fifteenth century, and from several positions of a saber of 1448 AC, as written on the blade of the saber (**Figure 6**), is presented in **Table 4**.

Figure 5. Meteorite alters Pendant of Umm El Marra.

Table 3. Analytical results (wt%) of meteorite alters Pendant of Umm El Marra (BDL = below detection limit).

	A	B	C	D	E	F
Balance	43.18	44.16	50.22	44.71	47.19	30.18
Fe	46.21	48.81	39.18	46.50	42.83	65.5
Ni	0.58	1.29	1.36	0.66	0.84	2.03
Co	0.21	0.25	0.27	0.19	0.26	0.31
S	3.67	0.62	1.35	3.04	0.83	0.22
Ni/Fe	0.013	0.026	0.035	0.014	0.02	0.03
Ni/Co	2.70	5.12	5.11	3.51	3.20	6.56
Total	93.85	95.13	91.32	95.10	91.95	97.79

Table 4. Analytical results of the Damask saber of **Figure 6**A (in a, b, c, and d locations), and for different sabers (B, C, D).

		Ni	Co	Fe	S	Ni/Co	Ni/Fe	Balance	Total
A	a	1.32	0.21	45.18	0.68	6.30	0.030	46.12	92.83
	b	1.16	0.25	41.12	1.12	4.64	0.028	48.53	91.06
	c	1.28	0.24	43.35	0.83	5.33	0.030	47.18	92.05
	d	1.41	0.29	44.16	1.18	4.86	0.032	46.23	91.99
B		1.34	0.22	47.18	0.62	6.10	0.030	45.17	93.93
C		1.29	0.23	45.43	2.04	5.61	0.028	46.61	93.56
D		1.23	0.19	48.19	0.42	6.47	0.025	45.29	94.90

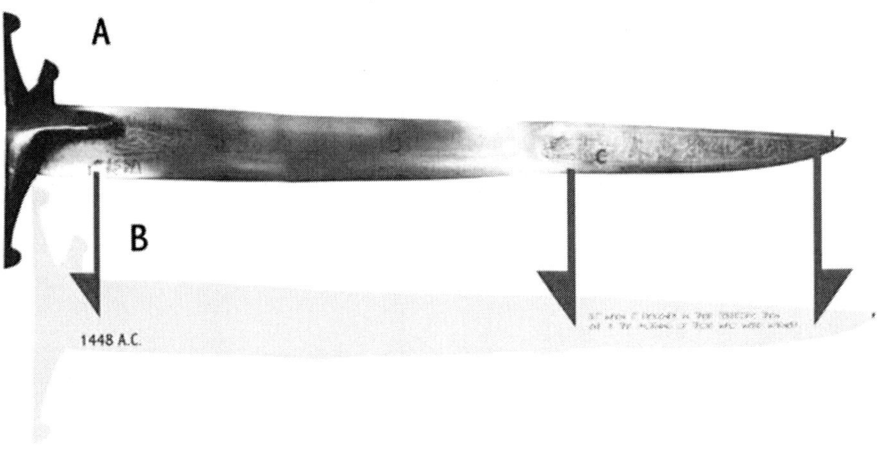

Figure 6. Photo a real Damask saber from 1448 AC (A), and a scheme showing the date, and the verse in Arabic, a, b, c, d, and locations of chemical analysis.

3.3. Origin of Damask Saber

These values, especially the average ratio Ni/Fe (=0.029), are in good similarity with the two upon analyzed meteoritic sample, from Umm El Marra, and Ras Shamra, taking in consideration the upon indicated contingencies factor of those samples. The meteoritic parental with the Damask saber is also indicated by the verse /177/ of Assafat, Surat /37/, written on the saber blade in Arabic, saying: But when it descends in their territory, then evil is the morning of those who were warned. Which descend, after this upon verse, are meteorites.

4. CONCLUSIONS

Ancient Arabs have an important contribution in the civilization of exploration, and manufacturing of metallic and mineral ores. They have discovered and used the most of the metallic ores, through more than /113/ ore mines, distributed in /13/ Arab countries.

During the Pre-Islamic and Islamic period they recognized, and exploited gold, silver, copper, lead, salt, etc. and they used advanced methods to exploit these ores. They used big stones to build citadels which made the Arab region the most famous archaeological reservoir in the world.

They especially known and used meteorites to manufacturing arms. The Damask saber, named Jaohar saber, is, by its mechanical properties and superelastic behavior, the most famous world saber. The parental meteoritic, proved in this paper, with this arm must be confirmed by more scientific investigations.

Ahmad Bilal This advanced role not to remember that ancient Arab role, but to incite us to feedback this role, and to contribute more in the modern civilization. Thus they can give to their ancient civilization the alive.

REFERENCES

1. Al Farabi (1953) Kitab Ihsa aluloum. The book of the Sciences Enumeration and translated by A. Gonzelez. 870-950.
2. Hakim, M.S. (1989) Al Birouni's book on mineralogy. Pakistan Hijra Council, Islamabad.
3. Bilal, A. (1978) Fluid inclusions in mantelic xenoliths— Geotectonic implications. Doctorat d'Etat Thesis, University of Paris, Paris.
4. Mohamadi, A. (1983) Techniques of rooms and pillars used in ancient Moroccan mines. Arab Mineral Resources Journal, n 3, 41-50.
5. Jambon, A. and Bilal, A. (2010-2012) Rapport de coopération. UMPC, Université de Damas sur les Météorites in Syrie, Paris.
6. Jamil, A.K. (1983) Metallurgy of ores and mineral separation in Iraq. Arab Mineral Resources Journal, n 3, 105- 113.
7. Ministry of Petrol and Metallic Resources (1974) Map occurrence of ores mines in the Arabic dorsal. Ministry Deputy for Metallic Resources, Saudi Arabia.
8. Adly, Ab. M. A. (1979) General outline of the geology and mineral occurrences of the Red Sea Hills. Bulletin No. 30, Ministry of Energy, Geological and Mineral Resources Department, Sudan.
9. Alomar, K. (1983) Ancient mines activities in Jordan. Journal Arab Mineral Resources, n 3, 95-102.
10. Bilal, A. (1983) Ancient Arab legacy in the exploitation and manufacturing of metallic ores. Arabe Minerales Resources Journal, n 3, 89-94.
11. Ibn Khaldun (1332-1406) Chronique général, procédé de prolégomènes. Ibn Khaldoun.

12. Afieh, M.S. (1983) Min activities in Egypt and Sudden during the successive civilization. Arab Mineral Resources Journal, n 3, 51-69.
13. Al Idrissi (1099-between 1165-1186) His book Nozhat al moshtak vi ikhtrak al afak, and his maps.
14. Al Saadi, M. (1983) Map occurrence of ancient ores mines in morocco. Ministry of Energy, Morocco.
15. Rosenberger, B. (1970) Les vieilles exploitations minières et les centres métallurgiques du Maroc. Essai de carte historique. Revue de Géographie du Maroc, no17 et 18.
16. Youssef, B. and Sassi, S. (1983) Metal manufacturing during the history in Tunisia. Arab Mineral Resources Journal, n 3, 133-139.
17. Gsells (1928) Ancient mines exploitation of Northern Africa. Morocco Institute of History High Studies Congress.
18. Reibold, M., Paufler, P., Levin, A.A., Kochm W., Patzke, N. and Meyer, C. (2006) Materials: Carbon nanotubes in an ancient Damascus saber. Nature, 284-286.

CHAPTER 16

The Role of Ultrahigh Resolution Fourier Transform Mass Spectrometry (FT-MS) in Astrobiology-Related Research: Analysis of Meteorites and Tholins

Árpád Somogyi [1,*], Roland Thissen [2], Francois-Régis Orthous-Daunay [2] and Véronique Vuitton [2]

[1]*Campus Chemical Instrument Center, Mass Spectrometry and Proteomics Laboratory, Ohio State University, Columbus, OH 43210, USA*

[2]*Université Grenoble Alpes, CNRS, IPAG, Grenoble F-38000, France*

ABSTRACT

It is an important but also a challenging analytical problem to understand the chemical composition and structure of prebiotic organic matter that is present in extraterrestrial materials. Its formation, evolution and content in the building blocks ("seeds") for more complex molecules, such as proteins and DNA, are key questions in the field of exobiology. Ultrahigh resolution mass spectrometry is one of the best analytical techniques that can be applied because it provides reliable information on the chemical composition and structure of individual components of complex organic mixtures. Prebiotic organic material is delivered to Earth by meteorites or generated in laboratories in simulation (model) experiments that mimic space or atmospheric conditions. Recent representative examples for ultrahigh resolution mass spectrometry studies using Fourier-transform (FT) mass spectrometers such as Orbitrap and ion cyclotron resonance (ICR) mass spectrometers are shown and discussed in the present article, including: (i) the analysis of organic matter of meteorites; (ii) modeling atmospheric processes in ICR cells; and (iii) the structural analysis of laboratory made tholins that might be present in the atmosphere and surface of Saturn's largest moon, Titan.

Keywords: origin of life; prebiotic material; ultrahigh resolution mass spectrometry; Fourier transform (FT); Orbitrap; ion cyclotron resonance (ICR); chemical compositions; isomeric structures

1. INTRODUCTION

The chemical composition and structure of prebiotic materials that are building blocks of more complex, biologically important molecules, including peptides, lipids, proteins, DNA, *etc.*, provide important information on their formation. Obviously, the creation of heavy elements (*i.e.*, heavier than He) is prerequisite for the formation of very small "pre-prebiotic" molecules and radicals, such as H_2O, CO, CH_4, $CH_2{=}O$, N_2, HCN molecules and OH, CN radicals/anions, *etc.* (for a recent review of the evolution of elements and organic matter in space, see [1]).

The early ingredients for organic matter were formed in interstellar clouds. Since the 1960s, microwave (MW) spectroscopy has provided strong evidence for the existence of these small molecules in interstellar clouds [2]. Almost 130 such molecules are known and identified by MW spectroscopy. Recently, high level quantum mechanical calculations allow further refinement of the observed spectra confirming these species more reliably (see, e.g., Reference [3]). The concentration of these molecules in interstellar clouds, however, is too small and the temperature is too cold to form more complex prebiotic molecules in large number and with reasonable rates (even on the cosmic timeframe). Nevertheless, shock waves generated by nearby supernovae swings the pendulum of molecular synthesis far away from the (cold) equilibrium and thus complex chemistry does occur. Over time, complex molecules such as amino acids, most notably glycine are formed and they can be delivered on the surfaces of rocky planets, ranging in size from nanometric dust to space rocks (asteroids), which can later be captured by the gravitational field of (newly formed) stars. It is well known that amino acids and other prebiotic molecules have been detected in meteorites that fell on Earth. From an astrobiological point of view, meteorites are of particular importance because we can study their chemical compositions in Earth based laboratories. Among other methods, such as nuclear magnetic resonance (NMR), ultrahigh resolution mass spectrometry (UHRMS) is a very effective tool to study the chemical composition of *individual* chemical compounds delivered from space. Below we will mention some of these studies.

The alternative to sample delivery by meteorites is to send laboratories to places where complex organic molecules can be formed. Unfortunately, present technologies prevent us from launching analytical instruments beyond the Solar System so we are restricted to a limited number of Sun bound "laboratories". Potential candidates include planets and/or moons with atmospheres. Besides Earth, Venus, Mars, and Pluto, only Saturn's largest moon, Titan has an atmosphere in the Solar System (in fact, a very thick one) consisting mostly of N_2, with 2% CH_4, 0.4% H_2 and trace amounts (ppm or ppb) of oxygen containing species such as CO. Analytical instruments, including mass spectrometers, have been sent to several Solar System bodies, including planets, comets, and moons. The success of the Cassini–Huygens mission provided us data on charged species in the atmosphere of Titan by using the "on-board" ion-neutral mass spectrometer (INMS) [4,5], Cassini plasma spectrometer (CAPS-IBS) [6], and the electron spectrometer (CAPS-ELS) [7,8].

An appealing and much less expensive alternative to sending spacecrafts (mass spectrometers) to celestial bodies is using Earth-linked laboratories to mimic, e.g., atmospheric processes and analyze their products. Although this approach does not give direct information on the "*in situ*" environment, it is flexible in the sense that we can easily change experimental conditions and we can use different instrumentations to analyze the outcome in great details. The products of these model reactions performed on Earth are very complex mixtures so mass spectrometry, especially ultrahigh resolution mass spectrometry occasionally coupled with separation techniques, such as gas or liquid chromatography (GC and LC, respectively) is a perfect tool to obtain information on chemical compositions and structures.

The research area on prebiotic materials is so broad and diversified that it cannot be adequately reviewed in a paper with limited length. Therefore, in the present paper, we show and discuss a few demonstrative examples of how useful ultrahigh resolution mass spectrometry (UHRMS) is in these astrobiologically related studies. We focus mostly on results obtained by the two best "performers" in terms of resolving power, both Fourier Transform: the FT-Orbitrap and FT-ICR mass spectrometers. The purpose of this paper is not to provide a comprehensive review of all mass spectrometry studies in the astrobiological field. Nevertheless, for detail-oriented readers, we recommend a recently published review on Titan tholins [9] and also the references throughout the paper. We will focus on the UHRMS analysis of: (i) organic contents of meteorites; (ii) the immediate, small molecule products of laboratory modeling of Titan's atmospheric reactions; and (iii) the structural investigation of higher molecular weight $C_xH_yN_z$ (>60 Da) species and their hydrolysis products.

2. RESULTS AND DISCUSSION

2.1. Determination of Organic Components in Meteorites by FT-MS

Obviously, from an astrobiological point of view, organic compounds containing C, H, N, O, S, and P, are of primary astrobiological interest [10,11,12,13,14]. Soluble organic matter (SOM) can be extracted from powdered meteorite by common solvents (e.g., methanol or dichloromethane) and these extracts can be ionized by electrospray ionization in both the positive and negative ion modes. Nevertheless, inorganic content, such as the presence of mass deficient $MgSO_4$, Na salts, and minerals can also be identified in otherwise organic rich chondrites [10].

For illustration, we present here some of our recent results using the ultrahigh resolution LTQ-Orbitrap (Grenoble, France) to analyze the soluble organic matter from a Martian meteorite (Figure 1). Small meteor rocks were ground and a methanol:toluene (1:2) solvent mixture was used for extraction. After centrifugation, direct infusion ESI was used in the positive ion mode to generate ions. The spectrum in Figure 1A contains a large number of peaks showing hundreds of components. Note that the vertical (intensity) axis is logarithmic in Figure 1A so the intensity range shown is 1–50,000. A mass defect *versus* exact mass (MDvEM) plot (for appropriate references, see Materials and Methods below) of the same and intricate MS is shown in Figure 1B. MDvEM analysis reveals immediately the very high degree of organization in the datasets, with obvious series in CH_2, but also addition of minerals, due to the interaction within the complex matrix (Figure 1B). Intensity dimension is color coded while the vertical axis plots the difference of

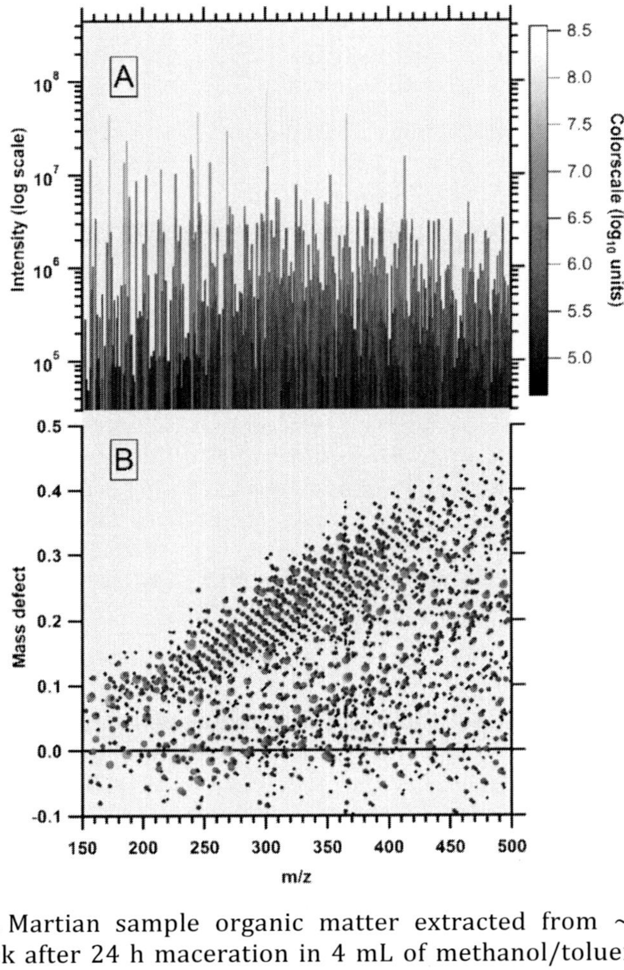

Figure 1. Martian sample organic matter extracted from ~20 mg of whole rock after 24 h maceration in 4 mL of methanol/toluene (1:2) at room temperature followed by storage in a dark place. (**A**) ESI produced positively charged mass spectrum. More than 2000 masses are detected in four decades in the 150–500 Da range. Average density in mass is 6.2 ions per Da. Neither a remarkable envelope nor peculiar organization can be seen. Colors indicate intensities on a logarithmic scale; (**B**) MDvEM representation of the mass spectrum. Each point represents one mass each. Mass defect is the difference between the mass coordinate and its closest integer so the value spans periodically and linearly within the −0.5 to +0.5 *m/z* range along the mass coordinate. This gives a modular space (like hours and minutes on a watch) where repetitive occurrences of mass difference draw lines. The visually remarkable alignment of points here is due to the CH_2 pattern repetition. Only singly charged ions are considered.

one exact mass to the nearest integer. Repetition of stoichiometric patterns generates lines in this diagram. Here, the most visible oblique lines are due to the multiple addition of CH_2.

The insoluble organic material (IOM) consists mostly of unsaturated polycyclic macromolecules that are difficult to ionize by ESI. Thus, the availability of dual MALDI/ESI sources (such as on the Bruker Apex Qh or SolarixR instruments) is incredibly useful. We detected organic molecules (ions) using simply laser desorption ionization (LDI) of Orgueil. In general, our results confirmed that ESI and LDI are complementary ionization techniques that probe different parts (organic content) of chondritic meteorites [10] (data not shown, but see other LDI results below).

Recent LTQ-Orbitrap studies also revealed that carbonaceous meteorites (such as Murchison and Lonewolf Nunatks 94,103 meteorites) contain a wide range extraterrestrial nucleobasis. [11] Tandem MS/MS fragmentation and comparison with reference standard provided unambiguous proof for the presence of purine, adenine, 2,6-diaminopurine and 6,8-diaminopurine that can be considered as components of an "extended genetic alphabet".

Note here that although FT-MS measurements (Orbitrap or ICR) provide reliable chemical compositions, it is difficult to distinguish between structural isomers based solely on (ultrahigh resolution) tandem MS/MS fragmentation. Other techniques, such as nuclear magnetic resonance (NMR) are extremely useful and improve the information content of analysis. A world leader in these combinative studies is a group led by Schmitt–Kopplin in the Helmholz–Zentrum in Munich, Germany [12,13]. Their recent studies include the one on Soltmany ordinary chondrite [12] and Murchison meteorite [13].

For correctness, we also note that conventional low resolution MS coupled with chromatography is still justified to use and provides useful information on organic content. An interesting recent example is a study when NWA, Gold Basin, Orgueil, and Al Haggounia meteorite powders were used to catalyze synthesis of nucleosides and other prebiotic compounds from formamide under proton irradiation [14]. Under these hostile conditions, about 30 different organic compounds have been detected by GC-MS. Because most of these compounds are not volatile enough derivatization was necessary before the analysis [14].

Based on the literature and our own experience, we can conclude that ultrahigh resolution MS (and tandem MS/MS) experiments are important to reveal detailed chemical information on extraterrestrial organic matter delivered to the Earth by meteorites. Although ESI is definitely a powerful ionization method, for the analysis of certain insoluble organic material we strongly recommend the use of laser desorption ionization (LDI) as a complementary ionization method (for other examples, related to laboratory "tholins", see below).

2.2. Modeling Reactions in Titan's Atmosphere and Analyzing the Products by FT-MS

Atmospheric processes have long been known to generate organic molecules. As mentioned in the Introduction, Titan's atmosphere is well known due to the successful Cassini–Huygens mission. Specifically, the INMS [4,5] mass spectrometer detected hydrocarbons and N-containing species up to m/z 100. Due to low (unit) resolution, however, the precise assignments are ambiguous. Heavier positively and negatively charged ions (with higher m/z up to several hundreds or even thousands) were detected by the CAPS-IBS [6] and CAPS-ELS [7,8] mass spectrometers, respectively. This is very important *in situ* information, but again, due to low resolution, it prevents reliable identification of the detected ions. Moreover, it is not clear how much and by what mechanism nitrogen is incorporated in Titan's organic aerosols, *i.e.*, the formation mechanisms of astrobiologically important organics are not fully understood.

To improve our understanding, in line with experiments, theoretical modeling of the kinetics of Titan's atmospheric processes is of primary importance. These studies are out of the scope of the current paper but we refer here to the work of Vuitton, Lavvas, Yelle, and others [15,16,17,18,19].

Obviously, for generating ion–molecule (or even molecule–molecule) interactions an energy source is required. On Titan (and, in general, other Solar System bodies with atmosphere), one of these sources are high enough energy photons in the UV range that are capable of ionizing/dissociating N_2 (and/or CH_4). Using tunable synchrotron radiation at the Advanced Light Source, Imanaka and Smith recently demonstrated the first evidence of nitrogenated organic aerosol production by extreme ultraviolet–vacuum ultraviolet irradiation of a N_2/CH_4 gas mixture [20]. The products of these UV-induced reactions were then analyzed by a low resolution quadrupole mass analyzer and by ultrahigh-mass-resolution FT-ICR (Bruker 9.4 T Apex Qh, Bruker, Billarica, MA, USA) mass spectrometry with LDI. The N_2/CH_4 photolytic solid products at 60 and 82.5 nm irradiation wavelengths indicate the predominance of highly nitrogenated compounds. ^{15}N labeling predicted that short lived nitrogen containing reactive species, such as the HC_2N radical might play a role in the formation of $C_xH_yN_z$ species. We note, however, that isotope labeling of the gas mixtures with ^{13}C clearly indicated the presence of unsaturated hydrocarbons at 60 nm irradiation [20].

FT-ICR cells can be used not only to detect large number of ions with high resolution and *m/z* precision but also to perform ion-molecule reactions in the cell. Maybe the simplest example is hydrogen/deuterium exchange (HDX) reactions of isomeric ions [21]. Different HDX rates indicate structural isomers of isomeric ions. In some cases, reaction products are also different (see, e.g., ND_3 addition to b_n peptide fragment ions, when $n > 5$ [22]).

A novel experiment relevant to modeling Titan's atmospheric chemistry has been performed inside the cell of a FT-ICR mass spectrometer [23]. The ICR cell served as a "reaction chamber" into which gas mixtures representative of Titan's atmospheric composition were injected. These gas mixtures were then irradiated with extreme ultraviolet radiation from the Elettra synchrotron. The influence of various gas mixtures and irradiation frequencies were studied on the ion evolution. This approach allowed the authors to compare the theoretical model predictions [15] to the laboratory experiments and to test for missing and/or poorly constrained pathways. Overall, the comparison between observed and modeled ion intensities validates the kinetic model. Unfortunately, the ICR cell is not infinite in size and its surface may initiate (catalyze) certain reactions implicating the involvement of heterogeneous chemistry on aerosols that might efficiently produce HCN and NH_3 in Titan's upper atmosphere.

In parallel to the above-mentioned comprehensive studies, we also performed simple experiments in the FT-ICR cell of an IonSpec 4.7 T instrument (IonSpec, Irvine, CA, USA) where 40 eV electrons were used as the energy source (unpublished results). The electrons were generated using a regular filament mounted outside the FT-ICR cell. Again, N_2/CH_4 gas mixtures were leaked into the ICR cell and the electron flux was varied as well as the reaction time (*i.e.*, ions evolving time). Typical spectra are shown in Figure 2a,b.

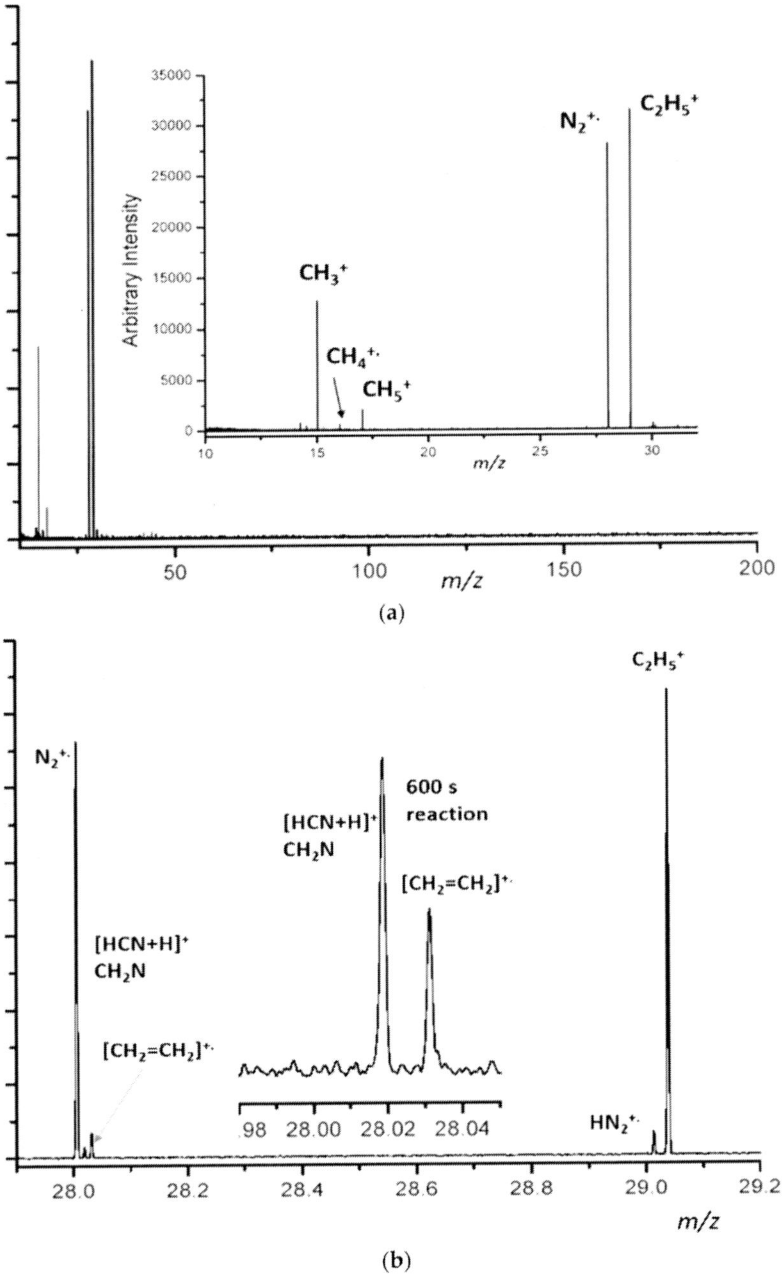

Figure 2. Detected reaction products obtained by a 15 s irradiation of a 95% N_2/5% CH_4 mixture at 5×10^{-5} torr after a 2 s (**a**) and 600 s (inset in (**b**)) reaction time in the ICR cell of a 4.7 T IonSpec instrument.

It is expected that 40 eV electrons ionize N_2 and CH_4. The irradiation time of 15 s of ionization with 40 eV electrons is sufficient time to ionize these molecules efficiently. A 2 s reaction time was allowed after the electron beam was turned off so other products are also formed (within the altogether 15 + 2 = 17 s time window). Not surprisingly, products that mimic chemical ionization (CI) when CH_4 is used as a CI gas (see, e.g., CH_5^+ and $C_2H_5^+$) have also been detected. Ultrahigh resolution allows us to take a "closer look" to identify $HCNH^+$ (most likely protonated HCN), N_2^+, and the ethylene molecular ion ($CH_2{=}CH_2^+$) all with m/z 28 nominal mass. When the gas mixture was allowed to react further for 600 s after the applied 15 s electron flux, the ionized N_2 disappeared and the ratio of CH_2N/CH_2CH_2 increased (see inset in Figure 2b). Note that the intensity of C_xH_y ions has been gradually increased and shifted to higher m/z masses (up to about m/z 180) and they overwhelmingly dominate the 600 s reaction time spectrum (full m/z range not shown). Ions with $C_xH_yN_z$compositions were detected at very low intensity (or not detected at all).

Unfortunately, albeit not surprisingly, similar to the Elettra experiments mentioned above [23], we also faced the problem of the involvement of the surface, as well as residual H_2O after 10 s of reaction time, after which the H_3O^+ ion becomes dominant. Evidence for surface chemistry was found in the negative ion mode by the detection of Cl^- and Br^- anions (spectrum not shown) that, most likely, are sputtered products from the mesh located between the filament and the ICR cell (that functions as an "on–off" switch to control the electron beam flux in and out of the ICR cell).

In spite of the shortcomings mentioned above, we believe that using the ICR cell as a gas phase "reactor" is justified and the ions produced are informative for the very early stage of ion formation generated by either high energy photons (in the UV range) or electrons (in the tens of eV energy range). In fact, surface chemistry, albeit a different one related, e.g., to microscopic dust in the atmosphere, may play a role in the formation of $C_xH_yN_z$ species that can grow to colloidal droplets responsible for the haze (non-transparent atmosphere). These droplets can rain down with liquid methane to Titan's surface and can be exposed to additional chemical reactions, most notably hydrolysis in ammonia/ice pounds (see discussion below).

2.3. Structural Analysis of Laboratory "Tholins", Analogues of Titan Aerosols by FT-MS

The first detailed ultrahigh resolution MS study on non-volatile laboratory tholin components (molecular weight > 120 Da) was published back in 2003 [24]. A homemade glow discharge (high voltage (HV)) reactor was built in the laboratory of Mark Smith (then at the University of Arizona) to generate laboratory tholin samples. This reactor was later modified and produced more pure tholin samples [25]. If methane is constantly replenished in the atmosphere, non-volatile aerosols have had billions of years to accumulate on Titan's surface. Thus, determining the chemical composition of its constituents is crucial to better understand and predict what might happen to them on the surface. Without ultrahigh resolution, significant chemical information would have been lost. Most notably, we realized that at almost every integer m/z there are several ions that are related to a $2CH_2/2N$ permutation that produce homolog series that grow and diminish with increasing m/z. (The mass difference between $2CH_2$ and 2N is 28.031300 – 28.006148 = 0.025152 Da, which is easily detectable using FT-MS even at higher m/z values.) These series become more unsaturated with increasing molecular mass. The success of this study inspired us as well as others to perform additional analysis by using ultrahigh resolution mass spectrometry [24,25]. Similar trends were observed by Pernot *et al.* [26] in a more detailed study.

It is well known that most of the prebiotic molecules on Earth contain oxygen. Because there is almost no oxygen containing compounds in Titan's atmosphere, the obvious question emerged: is there any way to incorporate oxygen into the $C_xH_yN_z$ molecules that are initially formed in a reductive atmosphere?

One avenue is to add the N_2/CH_4 mixtures with small amount of CO, perform the laboratory simulations and analyze the products by ultrahigh resolution mass spectrometry. A detailed study by Hörst *et al.* was published in 2011 [27]. The laboratory simulations were performed in the PAMPRE (Production d'Aerosols en Microgravité par Plasma Réactifs) apparatus [28], producing aerosols by cold plasma activation of a gaseous $N_2/CH_4/CO$ mixtures of 96.2%/2.0%/1.8% and 93.2%/5.0%/1.8%. Isotopically labeled $C^{18}O$ was used in these studies to further confirm the oxygen contribution from CO. The LTQ-Orbitrap mass spectrometer (Grenoble, France) was used to obtain the ultrahigh resolution mass spectra. Although the gas composition was significantly different from that of Titan's upper atmosphere, the study is important and clearly shows that when simple oxygen containing compounds (such as CO) are present, complex molecules containing oxygen are definitely formed. Most interestingly, several prebiotic molecules have been detected and confirmed, including adenine, cytosine, uracil, thymine, guanine, glycine and alanine [27].

An alternative way for oxygen incorporation is hydrolysis reactions of Titan aerosols on the surface of Titan. This sounds bizarre at first because the temperature is so low (about 95–100 K) on the surface. Obviously, water is frozen but water mixed with some ammonia has a significantly lower freezing point (e.g., 29.4% ammonia/water is liquid at about 193 K). Ammonia/water ponds can, furthermore, be heated up by meteorite impacts and they can stay melted for several thousands years.

We performed some qualitative hydrolysis reactions in H_2O and ammonia/water [29]. It was found that laboratory tholin components incorporate significant amount of oxygen within an hour at 373 K or within a few hours at 293 K. Exposure to air for several weeks also led to oxygen incorporation but the process was, as expected, much slower than in the liquid phase. In the same paper, we also proposed some characteristic fragmentation patterns using low resolution ion trap and high resolution (FT-ICR) tandem MS/MS measurements. In the positive ion mode, characteristic losses include the loss of NH_3, HCN, C_2H_4, CH_3CN, $CH_2{=}NH$, *etc.* Another importance of this work was to demonstrate the ESI ionization with methanol:acetonitrile (1:1) solvent combination is an inert mode to generate ions from tholin components, *i.e.*, when laboratory tholins are harvested carefully and anaerobically, oxygen contamination is negligible during the ESI process and the solutions are stable for at least 1–3 days allowing us more than enough time to gather ultrahigh resolution mass spectrometry data on the samples.

More comprehensive studies on laboratory hydrolysis have been carried out by Neish *et al.* in water and 14% ammonia/water solutions at different temperatures (–20, 0, 20, and 40 °C) [30,31,32]. Kinetic curves and rate constants have been determined at those temperatures for both decay processes (*i.e.*, when $C_xH_yN_z$ species decayed due to oxygen contribution) and for oxygen growth reactions (*i.e.*, the growth of $C_xH_yN_zO_n$ species). (Note that experiments with ^{18}O labeled water clearly proved that the source of oxygen was water and not contamination from moisture in air.) Activation energies were then calculated from Arrhenius plots for several species. The activation energies were found to be in the range of 50 ± 10 kJ/mol (decay) and 60 ± 10 kJ/mol (growth) [30]. Extrapolating down to about 100 K, these values encourages us to say that hydrolysis reactions can, indeed, occur within 3000–10,000 years, which is a very short time on the planetary time scale. Again, we note for correctness, that no phase transitions (*i.e.*, freezing of melted ponds) are assumed. But even then, meteorite impact can heat up local ammonia/water environment that might cool down within a couple of days or weeks but oxygen enrichment can occur within this relatively short time even though not at full scale

Similar to the work of Hörst *et al.* amino acids asparagine, glutamine, and histidine were detected among the hydrolysis products with 14% ammonia/water [32]. It is interesting to note here that narrow band experiments with resolution >200,000 clearly showed [31] that tholin components not only react with oxygen but also with NH_3 as was proved by using labeled $^{15}NH_3$. The power of using even higher resolution (up to almost 500,000) is demonstrated in Figure 3 and Table 1 that shows the unambiguous identification of additional species, e.g., the odd electron molecular ions for ions with the same *nominal* mass of *m/z* 192. The lower spectrum practically demonstrates an isotopic fine structure experiments that is widely used in metabolomics. Even though the bottom spectrum clearly shows ten different species, this spectrum is relatively simple compared to those obtained for species containing other mass deficient elements, such as S, P, and halogens.

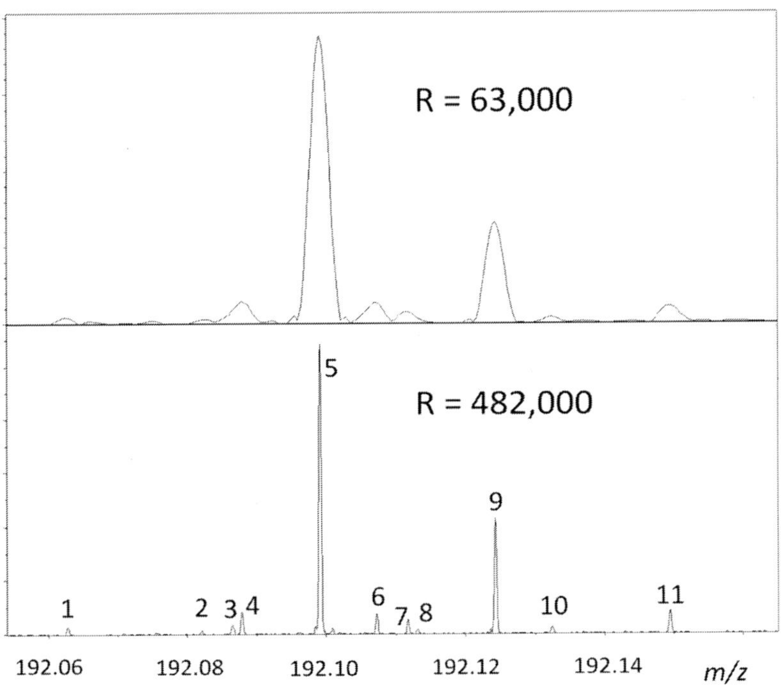

Figure 3. Comparison of information content obtained for LDI generated ions with the nominal mass of *m/z* 192 at two different resolutions (63,000 **upper**; 482,000 **lower** spectrum). The list of ions corresponding to the labeled peaks is collected in Table 1.

Table 1. Chemical compositions of the ions with nominal mass of *m/z* 192 at about 500,000 resolution (bottom spectrum in Figure 3).

Peak	Ionic Composition *	Calculated *m/z*	Measured *m/z*
1	$C_6H_6N_7O$	192.06283	192.06282
2	$C_6H_7N_8$ (^{13}C)	192.08217	192.08218
3	$C_6H_8N_8$	192.08664	192.08657
4	$C_8H_{10}N_5O$	192.08799	192.08799
5	$C_7H_{10}N_7$ *	192.09922	192.09922
6	$C_8H_{11}N_6$ (^{13}C)	192.10733	192.10732
7	$C_8H_{12}N_6$	192.11180	192.11179
8	$C_{10}H_{14}N_3O$	192.11314	192.11311
9	$C_9H_{14}N_5$	192.12437	192.12437
10	$C_{10}H_{15}N_4$ (^{13}C)	192.13248	192.13248
11	$C_{11}H_{18}N_3$	192.14952	192.14952

* denotes the ion used for internal calibration and ^{13}C indicates that one carbon atom is a ^{13}C isotope.

Data in Table 1 show that components 5, 9, and 11 are the well-characterized species related to the above-mentioned $2CH_2/2N$ replacements. In addition, it is expected to detect the ^{13}C isotope peaks for species with *m/z* 191 nominal mass (3, 6, 10). This sample is the same as in used in Reference [25] (denoted as UHVT_001) so it has been stored in a −80 °C freezer for years in a parafilmed bottle but has been opened a couple of times during these years so the appearance of small oxygen containing "contaminants" is not surprising (peaks 1, and 4). What is especially interesting and not reported before is the detection of odd electron molecular ions (peaks 3 and 7). Not shown in Table 1 but the largest error for accurate mass measurement that is possible to achieve with internal calibration and with a 15 T magnet is about 20 ppb. (First, the broadband spectrum is calibrated, ion compositions are identified and confirmed, then an ion in the narrow range is used to recalibrate the isotopic fine structure spectrum.) Obviously, the top spectrum in Figure 3 with a $\approx R = 63{,}000$ resolution is already informative but separation of peaks 3, 4 and 7, 8 is not adequate enough.

It is important to note here that the spectra shown in Figure 3 have been obtained on our Bruker SolarixR 15T FT-ICR instrument (Ohio State University, OH, USA) with an alternative ionization technique, laser desorption ionization (LDI). We first realized the effectiveness of LDI for laboratory tholin ionization back in 2005 [29] by generating LDI ions up to *m/z* 1000. That time LDI with FT-ICR capability was not available in our laboratory so only a time-of-flight (TOF) instrument (a Bruker Ultraflex III MALDI TOF-TOF, University of Arizona, Tucson, AZ, USA) was used. The resolution of TOF instruments (usually <50,000) is not always adequate to detect all "isobaric" ions, although they still can provide useful results just as demonstrated in [33] for tholin samples or de Marcellus *et al.* for aldehyde and carbohydrate molecules evolution in laboratory simulations by GC-TOF mass spectrometry [34].

Most of the studies mentioned above focused on positively charged ions (except with [33] that presented results for negatively charged ions). Recently, we performed a comparative study for positively and negatively charged ions by using both ESI and LDI ionization methods [25]. Characteristic differences have been observed: more saturated positively charged ions represent mostly amines and imines, meanwhile negatively charged ions are more unsaturated and represent compounds with nitrile (cyano) functionality (CN groups). In some cases, exotic C_xN_z ions are observed in the negative LDI spectra, such as the $C_{10}N_5^-$ anion that is most presumably a pentacyano cyclopentadienyl anion. Tandem MS/MS fragmentation confirmed the cyano functionality by losing HCN as a neutral molecule from several ions. Tandem MS/MS fragmentation and quantum chemical calculations suggested the formation of ion-molecule complexes, with a common anion of $C_2N_3^-$. (This ion at *m/z* 66 was first mentioned and studied in detail by Carrasco *et al.* [33].) Nevertheless, further experimental and theoretical studies are necessary to confirm the existence of stable ion-molecule complexes.

It is clear from the experimental and theoretical studies mentioned above that structural isomers do exist. Thus, even with the ultrahigh resolution power of FT-MS, isotope distinction is not always possible. Other than theoretical calculations, there are three different mass spectrometry related experiments that can be used to distinguish isomers: (i) ion mobility; (ii) gas-phase HDX; and (iii) variable wavelength infrared multiphoton dissociated spectroscopy (IRMPD). Although we have performed some ion mobility and HDX experiments in our laboratories, the result for laboratory tholins are not conclusive enough to be publishable. On the other hand, we successfully used IRMPD action spectroscopy in the CLIO laser facility [35] to obtain information on functional groups. The evaluation of the data and theoretical calculations are in progress and the results will be published elsewhere. For illustration, in Figure 4 we show one IRMPD action spectrum that was obtained for a $C_4H_3N_4^-$ anion (that we assume is an isomer (or isomers) of "deprotonated tetracyanide"). To our best knowledge, this is the first time a nitrile (CN stretching) band was detected at around 2180 cm^{-1}in the CLIO facility. This is an important result because it provides proof for the existence of –CN groups in highly unsaturated anions.

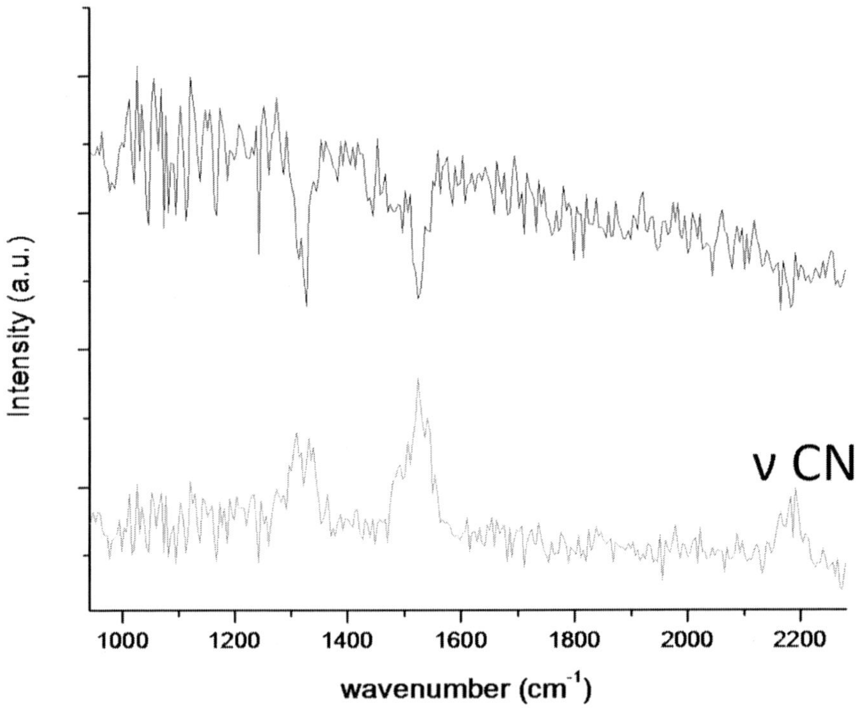

Figure 4. IRMPD action spectra of the anion $C_4H_3N_4^-$. The **upper** trace represents the trapped precursor ion *m/z* 117; and the **lower** trace represents all fragments of *m/z* 117 as a function of the IR radiation wavenumber. The CN stretching band is detected around 2180 cm^{-1}.

3. MATERIALS AND METHODS

The experimental details are quite diversified so we ask the readers to check the cited references below for details. Nevertheless, for guidance and easier reading, we summarize briefly the main features regarding model studies and instrumentation used in our laboratories. The individual references will be shown in the Results and Discussion Section.

Laboratory tholin samples were prepared by exposing a mixture of 2%–5% methane and 98%–95% nitrogen to a radio frequency electrical discharge under slow flow (6 atm L·h^{-1} and a pressure of 10^{-2} bar) at a temperature of 195 K in a recently designed ultrahigh vacuum (airtight) reaction chamber. The tholins were collected anaerobically in a glove box and stored in carefully sealed vials to avoid contamination so that tholin samples were produced with high purity (*i.e.*, only a trace amount of oxygenated species are detected, see discussion below).

The tholin samples were directly deposited on a conventional (stainless steel) matrix assisted laser desorption/ionization (MALDI) plate for laser desorption ionization (LDI) measurements that were carried out on a 9.4 T Bruker Daltonics Apex-Ultra FT-ICR instrument (at the University of Arizona, Tucson, AZ, USA) and a Bruker 15T SolariXR instrument (at the Ohio State University, Columbus, OH, USA). A Yag:Nd laser (352 nm) was used with variable laser power to reduce the possibility of laser induced fragmentation. Both FT-ICR instruments are equipped with a dual ESI/MALDI source so electrospray ionization (ESI) was also used for comparison. A Thermo Fisher Scientific LTQ-Orbitrap-XL instrument (University of Grenoble Alpes, Grenoble, France) was also used to perform comparative ultrahigh resolution MS and tandem MS/MS experiments. For electrospray ionizations, the tholin samples were dissolved in methanol and methanol:acetonitrile (1:1) with a ≈1–10 micromolar concentration. In all experiments, the instrument conditions were tuned and optimized for the detection of both positively and negatively charged ions in the mass range of m/z 50–400. Some experiments with electron ionization were also performed in the FT-ICR cell of a 4.7 T IonSpec instrument. Infrared multiphoton dissociation (IRMPD) action spectroscopy measurements were carried out at the Centre Laser Infrarouge d'Orsay (CLIO) facility [35] (Université Paris-Sud, Orsay, France). IRMPD action spectroscopy measures the trapped ion fragmentation efficiency as a function of the IR photons' wavenumber, *i.e.*, not the IR light absorption is recorded but the "action" (fragmentation) of the ion.

Before we show and discuss some recent results, we summarize briefly what kind of information can be obtained from ultrahigh resolution mass spectrometry (UHRMS) experiments and how data processing can simplify and make the information content more informative.

The ultrahigh resolution capability of Orbitrap (up to about 450,000, see Reference [36]) and ICR instruments (up to about 20,000,000 with a 21T magnet [37] and with the recently developed dynamically harmonized cell [38]) allow us:

(i)

To separate and determine major and minor components in complex mixtures, *i.e.*, significantly improve the chemical information. Without ultrahigh resolution, this information would be lost.

(ii)

To perform *accurate mass* measurements (*i.e.*, chemical composition determination) within <100 ppb error with internal calibration. Isotope fine structures [39] can be determined so that "isobaric" elements (e.g., ^{32}S (31.972071) and 2 O (31.989829)) are distinguished and used to confirm/eliminate certain chemical compositions. It has been shown that a mass resolution of 5,000,000 and mass accuracy of 100 ppb are sufficient to unambiguously ascribe the correct elemental composition for ions containing C, H, O, N, and S up to 500 Da [40].

(iii)

To determine fragmentation pattern of several precursor ions, *i.e.*, speeding up structural determination. Practically, this means data independent acquisitions (DIA) [41] that are widely used for complex metabolomics samples. Knowing the chemical compositions and characteristic fragments, e.g., loss of H_2O, NH_3, HCN, CH_3CN, *etc.*, one can easily generate a theoretical fragmentation matrix that can be compared to experimentally observed fragments to reveal structural features of individual components.

Instead of tabulating literally hundreds of chemical compositions, simple but informative representations, such as the Kendrick mass diagrams [42] (or mass defect *vs.* exact mass (MDvEM) plots [43]) or the (modified) van Krevelen diagrams [44,45] are commonly used for data interpretation. These diagrams are especially helpful in determining homolog series, such as molecules differing by CH_2 units, and chemical modifications, e.g., when 2 CH_2 units are replaced by "isobaric" (but, by FT-MS, easily distinguishable) 2 nitrogens. A total mass difference statistic algorithm was recently developed for the improved analysis of FT-ICR data obtained for natural organic matter that can be applied to organic samples of astrobiological importance [46].

4. CONCLUSIONS

Research on the formation of prebiotic, small organic molecules is a prerequisite to any further investigation about the origin of life in the Universe. Although we can now see far away and we can detect objects (galaxies) in the far remote corner of the observable Universe, we are very much bound to Earth and our Solar System to obtain detailed information on the existence (or absence) of prebiotic molecules. Planets have been detected around nearby stars (within about 100 light years) and improving instrumentation allows us to anticipate the possibility to detect "signatures of life" in the form of small organic molecules in planetary atmospheres. Nevertheless, detailed information on a wide variety of organic structures is not expected to be available for at least a couple of decades. For the time being, it is completely justified to study Solar System objects, such as meteoritic materials. Processes in the reductive atmosphere of the relatively close Titan need to be better understood for general chemical curiosity. The species formed in a reductive gas mixture under controlled laboratory conditions may provide insight for the origin of small, C, H, and N containing organic molecules.

The degree of information harvestable and its reliability strongly depend on the analysis methods and instrumentation. In this regard, ultrahigh resolution mass spectrometry (URHMS) is the best available method to obtain information on *individual* components of very *complex mixtures.* Nature tends to produce complex mixtures and not pure samples thus it is an obligation to use the best method that is available to unveil (certain) secrets of Nature.

We do hope that the illustrative examples shown above convince the readers about the importance, usefulness, reliability and necessity of using UHRMS in astrobiological research. Organic contents of meteorites, the products of atmospheric processes, and their further reaction (hydrolysis) products, can now be followed with incredible resolution (>500,000) and mass accuracy (<1 ppm). Isomeric structures can be distinguished by coupling UHRMS with additional techniques, such as IRMPD action spectroscopy, gas-phase HDX, and ion mobility.

Although the results obtained in laboratory environment on Earth are promising and revealing more and more structural details, *extrapolation* to other objects will always be risky. We do know a lot about the components and structure of laboratory-made tholins but we cannot be absolutely sure that these molecules are *really* there in the atmosphere and/or surface of Titan. Nevertheless, we strongly believe that the laboratory results will inspire the design of *ultrahigh resolution* mass analyzers that will, eventually, be sent to space, planets, moons, meteorites and comets. In fact, some of these design projects are in progress (multibounce TOF selected for future NASA mission at Europa [47] and Cosmorbitrap project to adapt the Orbitrap concept into a space borne instrument [48,49,50,51,52,53]). The technical challenges are enormous but physics allows such dreams that we the human species make to become reality.

ACKNOWLEDGMENTS

This work has been supported by a grant from Labex OSUG@2020 (Investissements d'avenir —ANR10 LABX56). The authors are grateful for the professional help by Vincent Steinmetz in the CLIO Facility (Paris Sud, Orsay, France) for detecting IRMPD action spectra. The authors are grateful for the Reviewers for their thorough reviews and valuable suggestions.

AUTHOR CONTRIBUTIONS

Árpád Somogyi and Roland Thissen conceived and designed most of the experiments; Árpád Somogyi, Roland Thissen, Francois-Régis Orthous-Daunay and Véronique Vuitton performed the experiments and analyzed the data; and all authors contributed to the writing of the paper.

CONFLICTS OF INTEREST

The authors declare no conflict of interest.

REFERENCES

1. Ehrenfreund, P.; Spaans, M.; Holm, H.G. The evolution of organic matter in space. Philos. Trans. R. Soc. 2011, 369, 538–554.
2. Irvine, W.M. Extraterrestrial organic matter. Orig. Life Evol. Bios. 1998, 28, 365–383.
3. Czakó, G.; Nagy, B.; Tasi, G.Y.; Somogyi, Á.; Šimunek, J.; Noga, J.; Braams, J.-B.; Bowman, J.M.; Császár, A.G. Proton affinity and enthalpy of formation of formaldehyde. Int. J. Quantum Chem. 2009, 109, 2393–2409.
4. Waite, J.H., Jr.; Niemann, H.B.; Yelle, R.V.; Kasprzak, W.T.; Cravens, T.E.; Luhmann, J.G.; McNutt, R.L., Jr.; Ip, W.-H.; Gell, D.; de La Haye, V.; et al. Ion neutral mass spectrometer results from the first flyby of Titan. Science 2005, 308, 982–986.
5. Cravens, T.E.; Robertson, I.P.; Waite, J.H., Jr.; Yelle, R.V.; Kasprzak, W.T.; Keller, C.N.; Ledvina, S.A.; Niemann, H.B.; Luhmann, J.G.; McNutt, R.L., Jr.; et al. Composition of Titan's ionosphere. Geophys. Res. Lett. 2006, 33, L07105.
6. Waite, J.H., Jr.; Young, D.T.; Cravens, T.E.; Coates, A.J.; Crary, F.J.; Magee, B.; Westlake, J. The process of tholin formation in Titan's upper atmosphere. Science 2007, 316, 870–875.
7. Coates, A.J.; Crary, F.; Lewis, G.; Young, D.T.; Waite, J.H., Jr.; Sittler, E.C.J. A new upper limit to the field-aligned potential near Titan. Geophys. Res. Lett. 2007, 34, L22103.
8. Coates, A.J.; Wellbrock, A.; Lewis, G.R.; Jones, G.H.; Young, D.T.; Crary, F.J.; Waite, J.H., Jr. Heavy negative ions in Titan's atmosphere: Altitude and latitude dependence. Planet. Space Sci. 2009, 57, 1866–1871.

9. Cable, M.L.; Hörst, S.M.; Hodyss, R.; Beauchamp, P.M.; Smith, M.A.; Willis, P.A. Titan tholins: Simulating Titan organic chemistry in the Cassini-Huygens era. Chem. Rev. 2012, 112, 1882–1909.
10. Orthous-Daunay, F.-R.; Thissen, R.; Vuitton, V.; Somogyi, Á.; Mespoulede, M.; Beck, P.; Bonnet, J.-Y.; Dutuit, O.; Schmitt, B.; Quirico, E. Orbitrap-MS and FT-ICR-MS of free and labile organic matter from carbonaceous chondrites. In Proceedings of the 42nd Lunar and Planetary Science Conference, The Woodlands, TX, USA, 7–11 March 2011; p. 2654.
11. Callahan, M.P.; Smith, K.E.; Cleaves, H.J., II; Ruzicka, J.; Stern, J.C.; Glavin, D.P. Carbonaceous meteorites contain a wide range of extraterrestrial nucleobases. Proc. Natl. Acad. Sci. USA 2011, 34, 13995–13998.
12. Schmitt-Kopplin, P.; Harir, M.; Kanawati, B.; Tziozis, D.; Hertkorn, N.; Gabelica, Z. Chemical footprint of the solvent soluble extraterrestrial organic matter occluded in Sołtmany ordinary chondrite. Meteorites 2012, 2, 79–92.
13. Schmitt-Kopplin, P.; Gabelica, Z.; Gougeon, R.G.; Fekete, A.; Kanawati, B.; Harir, M.; Gebefuegi, I.; Eckel, G.; Hertkorn, N. High molecular diversity of extraterrestrial organic matter in Murchison meteorite revealed 40 years after its fall. Proc. Natl. Acad. Sci. USA 2010, 107, 2763–2768.
14. Saladino, R.; Carota, E.; Botta, G.; Kapralov, M.; Timoshenko, G.N.; Rozanov, I.Y.; Krasavin, E.; di Mauro, E. Meteorite-catalyzed syntheses of nucleosides and of other prebiotic compounds from formamide under proton irradiation. Proc. Natl. Acad. Sci. USA 2015, 112, E2746–E2755.
15. Vuitton, V.; Yelle, R.V.; McEwan, M.J. Ion chemistry and N-containing molecules in Titan's upper atmosphere. Icarus2007, 191, 722–742.
16. Yelle, R.V.; Vuitton, V.; Lavvas, P.; Klippenstein, S.S.; Smith, M.A.; Hörst, S.M.; Cui, J. Formation of NH3 and CH2NH in Titan's upper atmosphere. Faraday Discuss. 2010, 147, 31–49.
17. Lavvas, P.; Yelle, R.V.; Vuitton, V. The detached haze layer in Titan's mesosphere. Icarus 2009, 201, 626–633.
18. Lebonnois, S.; Bakes, E.L.O.; McKay, C.P. Transition from gaseous compounds to aerosols in Titan's atmosphere. Icarus 2002, 159, 505–517.
19. Magee, B.A.; Waite, J.; Mandt, K.E.; Westlake, J.; Bell, J.; Gell, D.A. INMS-derived composition of Titan's upper atmosphere: Analysis methods and model comparison. Planet. Space Sci. 2009, 57, 1895–1916.
20. Imanaka, H.; Smith, M.A. Formation of nitrogenated organic aerosols in the Titan upper atmosphere. Proc. Natl. Acad. Sci. USA 2010, 107, 12423–12428.
21. Somogyi, Á. Probing peptide fragment ion structures by combining sustained off-resonance collision-induced dissociation and gas-phase H/D exchange (SORI-HDX) in fourier transform ion-cyclotron resonance (FT-ICR) instruments. J. Am. Soc. Mass Spectrom. 2008, 19, 1771–1775.
22. Somogyi, Á.; Harrison, A.G.; Paizs, B. Using gas-phase guest-host chemistry to probe the structures of b ions of peptides. J. Am. Soc. Mass Spectrom. 2012, 23, 2055–2058.
23. Thissen, R.; Vuitton, V.; Lavvas, P.; Lemaire, J.; Dehon, C.; Dutuit, O.; Smith, M.A.; Turchini, S.; Catone, D.; Yelle, R.V.; et al. Laboratory studies of molecular growth in the Titan ionosphere. Phys. Chem. A 2009, 113, 11211–11220.
24. Sharker, N.; Somogyi, Á.; Lunine, J.I.; Smith, M.A. Titan aerosol analogues: Analysis of the nonvolatile tholins. Astrobiology 2003, 3, 719–726.
25. Somogyi, Á.; Smith, M.A.; Vuitton, V.; Thissen, R.; Komáromi, I. Chemical ionization in the atmosphere? A model study on negatively charged "exotic" ions generated from Titan's tholins by ultrahigh resolution MS and MS/MS. Int. J. Mass Spectrom. 2012, 316–318, 157–163.
26. Pernot, P.; Carrasco, N.; Thissen, R.; Schmitz-Afonso, I. Tholinomics-chemical analysis of nitrogen-rich polymers. Anal. Chem. 2010, 82, 1371–1380.
27. Hörst, S.M.; Yelle, R.V.; Buch, A.; Carrasco, N.; Cernogora, G.; Dutuit, O.; Quirico, E.; Sciamma-O'Brien, E.; Smith, M.A.; Somogyi, Á.; et al. Formation of prebiotic

molecules in a Titan atmosphere simulation experiment. Astrobiology2012, 12, 809–817.
28. Szopa, C.; Cernogora, G.; Boufendi, L.; Correia, J.-J.; Coll, P. PAMPRE: A dusty plasma experiment for Titan's tholins production and study. Planet. Space Sci. 2006, 54, 394–404.
29. Somogyi, Á.; Oh, C.-H.; Lunine, J.I.; Smith, M.A. Organic environments on Saturn's moon, Titan: Simulating chemical reactions and analyzing products by FT-ICR and ion-trap mass spectrometry. J. Am. Soc. Mass Spectrom. 2005, 16, 850–859.
30. Neish, C.D.; Somogyi, Á.; Imanaka, H.; Lunine, J.I.; Smith, M.A. Rate measurements of the hydrolysis of complex macromolecules in cold aqueous solutions: Implications for prebiotic chemistry on the early Earth and Titan. Astrobiology 2008, 8, 273–287.
31. Neish, C.D.; Somogyi, Á.; Lunine, J.I.; Smith, M.A. Low temperature hydrolysis of laboratory tholins in ammonia-water solutions: Implications for prebiotic chemistry on Titan. Icarus 2009, 201, 412–421.
32. Neish, C.D.; Somogyi, Á.; Smith, M.A. Titan's primordial soup: Formation of amino acids via low-temperature hydrolysis of tholins. Astrobiology 2010, 10, 337–347.
33. Carrasco, N.; Schmitz-Afonso, I.; Bonnet, J.-Y.; Quirico, E.; Thissen, R.; Dutuit, O.; Bagag, A.; Laprévote, O.; Bush, A.; Giulani, A.; et al. Chemical characterization of Titan's tholins: Solubility, morphology and molecular structure revisited. J. Phys. Chem. A 2009, 113, 11195–11203.
34. De Marcellus, P.; Meinert, C.; Myrgorodska, I.; Nahon, L.; Buhse, T.; Le Sergeant d'Hendecourt, L.; Meierhenrich, U.J. Aldehydes and sugars from evolved precometary ice analogs: Importance of ices in astrochemical and prebiotic evolution. Proc. Natl. Acad. Sci. USA 2015, 112, 965–970.
35. Lemaire, J.; Boissel, P.; Heninger, M.; Mauclaire, G.; Bellec, G.; Mestdagh, H.; Simon, A.; le Caer, S.; Ortega, J.M.; Glotin, F.; et al. Gas Phase infrared spectroscopy of selectively prepared ions. Phys. Rev. Lett. 2002, 89, 273002.
36. Makarov, A.; Denisov, E.; Kolomeev, A.; Balschun, W.; Lange, O.; Strupat, K.; Horning, S. Performance evaluation of a hybrid linear ion trap/orbitrap mass spectrometer. Anal. Chem. 2006, 78, 2113–2120.
37. Hendrickson, C.L.; Quinn, J.P.; Kaiser, N.K.; Smith, D.F.; Blakney, G.T.; Chen, T.; Marshall, A.G.; Weisbrod, C.R.; Beu, S.C. 21 Tesla fourier transform ion cyclotron resonance mass spectrometer: A national resource for ultrahigh resolution mass analysis. J. Am. Soc. Mass Spectrom. 2015, 26, 1626–1632.
38. Kostyukevich, Y.I.; Vladimirov, G.N.; Nikolaev, E.N. Dynamically harmonized FT-ICR cell with specially shaped electrodes for compensation of inhomogeneity of the magnetic field. computer simulations of the electric field and ion motion dynamics. J. Am. Soc. Mass Spectrom. 2013, 23, 2198–2207.
39. Loos, M.; Gerber, C.; Corona, F.; Hollender, J.; Singer, H. Accelerated isotope fine structure calculation using pruned transition trees. Anal. Chem. 2015, 87, 5738–5744.
40. Kim, S.; Rogers, R.P.; Marshall, A.G. Truly "exact" mass: Elemental composition can be determined uniquely from molecular mass measurement at ≈0.1 mDa accuracy for molecules up the ≈500 Da. Int. J. Mass Spectrom. 2006, 251, 260–265.
41. Doerr, A. DIA mass spectrometry. Nat. Methods 2015, 12, 35.
42. Kendrick, E. A Mass scale based on CH2=14.000 for high resolution mass spectrometry of organic compounds. Anal. Chem. 1963, 35, 2146–2154.
43. Danger, G.; Orthous-Daunay, F.-R.; de Marcellus, P.; Modica, P.; Vuitton, V.; Duvernay, F.; Flandinet, L.; d'Hendecourt, L.L.; Thissen, R.; Chiavassa, T. Characterization of laboratory analogs of interstellar/cometary organic residues using very high resolution mass spectrometry. Geochim. Cosmochim. Acta 2013, 118, 184–201.
44. Van Krevelen, D.W. Graphical-statistical method for the study of structure and reaction process of coal. Fuel 1950, 29, 269–284.
45. Kim, S.; Kramer, R.W.; Hatcher, P.G. Graphical method for analysis of ultrahigh-resolution broadband mass spectra of natural organic matter, the van Krevelen diagram. Anal. Chem. 2003, 75, 5336–5344.

46. Kumenkov, E.V.; Kononikhin, A.S.; Perminova, R.V.; Hertkorn, N.; Gaspar, A.; Schmitt-Kopplin, P.; Popov, I.A.; Garmash, A.V.; Nikolaev, E.N. Total mass difference statistics algorithm: A new approach to identification of high-mass building blocks in electrospray ionization Fourier transform ion cyclotron mass spectrometry data of natural organic matter. Anal. Chem. 2009, 81, 10106–10115.
47. Hässig, M.; Libardoni, M.; Mandt, K.; Miller, G.; Blasé, R. Performance evaluation of a prototype multi-bounce time-of-flight mass spectrometer in linear mode and applications in space science. Planet Space Sci. 2015, 117, 436–443.
48. Briois, C.; Lethuillier, A.; Thirkell, L.; Boukrara, A.; Lebreton, J.-P.; Carrasco, N.; Szopa, C.; Chapuis, C.; Bouabdellah, A.; Thissen, R.; et al. A high mass-resolution orbitrap mass spectrometer for in Situ analysis in planetary science. In Proceedings of the JFSM Conference, Orléans, France, 17–20 September 2012.
49. Briois, C.; Lebreton, J.-P.; Szopa, C.; Thirkell, L.; Aradj, K.; Bouabdellah, A.; Boukrara, A.; Carrasco, N.; Chalumeau, G.; Chapelon, O.; et al. High resolution mass spectrometry for future space instrumentation: Current development within the french space orbitrap consortium. In Proceedings of the Geophysical Research, 16 2014 EGU2014-1639, Vienna, Austria, 27 April–2 May 2014.
50. Carrasco, N.; Szopa, C.; Chapuis, C.; Bouabdellah, A.; Thissen, R.; Cornelli, S.; Puget, P.; Briois, C.; Thirkell, L.; Boukrara, A.; et al. Ultra high mass-resolution for space exploration: The orbitrap mass analyzer. In Proceedings of the EPSC 2012, Madrid, Spain, 23–28 September 2012.
51. Thissen, R.; Thirkell, L.; Makarov, A.; Puget, P.; Briois, C.; Szopa, C.; Carrasco, N.; Berthelier, J.-J.; Cottin, H.; Grand, N.; et al. Ultrahigh resolution fourier transform mass analyzer for space exploration: Orbitrap. In Proceedings of the DPS-EPSC 2011, Nantes, France, 2–7 October 2011.
52. Vuitton, V.; Briois, C.; Thirkell, L.; Aradj, K.; Bouabdellah, A.; Boukrara, A.; Carrasco, N.; Chalumeau, G.; Chapelon, O.; Colin, F.; et al. The orbitrap mass analyzer as a space instrument for the understanding of prebiotic chemistry in the solar system. In Proceedings of the 40th COSPAR Scientific Assembly, Moscow, Russia, 2–10 August 2014.
53. Kempf, S.; Srama, R.; Grün, E.; Mocker, A.; Postberg, F.; Hillier, J.K.; Horany, M.; Sternovsky, Z.; Abel, B.; Beinsen, A.; et al. Linear high resolution dust mass spectrometer for a mission to the Galilean satellites. Planet. Space Sci. 2012, 65, 10–20.

CHAPTER 17

The Formation of Jupiter, the Jovian Early Bombardment and the Delivery of Water to the Asteroid Belt: The Case of (4) Vesta

Diego Turrini [1,*] and Vladimir Svetsov [2]

[1]*Istituto di Astrofisica e Planetologia Spaziali (INAF-IAPS), Istituto Nazionale di Astrofisica, Via del Fosso del Cavaliere 100, Rome I-00133, Italy*

[2]*Institute for Dynamics of Geospheres, Russian Academy of Sciences, Leninskiy Prospekt 38-1, Moscow 119334, Russia*

ABSTRACT

The asteroid (4) Vesta, parent body of the Howardite-Eucrite-Diogenite meteorites, is one of the first bodies that formed, mostly from volatile-depleted material, in the Solar System. The Dawn mission recently provided evidence that hydrated material was delivered to Vesta, possibly in a continuous way, over the last 4 Ga, while the study of the eucritic meteorites revealed a few samples that crystallized in presence of water and volatile elements. The formation of Jupiter and probably its migration occurred in the period when eucrites crystallized, and triggered a phase of bombardment that caused icy planetesimals to cross the asteroid belt. In this work, we study the flux of icy planetesimals on

Vesta during the Jovian Early Bombardment and, using hydrodynamic simulations, the outcome of their collisions with the asteroid. We explore how the migration of the giant planet would affect the delivery of water and volatile materials to the asteroid and we discuss our results in the context of the geophysical and collisional evolution of Vesta. In particular, we argue that the observational data are best reproduced if the bulk of the impactors was represented by 1–2 km wide planetesimals and if Jupiter underwent a limited (a fraction of au) displacement.

Keywords: comets; Vesta; impacts; asteroids; Jupiter; water; Solar System formation

1. INTRODUCTION

The observations of the Dawn mission [1,2] recently confirmed the genetic link, suggested more than 40 years ago based on spectroscopic measurements [3], between the asteroid (4) Vesta and the Howardite-Eucrite-Diogenite (HED) class of meteorites. From the HED meteorites we know that Vesta was a mostly volatile-depleted body (see e.g., [4] and references therein) and, based on the crystallization ages of the oldest eucrites [5] and diogenites [6], we know that it formed and differentiated in the first 3 Ma of the life of the Solar System. The observational data provided by the Dawn mission [2,7,8] and the laboratory studies of the HED meteorites [4], however, also revealed that water and/or hydrated materials were present on Vesta both in the most ancient and in the more recent past. The instruments on-board the Dawn spacecraft detected the presence of hydroxyls (OH) [7] and hydrogen (H) [2], mainly associated with the presence of dark material [2,7,9], and revealed the existence of pitted terrains [8], interpreted as the results of impact-triggered degassing events of buried volatile-rich materials. The dark material was argued to be an exogenous contaminant delivered to Vesta by the impacts of carbonaceous chondrite-like asteroids [9,10,11] and was suggested to be possibly linked to the existence of the pitted terrains [8]. Alongside these evidences for the delivery of hydrated material over the last 4 Ga (see [11] for a discussion), HED meteorites revealed the presence of water and other volatile elements (e.g., clorine and fluorine) in the vestan magma during the crystallization of some eucritic samples [4]. As a consequence, volatile materials should have been present, or delivered, to Vesta before the complete solidification of its eucritic crust.

According to [12], the thickness of the crust from which eucrites and diogenites originated should have ranged between 40 km and 80 km. If we follow [13] and assume an initial composition of Vesta similar to CI meteorites (olivine-rich carbonaceous chondrites named after the Ivuna meteorite), the results of [12] would indicate that the eucritic and diogenitic layers originally had a thickness of 26 km and 13 km, respectively. The results of thermal models (see e.g., [14] and references therein for previous works on the subject by other authors) and geophysical models [15] indicate that the eucritic crust of Vesta solidified quite early in the life of the asteroid. As discussed by [14], depending on the accretion time and the initial porosity of Vesta, the thickness of the solid crust between 3 and 5 Ma from the condensation of the CAIs (the calcium-aluminum inclusions, which are the oldest known solids in the Solar System; see [16] and references therein) could have ranged from a minimum of 7–9 km to a maximum of 20–30 km. The independent study by [15], using a more complete physical model, gives thickness values for the solid crust varying between 7 km and at least 10 km ([15], Supplementary Information; G. Golabek, personal communication) between 3 Ma and 5 Ma from CAIs, and up to 20–30 km at about 9 Ma [15]. Therefore, the basaltic (eucritic) crust should have completely solidified somewhere between the first 3–10 Ma. In order to reach the molten layer toward the end of this temporal interval (e.g., when the solid crust was about 15 km thick) an impact would cause the formation of a ~200 km wide crater, with major implications for the survival of the basaltic crust (e.g., the effusion of diogenite-rich magma and the formation of diogenitic terrains units, [17,18]) that have not been observed by the Dawn mission [1,2]. As a consequence, whatever mechanism delivered the volatile elements incorporated in the eucritic samples studied by [4], it should have acted during the first few Ma of the life of Vesta and it should have preserved the basaltic crust of the asteroid, which we know survived until present time [1,2].

At the time Vesta was forming its basaltic crust, the Solar System was in the phase of its evolution known as the Solar Nebula [16], *i.e.*, it was a circumsolar disk of gas and dust where the first generations of planetary bodies were forming. The beginning of the Solar Nebula is generally assumed to coincide with the condensation of the CAIs about 4.6 Ga ago (the oldest CAI currently known dates *4568.2−0.4+0.2* Ma ago [19]), while its duration is indirectly constrained by observations and theoretical studies of circumstellar disks, whose median (inner disk) lifetime is about 3 Ma with the range of observed values spanning between 1–10 Ma [20]. The giant planets of the Solar System should have formed across this timespan (see [21] for an overview on the processes and the timescales governing the formation of giant planets) and, in particular, theoretical [22,23] and observational [24] arguments suggest that Jupiter formed 3–5 Ma after the condensation of CAIs. The formation of Jupiter has been shown by different authors [17,18,25,26,27,28] to trigger a sudden spike in the flux of impactors in the early history of the Solar System. This event, named the Jovian Early Bombardment ([17,28], JEB in the following), is caused by the scattering of ice-rich planetesimals from the outer Solar System due to the gravitational perturbation of the giant planet [17,25,26,27,28] and by the appearance of the Jovian mean motion resonances in the asteroid belt, in particular the 3:1 and 2:1 resonances [17,18,27,28]. The duration of the JEB is limited to about 1 Ma [17,26,28], with the bulk of the impacts taking place in the first (3–5) $\times 10^5$ years [17]. The flux of impactors due to the Jovian resonances is the dominant one in the inner Solar System [17,18] and is the one shaping the early collisional evolution of the asteroid belt [17,18,28].

Turrini *et al.* [17] estimated the fluxes of impactors coming from the outer Solar System and from the Jovian resonances during the JEB, the crater populations they produce and the probability of Vesta being destroyed during the bombardment using different size-frequency distributions (SFDs in the following) of the impactors. Their results showed that the probability of Vesta undergoing a catastrophic impact are negligible, but suggested that cratering erosion could play an unexpectedly significant role due to the higher, pre-depletion population of planetesimals inhabiting the asteroid belt at the time. Turrini *et al.* [28] further investigated the subject of asteroidal erosion during the JEB and, using a more detailed physical description of the mass loss processes, showed that cratering erosion indeed played a much more relevant role than catastrophic disruption in determining the fate of primordial asteroids. Turrini *et al.* [28] showed that cratering erosion is a function of the extent of Jupiter's migration and of the position of the target body in the asteroid belt. Depending on the considered scenario and SFD of the impactors, planetesimals the size of Vesta could lose from a few times 1% to a few times 10% of their original mass [28]. Turrini [18] re-evaluated the collisional evolution of Vesta due to the impactors from the inner Solar System by reprocessing the results of the simulations from [17] with an improved version of the collisional model detailed by [28] and by investigating the roles of crater saturation, surface excavation and cratering erosion. According to the results of [18], the survival of the basaltic crust of Vesta would favor scenarios where Jupiter underwent a limited (no more than 0.25 au) migration and the protoplanetary disk was dominated, in terms of population, by planetesimals with diameters of 1–2 km while most of the mass of the disk was in the form of 1000 km wide or larger planetary embryos. However, Turrini [18] also pointed out that, if an undifferentiated crust survived on top of the molten interior of Vesta after the differentiation process as suggested by the results of [14,15], the favored scenarios would be those where migration was more significant (0.50–1.00 au) to allow for the removal of said undifferentiated crust.

The aim of this work is to evaluate the implications of the fluxes of impactors originating from the outer Solar System in the simulations of [17] for the surface evolution of Vesta and the delivery of volatile materials to the asteroid. To achieve this goal, we will proceed in a way similar to what has been done by [18] for the impactors originating from the inner Solar System. We will use the hydrodynamic method developed by [29] to assess the erosion caused by the impacts of icy planetesimals at different velocities and, at the same time, to evaluate the fraction of the mass of these icy impactors that would be retained by Vesta. Using these data, we will adapt the collisional model developed by [18] to the case of icy impactors, we will evaluate their effects in terms of crater saturation, surface erosion and excavation for Vesta, and we will assess whether the JEB could be the source of the water and volatile elements observed in the eucritic samples studied by [4]. Finally, we will discuss the results obtained in this study together with those obtained by [18] to constrain the migration of Jupiter and the size-frequency distribution of the planetesimals in the Solar Nebula.

2. THE JOVIAN EARLY BOMBARDMENT AND THE COLLISIONAL MODEL

The main input data we used in this study of the delivery of water to Vesta across the JEB are the fluxes of impactors described in [17], hereafter Paper I. To assess the effects of the JEB in terms of the collisional evolution of Vesta (e.g., cratering erosion, crater saturation, excavation of the crust) we based on the collisional model described in [28], hereafter Paper II, and [18], hereafter Paper III. As the dynamical, physical and numerical details of the model used to study the JEB have been extensively described in Papers I, II and III, here we will discuss the main aspects of the simulations and highlight the differences with respect to the previous works. The interested readers are referred to Paper I for more details on the fluxes of impactors on Vesta during the JEB and to Paper III for an extensive discussion of the effects of the impactors originating from the inner Solar System on the collisional and geophysical evolution of the asteroid. The numerical model used to assess the amount of cometary material surviving each impact and the associated erosion of the vestan surface is based on the one described in [29]: in the following, the improvements and the modification to the original model will be described in more detail.

2.1. The Solar Nebula

The template of the Solar System at the beginning of the simulations of Paper I was composed of the Sun, the forming Jupiter, Vesta, Ceres and a disk of planetesimals modeled as massless particles. The evolution of this template of the Solar System was followed for 2×10^6 years, centered on the time Jupiter started to accrete the nebular gas.

2.1.1. The Protoplanetary Disk

In the simulations of Paper I the disk of planetesimals extended between 2 au and 10 au. The disk was divided into annular regions of width equal to 1 au, each containing 10^4 massless particles. The planetesimals had initial values of the orbital eccentricity and of the inclination ranging between $0 \leq e_i \leq 3 \times 10^{-2}$ and 0 rad $\leq i_i \leq 3 \times 10^{-2}$ rad, respectively. Planetesimals formed in the inner Solar System (2–4 au, ISS in the following) were considered rocky bodies with mean density $\rho_{iss} = 3.0$ g cm^{-3}, while planetesimals formed in the outer Solar System (4–10 au, OSS in the following) were considered volatile-rich bodies with mean density $\rho_{oss} = 1.0$ g cm^{-3}. In Paper I the transition between ISS and OSS impactors, assumed to coincide with the location of the Snow Line, took place at $r_{SL} = 4.0$ au or at $r_{SL} = 3.0$ au depending on the assumed size-frequency distribution of the impactors. In this work, the transition between ISS and OSS impactors will always take place at $r_{SL} = 4.0$ au. The planetesimals were removed from the simulations of Paper I if their semimajor axes became smaller than 1 au, or larger than 30 au, or if they impacted the Sun or Jupiter. Each massless particle could impact only once with each of the target bodies considered in Paper I (*i.e.*, Vesta and Ceres) but was not removed from the simulations unless it impacted both targets or one of the previously listed conditions occurred.

2.1.2. Vesta

In the simulations of Paper I the target bodies considered were Vesta and Ceres. The semimajor axis of Vesta was assumed $a_v = 2.362$ au and the asteroid was initially on a planar, circular orbit. Its mean radius was assumed $r_v = 258$ km based on [30] and its mass value $m_v = 2.70 \times 10^{23}$ g was derived from [31]. Note that the values of the mean radius and of the mass adopted for Vesta in Paper I are slightly different from the ones recently estimated by the Dawn mission (262.7 km and 2.59×10^{23} g respectively, [32]). However, these values were only used to evaluate the collisional cross-section of Vesta, and the differences between the two sets of values are of the order of only 2%–4%, so they do not significantly affect the fluxes estimated in Paper I.

2.1.3. Jupiter

At the beginning of the simulations, Jupiter was a planetary embryo with mass $M_0 = 0.1\ M_\oplus$ (where $M_\oplus = 5.9726 \times 10^{27}$ g is the Earth's mass) that grew to the critical mass $M_c = 15\ M_\oplus$ in $\tau_c = 10^6$ years (where τ_c can be interpreted as the oligarchic growth timescale, [21]) as:

$$M = M_0 + \left(\frac{e}{e-1}\right)(M_c - M_0) \times \left(1 - e^{-t/\tau_c}\right) \quad (1)$$

As shown in Papers I and II, the effects of Jupiter on the cometary flux on Vesta and the inner asteroid belt during this phase are limited. A few OSS planetesimals can have close encounters with Jupiter, be injected on orbits crossing the asteroid belt and impact on Vesta or one of the largest asteroids, but they are not considered in this analysis as their contribution to the total flux on Vesta is smaller than that of OSS planetesimals during the JEB (see Paper I). As shown in Paper I, however, this approximation would not be correct in the case of Ceres and, more generally, a larger statistics (in terms of simulations) is needed in order to properly assess their role in the flux of volatile-rich impactors to the asteroid belt.

When the critical mass value M_c is reached, the nebular gas surrounding Jupiter is assumed to be rapidly accreted by the planet, whose mass then grows as:

$$M = M_c + (M_J - M_c) \times \left(1 - e^{-(t-\tau_c)/\tau_g}\right) \quad (2)$$

where $M_J = 1.8986 \times 10^{30}\ g = 317.83\ M_\oplus$ is the final mass of the giant planet. The e-folding time $\tau_g = 5{\times}10^3$ years is derived from the hydrodynamical simulations described in [33,34]. While accreting the nebular gas, Jupiter migrates inward due to disk-planet interactions as:

$$r_p = r_0 + (r_J - r_0) \times \left(1 - e^{-(t-\tau_c)/\tau_r}\right) \quad (3)$$

where r_p is the instantaneous orbital radius of Jupiter, r_0 is its orbital radius at the beginning of the simulation, r_J is its present orbital radius and $\tau_r = \tau_g = 5 \times 10^3$ years. Paper I considered four different migration scenarios: no displacement and 0.25 au, 0.50 au and 1.00 au displacements. The initial position of Jupiter is chosen in the different scenarios so that its final position is always the present one, consistently with the original Nice model ([35], note that in the most recent developments of the Nice model [36] Jupiter is assumed to be initially located at about 5.4 au). With the chosen values of the e-folding time τ_r, the migration scenarios are equivalent to assume values of $a/\dot{a}$ (*i.e.*, the radial migration timescale, see [37]) respectively equal to 3.2 × 10^5 years, 1.6 × 10^5 years and 8 × 10^4 years, consistently with the results of theoretical studies [37]. As a test, Paper I modified the values of τ_g and τ_r to 2.5 × 10^4 years in the scenario were Jupiter migrates by 1 au. Neither the fluxes of impactors nor the impact velocities changed in any significant way on Vesta. The same was not true for Ceres, which received a flux of OSS impactors a factor of 4 higher.

2.2. Size-Frequency Distributions of Planetesimals

In this study we will consider a total of 4 SFDs of the primordial planetesimals populating the Solar Nebula, whose selection is based on the results of Papers II and III, and on the criterion of the survival of Vesta and its basaltic crust [38]. The first two SFDs considered [39,40] supply the post-formation, non-collisionally evolved average size of the primordial planetesimals as a function of their distance from the Sun. As a consequence, these SFDs can be be applied directly to both ISS and OSS impactors. The last two SFDs [41,42] provide instead the collisionally evolved size distribution of the planetesimals after 3 Ma, assumed to be the formation time of Jupiter, but were derived exclusively for the orbital region of the asteroid belt. In principle, therefore, they cannot be applied directly to the OSS impactors that are the focus of this work.

The results of [42], however, showed that for diameters down to 1 km the collisionally evolved SFD of the primordial planetesimals varies little with distance in the orbital region between 1.5 au and 4 au (see Figure 14 from [42]). According to [42], the collisionally evolved SFD also shows little sensitivity to the assumed planetesimal density (see Figure 13 from [42]) and to the surface density of the swarm (see Figure 19 from [42]). Finally, Weidenschilling [43] showed that, for diameters smaller than 100 km, the SFD of planetesimals that underwent between 1 Ma and 5 Ma of collisional evolution is essentially similar between 0.5 and 10 au (see Figure 4 from [43]). It must be noted, however, that [43] considered a initial size of the planetesimals larger than that of [42], *i.e.*, 1 km *versus* 100–500 m respectively.

Because of the previous results, the SFDs by [41,42] were included in this work to investigate the cometary bombardment in the asteroid belt in a collisionally evolved disk of planetesimals. This is equivalent to assuming that the invariance of the SFD with distance, density of the planetesimals and surface density of the swarm observed by [42] between 1.5 and 4 au also holds between 4 au and 10 au. Based on the results of [43] this assumption appears to be reasonable, especially for the plausible temporal range of the formation of Jupiter (3–5 Ma) and size range of the planetesimals (1–10 km) that constitutes the bulk of the JEB based on the results of Paper III. As this assumption may nevertheless be not correct, the results that will be obtained using these SFDs should be regarded as an indication and should be considered with caution.

From the results of [43], however, we can extrapolate that the main effect (for the JEB) of the uncertainties on the SFD of the OSS planetesimals would be on the relative abundances of the impactors between 1 km and 10 km in diameters. This effect would be larger at about 1 Ma after the beginning of the planetary formation process, but would become significantly less important by about 5 Ma [43]. Depending on the actual value of the slope of the differential size distribution of the impactors [44], the uncertainty on the slope can result in either an increase or a decrease in the OSS mass flux on Vesta. Based on the results of Paper III when considering the SFDs from [39,41,42] for what it concerns the magnitude of the effects linked to the mass flux and the size of the impactors (e.g., the surface erosion), we can expect this uncertainty to change only quantitatively (by a factor of a few) but not qualitatively the results of this work.

2.2.1. Planetesimals Formed in a Quiescent Disk

The first SFD considered is that of a disk of planetesimals formed by gravitational instability of the dust in the mid-plane of a non turbulent protoplanetary nebula [25,45,46]. The protoplanetary nebula is assumed to have a mass M_{neb} = 0.02 $M_\odot$(where $M_\odot$ = 1.9891 × 10^{33} g is the solar mass) distributed between 1–40 au, dust-to-gas ratio ξ = 0.01 inside the Snow Line and ξ = 0.02 beyond the Snow Line, and density profile

$$\sigma = \sigma_0 \left(\frac{r}{1\,au}\right)^{-n_s}, \text{ where } \sigma_0 = 2700\text{g cm}^{-2}$$

where σ_0 = 2700g cm^{-2} is the surface density at 1 au, r is the heliocentric distance expressed in cm, the symbol 1 au indicates the value of the astronomical unit expressed in cm, and n_s = 1.5. The initial mass of solids contained in the region comprised between 2 au and 3 au (*i.e.*, the reference region considered also by [41,42]) is about 4 $M_\oplus$. For such a nebula [39] showed that the average mass of the planetesimals $\bar{m}_p$ would follow the semi-empirical relationship

$$\overline{m}_p = m_0 \left(\frac{r}{1\,au}\right)^{\beta} \tag{4}$$

where $\bar{m}_p$ and m_0 are expressed in g, and β = 1.68. The value m_0 is the average mass of planetesimals at 1 au, *i.e.*, 2 × 10^{17} g [39]. Paper I showed that, assuming that the mass dispersion of the planetesimals about the average values of Equation (4) is governed by a Maxwell-Boltzmann distribution, a mass value can be associated to each test particle by means of a Monte Carlo method where the uniform random variable *Y* varying in the range [0, 1] is

$$Y = \frac{2\gamma\left(3/2, y^*\right)}{\sqrt{\pi}} = P\left(3/2, y^*\right) \tag{5}$$

where $P\left(3/2, y^*\right)$ is the lower incomplete Gamma ratio. The inverse of the lower incomplete Gamma ratio can be computed numerically and, by substituting y^* back with $m^*/\bar{m}_p(r)$ we obtain

$$m(r) = \overline{m}_p\, inv\left(P\left(3/2, Y\right)\right) \tag{6}$$

Since the use of massless particles assures the linearity of the processes investigated over the number of considered bodies, the number of impacts expected in such a disk of planetesimals is extrapolated by multiplying the number of impacts recorded in the simulations of Paper I by a factor γ where

$$m^*/\overline{m}_p(r) \text{ we obtain} \tag{7}$$

where n_{mp} = 8 × 10^4 and N_{tot} is given by

where $n_{mp} = 8 \times 10^4$ and N_{tot} is given by

$$\begin{aligned} N_{tot} &= \int_{r_{min}}^{r_{max}} 2\pi r n^*(r) dr \\ &= \pi^{3/2} \frac{\xi\sigma_0}{m_0} (1\,au)^2 \left(\frac{1}{2 - n_s - \beta}\right) \times \left(\left(\frac{r_{max}}{1\,au}\right)^{2-n_s-\beta} - \left(\frac{r_{min}}{1\,au}\right)^{2-n_s-\beta}\right) \end{aligned} \tag{8}$$

where r_{min} = 2 au, r_{max} = 10 au. As in this study we are focusing on the role of OSS planetesimals, we will consider only the outer part of the protoplanetary disk, *i.e.*, r_{min} = 4 au and n_{mp} = 6 × 10^4.

2.2.2. Planetesimals Formed in a Turbulent Disk

The second SFD considered is that of planetesimals formed by concentration of dust particles in low vorticity regions in a turbulent protoplanetary nebula [47,48]. Following [40], the protoplanetary nebula is characterized by a surface density σ'_0= 3500 g cm^{-2} at 1 au, a nebular density profile with exponent $n'_s = -1$ and a dust-to-gas ratio $\xi' = 0.01$ beyond the Snow Line and $\xi' = 0.005$ inside the Snow Line (see Figure 14 from [40], gray dot-dashed line). Differently from Papers I and II and from [40], here the Snow Line will be placed at 4.0 au. As in the case of the SFD by [39], the initial mass of solids contained in the region comprised between 2 and 3 au is about 4 $M_\oplus$. The results of [40] supply the average diameter of planetesimals as a function of heliocentric distance (see Figure 14 from [40], gray dot-dashed line), from which Paper I derived the following semi-empirical relationship analogous to Equation (4):

$$\overline{m'}_p = \frac{\pi}{6}\rho D_0^3 \left(\frac{r}{1\,au}\right)^{3\beta'} \quad (9)$$

where $\beta' = 0.4935$ and $D_0 = 70$ km is the average diameter of the planetesimals at 1 au. By substituting the primed quantities to the original ones in Equations (6) and (8), the mass and the normalization factor for each massless particle can be obtained through the same approach described previously.

2.2.3. The "Asteroids were Born Big" Scenario

The third SFD considered is derived from the results of [41]. Morbidelli *et al.* [41] did not explore a specific model of planetesimal formation in quiescent or turbulent disks but instead tried to constrain the initial size-frequency distribution of planetesimals in the orbital region of the asteroid belt, assuming an initial mass of 1.6 $M_\oplus$ in the region comprised between 2 and 3 au. Their results suggest that the best match with the present-day SFD of the asteroid belt is obtained for planetesimal sizes initially spanning 100–1000 km (see Figure 8 from [41]), a range consistent with their formation in a turbulent nebula. Accretion and break-up of these primordial planetesimals extended the size distribution between 5–5000 km (see Figure 8a from [41], black solid line) in about 3 Ma, assumed by [41] to be the formation time of Jupiter respect to CAIs.

The number of planetesimals in the orbital range between 4 and 10 au was estimated for this SFD by assuming a protoplanetary nebula similar to that described for the SFD of [39] but with a surface density σ_0 at 1 au chosen so that the total mass of solids between 2 and 3 au is equal to 1.6 $M_\oplus$. The population of planetesimals supplied by [41] was then multiplied by the ratio between the mass contained in the 4–10 au orbital region to that of the 2–3 au orbital region. For each OSS impact event in the simulations of Paper I the mass of the impacting planetesimal was then estimated through a simple Monte Carlo extraction based on the cumulative probability distributions of the SFD supplied by [41]. The normalization factor was estimated through Equation (7) using the scaled number of planetesimals.

2.2.4. The "Asteroids were Born Small" Scenario

The final SFD considered is derived from the results of [42], who studied the accretion of primordial planetesimals in an annular region comprised between 1.5 and 4 au and containing 4.9 $M_{\oplus}$ (2 $M_{\oplus}$ in the region between 2 and 3 au). Using an approach analogous to the one also used by [41] but differing in the algorithms governing the computation of the collisional probabilities, [42] showed that a primordial SFD of the asteroid belt capable of reproducing the features of the present day SFD can be obtained also from disks initially populated by planetesimals as small as 50–200 m. Planetesimals of 500 m in diameter succeed only partially in producing a satisfactory SFD. This study focuses on the SFD of the asteroid belt that [42] refers to as the "standard case", *i.e.*, the one produced from a disk initially populated by planetesimals with a diameter of 100 m (see Figure 8 from [42]).

As in the case of the SFDs by [41], the number of planetesimals in the orbital range between 4 and 10 au was estimated for this SFD by assuming a protoplanetary nebula similar to that described for the SFD of [39] but with a surface density $\sigma_0$at 1 au chosen so that the total mass of solids between 2 and 3 au is equal to 2 $M_{\oplus}$. The mass of the impacting planetesimals was then estimated through a simple Monte Carlo extraction and the normalization factor γ was computed using the scaled number of planetesimals. In the following, when using this SFD only planetesimals whose size is greater than or equal to 1 km will be considered.

2.3. Collisional Model: Cratering, Erosion and Water Delivery

To estimate the fluxes of impactors on Vesta, Paper I opted for a statistical approach based on solving the ray–torus intersection problem between the orbital torus of Vesta and the linearized path of a massless particle across a time step. The method is similar to the analytical method developed by [49], but does not require averaging over orbital angles other than the mean anomaly. Interested readers are referred to Papers I and II for details on the algorithm.

We used the fluxes estimated in Paper I as the basis for an improved assessment of the collisional evolution of Vesta during the JEB. A set of 10^4 Monte Carlo simulations was run for each of the planetesimal SFDs described in Section 2.2. In each run a new mass value was extracted for each impact event recorded in the simulations of Paper I and, using the corresponding impact velocity, we computed the diameter and depth of the produced crater, the energy of the impact event, and the eroded mass. Averaging over each set of 10^4 Monte Carlo simulations, we then computed for each SFD of the primordial planetesimals the total eroded mass, the cumulative probability of Vesta undergoing catastrophic disruption, and the fraction of the Vestan surface affected by the impacts.

The diameter of the craters produced by the flux of impactors was estimated, as in [18], using the following scaling law for rocky targets by [50]:

$$\frac{R_c}{r_i} = 0.93\left(\frac{g\,r_i}{v_i^2}\right)^{-0.22}\left(\frac{\rho_i}{\rho_v}\right)^{0.31} + 0.93\left(\frac{Y_v}{\rho_v v_i^2}\right)^{-0.275}\left(\frac{\rho_i}{\rho_v}\right)^{0.4} \tag{10}$$

where R_c is the final radius of the crater, r_i is the radius of the impactor, g = 0.25 m s^{-2} is the surface gravity of Vesta, v_i is the impact velocity, Y_v = 7.6 MPa is the strength of the material composing the surface of Vesta (assumed to behave as soft rock, [51]), ρ_i = 1000 kg/m^3 is the average density of the OSS impactors, and ρ_v = 3090 kg/m^3 is the density of the basaltic surface of Vesta [32,52].

From the results of Equation (10) we then estimated the excavation depths of the crater populations in the same way as [18]. According to [53], the transition from simple to complex craters on Vesta seems to occur at diameters of about 30 km. For craters smaller than this value we assumed a constant depth-to-diameter ratio of 0.168, *i.e.*, equal to the average value measured on Vesta by the Dawn mission [53]. For larger craters we used the conservative relation from [54]

$$d_{exc} = 0.1\,D_t = 0.1\,(D/1.3) = 0.077\,D \tag{11}$$

where d_{exc} is the depth of excavation, D_t is the diameter of the transient crater and D is the final diameter of the crater. The factor 1.3 is used to scale the diameter of the transient crater to that of the final crater [51,54].

To assess how energetic the different impact events are and to which mass loss regime they are associated to, for each impact event we computed its specific kinetic energy per unit of the vestan mass Q_D and compared it with the catastrophic disruption threshold *QD** of Vesta*QD** is defined as the energy needed to destroy a target body, leaving a largest fragment with a mass equal to half the mass of the original body [55]. We evaluated the catastrophic disruption threshold *QD** of Vesta using Equation (6) from [55] and the coefficients for basaltic targets computed by these authors (see Table 3 from [55]). As in Papers II and III, the coefficients of the case v_i = 5 km s^{-1} were used for all impact events with a velocity greater than or equal to 5 km s^{-1}, and those of the v_i = 3 km s^{-1} were used for all the other impact events. For high-energy impacts (*0.1≤QD/QD*<1*), we computed the mass loss of Vesta using Equation (8) from [55] expressed in terms of the eroded mass m_e:

$$(0.1 \leq Q_D/Q_D^* < 1), \tag{12}$$

where m_t is the mass of the considered target body, s = 0.5 for v_i < 5 km s^{-1} and s = 0.35 for $v_i \geq$ 5 km s^{-1}. The effects of catastrophic impacts

$$(Q_D/Q_D^* \geq 1)$$

were not directly accounted for in the estimates of the eroded mass, as they would always result (by definition) in the loss of half the mass of Vesta. To properly treat them and include them in the collisional model, we would need to simulate the following dynamical evolution of the fragments and to estimate their re-accretion efficiency. The cumulative number of catastrophic impacts was used instead to assess the probability of Vesta surviving the JEB intact and to reject those scenarios where Vesta has a large chance of undergoing such events. Finally, the mass loss associated to cratering erosion caused by low-energy impacts (*QD/QD*<0.1*) was calculated based on the results of the hydrodynamic simulations we will now detail.

In order to determine the amount of water and volatile elements delivered to Vesta by OSS impactors, we carried out a set of numerical simulations of the impacts of icy planetesimals on the asteroid. Numerical simulations of impacts on Vesta, treated as a planar target, have been previously performed by [29], using a hydrodynamic approach and assuming that both the projectile and the target consisted of dunite with a density of 3.32 g/cm^3.

With respect to [29], in this study we made some modifications and improvements of the numerical model. As in [29], we used the numerical hydrodynamic method SOVA ([56], SOVA is an acronym for Solid-Vapor-Air, as the code is designed for simulations of multi-material, multi-phase flows), but we adapted the hydrodynamic equations and the method to a 3D spherical system of coordinates. This choice seems to be more convenient to study impacts that are not as large as the high-energy impacts that can shatter the target but at the same time are not so small that the sphericity of the target could be neglected. The gravitational field was assumed central and constant in time. A similar approach has been developed in 2D spherical geometry for simulations of large vertical impacts on the Earth [57]. Furthermore, we added the effects of dry friction as described in [58]. The equations with friction are similar to the Navier-Stokes equations and dry friction does not violate the hydrodynamic similarity: the results, however, can depend on the dimensionless coefficient of friction. For the latter we adopted a value of 0.7, which is typical for rocks and sand. The numerical grid consisted of 250 × 100 × 225 cells over azimuth, polar angle and the radial distance respectively. We assumed bilateral symmetry, which allowed us to model only the half-space in the zenith direction (we assumed that the impact velocity vector lied in the reference plane that passes through the origin of coordinates and is orthogonal to the zenith). Cell sizes were 1/40 of the projectile's diameter around the impact point and increased to the antipodal point and to the radial boundaries located at distances of about 10 vestan radii. The minimum cell sizes varied and became larger when the crater grew.

Vesta was assumed to be in equilibrium and at a low temperature. Its radius was assumed equal to 260 km, the core radius equal to 110 km [32,52], the thickness of the crust equal to 23 km, and the mass of Vesta equal to 2.59×10^{23} g. We used the available ANEOS equations of state [59] for the mantle and the crust with input data from [60] and Tillotson's equation of state [61] for the iron core. ANEOS is an acronym for Analytic Equations of State: however, ANEOS does not have a simple analytic form but is instead a code with a large number of input parameters, the choice of which for a definite material is a difficult task. We assumed that the mantle material consists of dunite and the crust consists of granite. Usually just these materials are used in simulations of impacts because they are the only ones with proper equations of state that fit for the mantle and the crust. However, we should elucidate that various equations of state appropriately describe the state of materials compressed by the shock wave. The difficulties arise if one wants to obtain the amount of melted or vaporized material after the release phase. The impact velocities on Vesta are fairly low and the mass of escaped material, which is determined by the flow at rather low velocities, can be calculated with a reasonable accuracy for the chosen equations of state. We assumed that the icy planetesimals consisted of water and had a spherical shape. The diameter of impacting bodies in the simulations was set to 1 km: hydrodynamic similarity, however, holds valid when the diameters of the projectiles are much smaller than the diameter of Vesta and smaller than the thickness of the vestan crust. Therefore, the results of our simulations are valid for diameters of the impactors up to about 10–20 km. The impact velocities ranged from 1 to 10 km/s and the impact angle was taken to be 45°, which is the most probable impact angle [54].

During the impacts some portion of material ejected from the craters gained velocities higher than the escape velocity of Vesta (0.35 km/s). Obviously, the mass of the escaped material grows with the impact velocity. The retained material of the impacting planetesimals concentrates mainly within and around the impact craters, but small portions of the retained material are ejected and fall over the whole surface of Vesta as shown in Figure 1 and Figure 2. The relative amounts of retained mass of the impactors and eroded mass of the target are reported in Table 1 and shown in Figure 3. The simulations of the impacts of icy bodies have been made both with and without friction. Friction consolidates the projectile's material and slightly accelerates the target's material. For probable impact velocities of 4 km/s and 8 km/s (see Paper I) the fraction of retained icy material is 0.25 and 0.03 of the projectile's mass. If friction is neglected, these retained masses become 0.42 and 0.13 respectively. We used the values tabulated in Table 1 and shown in Figure 3 also to estimate the erosion caused by the (low-energy) impacts of icy planetesimals. To do this, for each impact event recorded in the simulation of Paper I we derived the erosion efficiency associated to its impact velocity by linearly interpolating between the values tabulated in Table 1.

Table 1. Erosion and retention efficiencies of Vesta for cometary impacts, expressed in units of the mass of the impactors. The erosion and retention factors for the different impact velocities are computed both including and neglecting the effects of dry friction.

Impact Velocity (km/s)	Erosion (with friction)	Erosion (without friction)	Retention (with friction)	Retention (without friction)
1	0.19	0.23	0.45	0.54
2	0.49	0.82	0.33	0.51
4	3.80	2.84	0.25	0.42
6	6.50	6.30	0.09	0.28
8	11.5	9.70	0.03	0.13
10	15.0	13.3	0.03	0.09

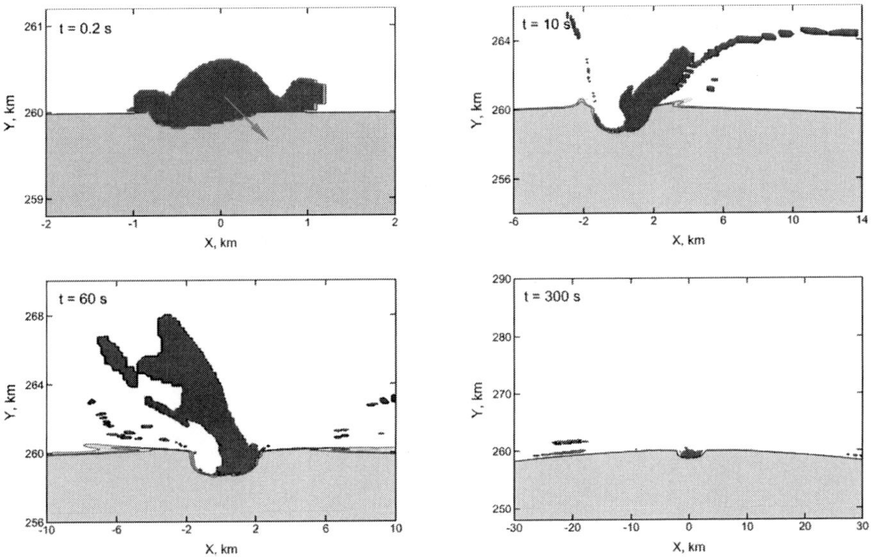

Figure 1. Formation of the crater and evolution of the projectile's material after the impact on Vesta of a OSS planetesimal with a velocity 4 km/s and diameter of 1 km. Isolines of density are shown in the plane XY passing through the vector of impact velocity and the center of Vesta. The crustal material of Vesta with density above 0.5 g/cm^3 is shown by the grey color. The material of the impactor is indicated in blue. The red arrow in the first panel shows the direction of the impact (45° respect to the local vertical). Note the different scales on the axes in the four panels.

Before proceeding, it must be noted that the ultimate fate of the retained water can depend on several processes and on their balance. First, water deposited in the crater can penetrate into the brecciated floor and be buried under the surface. Second, subsequent impacts can remove the water from the surface, vaporizing it (as suggested by [8] as a possible origin of the pitted terrains on Vesta) or ejecting it with velocities higher than the vestan escape velocity. Third, material accreted by Vesta with low collision speeds as well as low-velocity ejecta from the craters of subsequent impacts can cover the retained water that remained on the surface and mix it with the target material. Forth, water lying or exposed on the surface will evaporate. Finally, in the temporal interval we are considering impacts can penetrate the solid crust of Vesta and deliver the water to the molten interior of the asteroid. Alternatively, impact-triggered effusive phenomena can cause the incorporation of part of the water on the surface into the solidifying magma. In this study we will limit ourselves to the estimation of the maximum amount of water delivered to Vesta, neglecting the loss processes previously discussed.

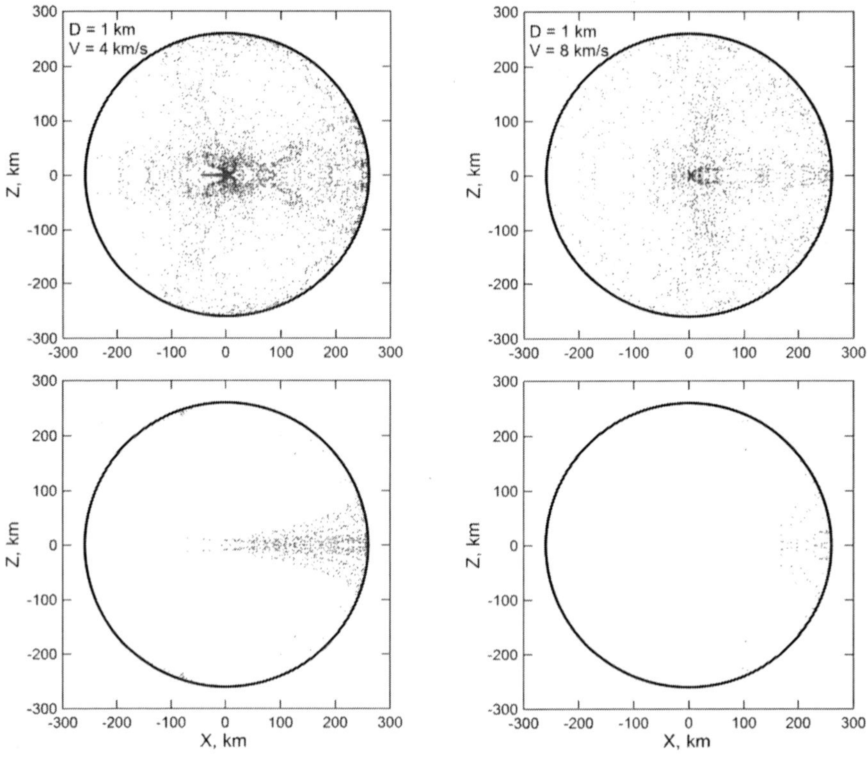

Figure 2. Distributions of the material of the impactors over the vestan surface after the impacts of OSS planetesimals with diameter of 1 km at 4 km/s (**left** plots) and at 8 km/s (**right** plots). The upper panels show the hemisphere where the impacts occurred, with the center of the craters in the middle of the hemisphere. The lower panels show instead the opposite hemisphere. The impacting bodies strike Vesta along the positive direction of the X axis at 45° respect to the local vertical.

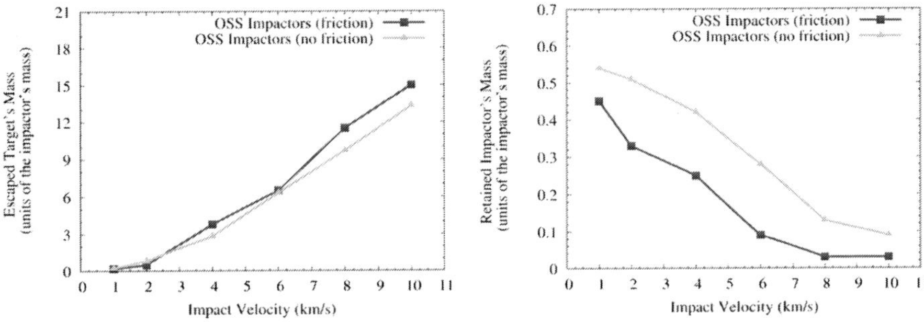

Figure 3. Erosion (**left** plot) and accretion (**right** plot) efficiencies of cometary impactors on Vesta, expressed in units of the mass of the impactors. The plotted values are from Table 1.

3. RESULTS

To investigate the delivery of volatile elements to Vesta by the JEB, we will first characterize the flux of OSS impactors on the asteroid based on the results of Paper I. We will estimate the possibility of Vesta undergoing catastrophic impacts and we will quantify the erosion of its basaltic surface, to rule out implausible scenarios. We will then assess the amount of volatile elements that OSS impactors would deliver to the asteroid and, finally, the excavation of the surface due to the impacts, to assess whether the delivered volatile materials would be confined on the surface of the asteroid or could penetrate the solid crust and mix with the molten material.

3.1. Characterization of the Jovian Early Bombardment

The fluxes of OSS impactors hitting Vesta in the four migration scenarios we considered are quite different, as discussed in Paper I. The first difference is due to the distribution of the orbital elements of the OSS planetesimals in the semimajor axis–eccentricity (a–e) plane, as shown in Figure 4, and divides the scenarios with limited (0.25 au) or no migration of Jupiter from the scenarios with a more significant displacement of the giant planet (0.5 au and 1 au). The different values of the orbital eccentricity of the impactors on Vesta between the migration scenarios (see Figure 4 and Figure 1 of Paper I) translate into a different distribution of their impact velocities, as shown in Figure 5. In the scenarios with limited or no migration the impact velocities range from 2 km/s to 12 km/s, while in the scenarios with larger displacements of Jupiter the impact velocities mostly range between 3 km/s and 7 km/s. As reported already in Paper I, in the scenario where Jupiter migrates by 1 au about 2% of the impacts occur at ~40 km/s. For these impacts we considered that no fragment of the impactor survived on Vesta.

The second difference is in the number of OSS impactors on Vesta. As discussed in Paper I and here shown in Figure 6, the scenario with no migration of Jupiter is the one with the highest flux of OSS impactors hitting the asteroid, the flux being 1–2 orders of magnitude higher than in the other cases. The number of OSS impactors on Vesta in the 0.25 au and 0.50 au migration scenarios decreases with the increase of the Jovian displacement, *i.e.*, the larger the migration the lower the flux. The flux of OSS impactors then increases in the case of Jupiter migrating by 1 au: even if it still is 1 order of magnitude lower than in the case of no migration, the flux is the second highest among the scenarios we considered. Note that, because of this trend in the number of OSS impactors, the SFD by [40] can produce (cumulatively) 1 impact only in the case where Jupiter does not migrate.

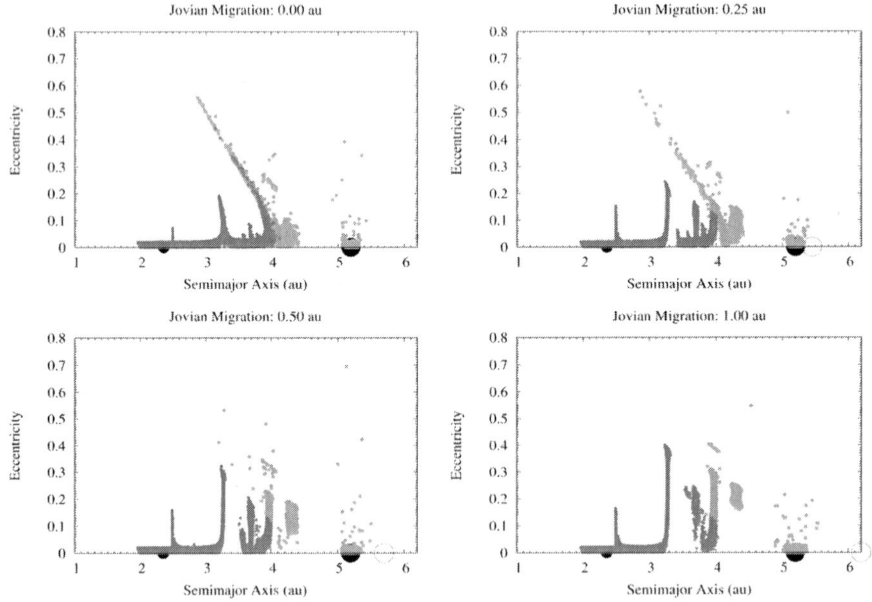

Figure 4. The orbital structure of the protoplanetary disk 2×10^5 years after Jupiter started to accrete the nebular gas (*i.e.*, 1.2×10^6 years from the beginning of the simulations). The black symbols at 2.36 au and at 5.2 au indicate respectively Vesta and Jupiter. The open circles in the cases of 0.25 au, 0.50 au and 1.00 au migration indicate the initial position of Jupiter. ISS impactors are indicated in red, OSS impactors in blue.

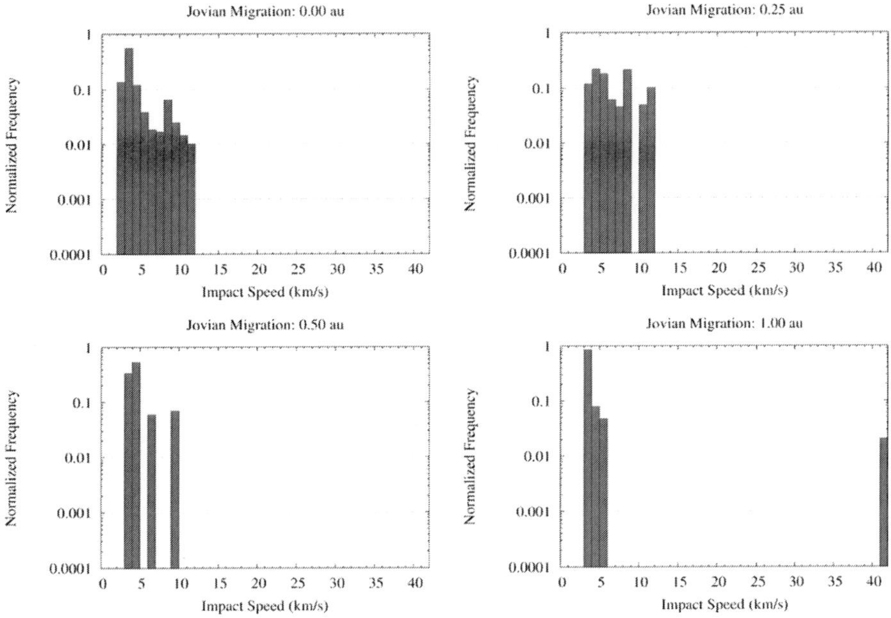

Figure 5. Normalized distribution of the impact velocities of OSS impactors reported in Paper I in the four migration scenarios. The distribution of the impact velocities is common to all SFDs of the impactors.

As can be seen in Figure 6, the OSS impactors are planetesimals whose diameter is about 1–2 km in the case of the SFD by [42], about 5–10 km in the case of the SFD by [41], and about 10–30 km in the case of the SFD by [39]. In the case of the SFD by [40], the OSS impactors that could hit Vesta would range in diameter between 100 km and 200 km but, as we discussed previously, this SFD can produce at most 1 impact on Vesta only in the case where Jupiter did not migrate.

3.2. Catastrophic and High-Energy Impacts

As shown in Table 2 and in agreement with the results of Paper I, the chances of Vesta being destroyed by a catastrophic impact during the JEB are always lower than 0.1%, mostly due to the limited duration of the JEB. The same holds true for high-energy impacts ($0.1QD^* \leq QD < QD^*$) in the cases of the SFDs by [39,41,42], due to the relatively small size of the impactors (as discussed in Section 3.1 and shown in Figure 6). In the case of the SFD by [40], however, in the scenario where Jupiter does not migrate there is a fairly large chance (~13%) of Vesta undergoing a high-energy impact, which would have major consequences for the survival of its crust.

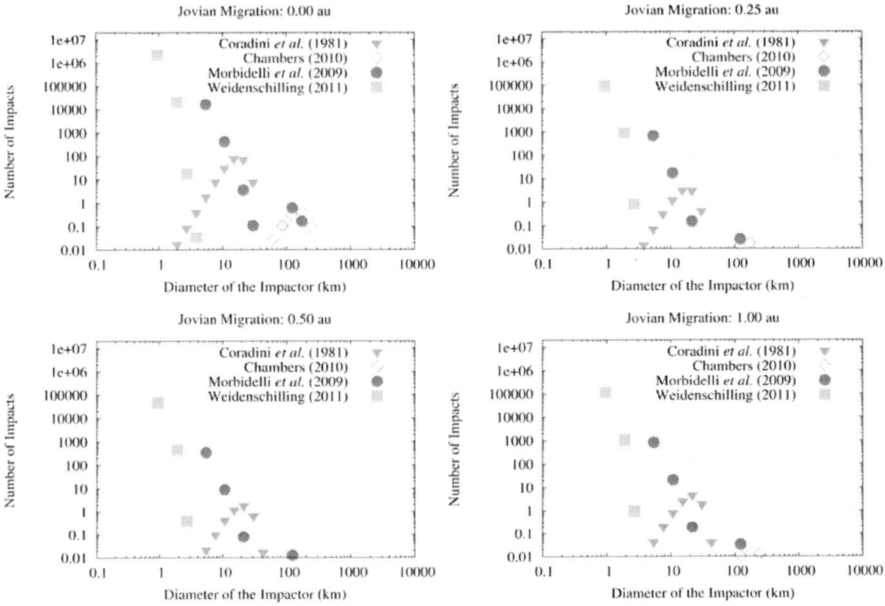

Figure 6. Size-frequency distributions of the OSS impactors on Vesta in the four migration scenarios for the different SFDs of the primordial asteroids considered in this work. Numbers of impacts lower than 1 indicate stochastic events (even when considered cumulatively): these cases are not considered in the following analysis.

Table 2. Number of catastrophic impacts on Vesta across the JEB in the different migration scenarios and for the different SFDs considered.

Migration (au)	Coradini *et al.* (1981)	Chambers (2010)	Morbidelli *et al.* (2009)	Weidenschilling (2011)
0.00	0	$<10^{-3}$	$<10^{-3}$	$<10^{-3}$
0.25	0	$<10^{-3}$	$<10^{-3}$	$<10^{-3}$
0.50	0	$<10^{-3}$	$<10^{-3}$	0
1.00	0	$<10^{-3}$	$<10^{-3}$	0

3.3. Erosion of Vesta due to OSS Impactors

Now that we have verified that Vesta would survive the flux of OSS impactors triggered by the JEB without being shattered or destroyed, we need to assess the degree of erosion that these impactors would cause to its surface and test whether any of our scenarios is inconsistent with the survival of the basaltic crust of Vesta. In Figure 7 we show the mass loss of Vesta due to cratering erosion: as we expected following our discussion in Section 3.2, the case of SFD by [40] and no migration of Jupiter results in a significant mass loss of the asteroid (~4%), about twice as large as the one caused by ISS impactors as estimated in Paper III.

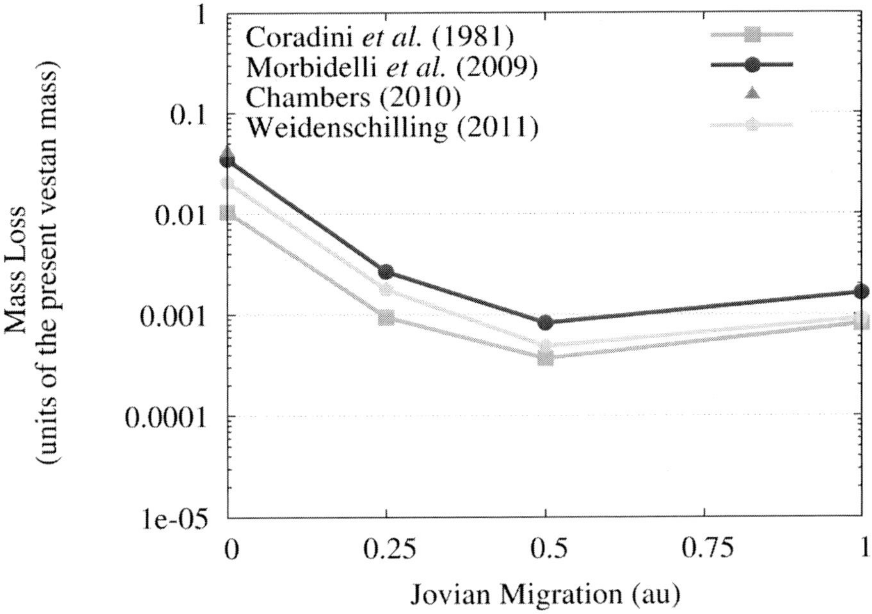

Figure 7. Mass loss of Vesta in the different migration scenarios and for the different SFDs considered. The mass loss is expressed in units of the present Vestan mass ([32,52]).

Also in the case of the SFDs by [39] the erosion caused by OSS impactors in the no migration scenario is twice as large as that caused by ISS impactors and estimated in Paper III, reaching about 1% of the present vestan mass. In the cases of the SFDs by [41,42], the erosion caused by the OSS impactors is respectively about six and twenty times larger than the one caused by ISS impactors in Paper III. The eroded mass amounts to about 3% of the present vestan mass for the SFD by [41] and to about 2% of the present vestan mass for the SFD by [42]. In all the other migration scenarios, the cratering erosion caused by OSS impactors is significantly lower than the one due to ISS impactors and never rises above 0.3% of the present vestan mass. As a consequence, the results of Paper III are not affected by the OSS planetesimals in those scenarios where Jupiter migrates.

To better understand the effects of the OSS impactors during the JEB for the survival of the basaltic crust of Vesta, we converted the mass loss values shown in Figure 7 to the thickness ΔR of the shell, extending outward from the present surface of Vesta, whose mass matches the mass ΔM lost by the asteroid, *i.e.*,

$$\Delta R = \left(R_V^3 + \frac{3\Delta M}{4\pi\rho_V}\right)^{1/3} - R_V \qquad (13)$$

where $R_V = 262.7$ km and $\rho_V = 3090$ kg m^{-3} are respectively the mean radius and the mean crustal density of Vesta measured by the Dawn mission ([32,52]).

The values of the thickness of the eroded crust we obtained are shown in Figure 8, where we considered both cases of erosion discussed in Paper III. The first one is the case of uniform erosion, which assumes that impacts are distributed isotropically on the surface of Vesta. The second case instead assumes, based on the compact vertical extension of the protoplanetary disk discussed in Section 2.1, that impacts on Vesta takes place mostly on the ecliptic plane and therefore they are distributed as a function of the cross-sectional area of the different regions of the asteroid. As a consequence, the region comprised between 45° and −45° of latitude should receive about 70% of the impacts and erode more than the regions between −45° and −90° and between 45° and 90°. As discussed in Paper III, this case results in an erosion of this "equatorial" belt twice as high as the one we obtain in the isotropic case.

As we can see from Figure 8, the erosion of the vestan crust in the isotropic case is of about 100–300 m or less for all the SFDs in all the scenarios where Jupiter migrates while accreting the gas. In the scenario of no migration of Jupiter, the surface erosion of Vesta rises instead to 1–2 km (see Figure 8) for the SFDs from [39,42] and to 3 km for the SFDs from [40,41]. Once we sum the erosion due to OSS impactors to the one we estimated for ISS impactors in Paper III, we obtain the erosion pattern discussed in Papers I and II. The erosion of Vesta decreases when we move from the scenario of no migration to that of limited (0.25 au) migration due to the drop in the flux of OSS impactors (characterized by higher velocities), which is not compensated by the increase in the ISS impactors (characterized by lower velocities). However, if Jupiter undergoes more extensive migration (0.50–1.00 au) the new ISS impactors due to the 2:1 resonance cause the erosion of Vesta to grow again, reaching values higher than those due to the ISS and OSS impactors in the no migration scenario.

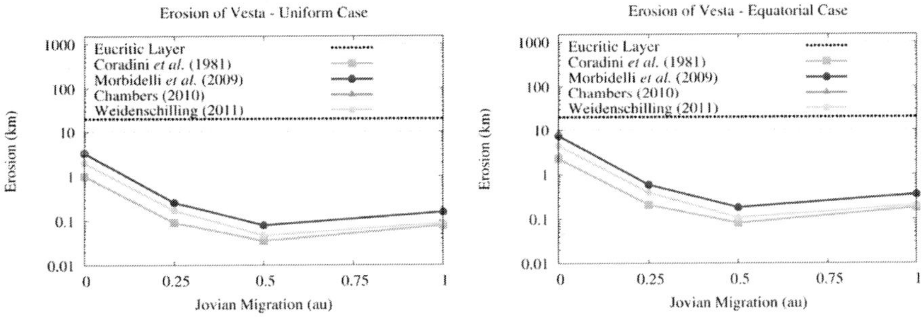

Figure 8. Surface erosion of Vesta in the different migration scenarios and for the different SFDs considered. Surface erosion is expressed as the thickness of a shell, with density equal to the present crustal one of Vesta and mass equal to the eroded mass, extending from the present radius of Vesta outwards. The plot on the left shows the case of a uniform erosion, the plot on the right shows instead the case of erosion proportional to the cross-sectional area of Vesta. The dotted line indicates the thickness of the eucritic layer according to [12].

Finally, in order to assess whether erosion would be a global effect on Vesta or not, we built the R-plot of the crater populations produced by the OSS impactors for the different SFDs of the planetesimals and the four migration scenarios. Following [54], we took as the smallest crater diameter the value of 1 km and we divided the crater population associated to each SFD in bins where *Di+1=2Di* and the central diameter is the geometric mean *D¯i=DiDi+1*. We computed the R-value as

$$\overline{D_i} = \sqrt{D_i D_{i+1}}. \tag{14}$$

where f_i is the fraction of the vestan surface covered by the craters in the relevant bin. The value of f_i of each bin is obtained simply by summing the surface areas A_j covered by the N_i craters in the bin and dividing it by the surface of Vesta S_V :

$$f_i = \left(\Sigma_{j=0}^{N_i} A_j\right) / S_V \tag{15}$$

where A_j is the geometrical area of each crater of diameter D_j:

$$A_j = \frac{\pi}{4} D_j^2 \tag{16}$$

Note that, when the condition $D_{i+1} = \sqrt{2} D_i$ is satisfied and the average diameter of each bin is computed as the geometric mean $\overline{D_i} = \sqrt{D_i D_{i+1}}$, this definition of the R-value is equivalent to the one given by [62], *i.e.*,

$$R_i = \frac{N_i \overline{D_i}^3}{S\left(D_{i+1} - D_i\right)} \tag{17}$$

where N_i is the number of craters in the relevant bin and S is the surface area whose population of craters is under study, which in our case is the whole surface of Vesta ($S = S_V$). In building the R-value distributions, we considered only the effects of low-energy impacts ($(Q < 0.1\, Q_D^*)$): high-energy and catastrophic impacts were not included. Moreover, we considered only those bins where the cumulative impact probability produced at least 1 impact once normalized to the real population of the disk. As a consequence, we did not consider the SFD by [40].

The R-plots for the SFDs by [39,41,42] are shown in Figure 9, where we also showed the R-value that would be associated to a 5% and a 13% saturation of the surface of the asteroid. These two threshold levels represent respectively the minimum R-value for which a crater population can reach equilibrium [54] and the R-value estimated for Mimas, whose surface is the most densely cratered in the Solar System [54]. As a reference, in Figure 9 we also showed the R-values of the crater population produced on Vesta by asteroidal impactors over the last 4 Ga estimated by [11].

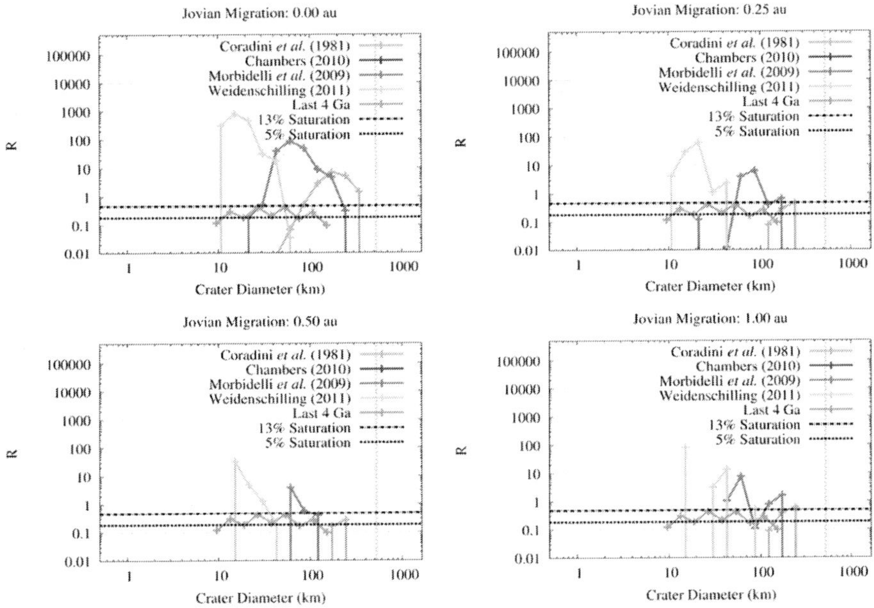

Figure 9. R-plots of the crater populations produced by the different SFDs of the OSS planetesimals in the four migration scenarios. The 5% and 13% saturation levels and the crater population caused by the last 4 Ga of collisional evolution of Vesta are also shown for reference. The light blue vertical dashed line indicates the present diameter of Vesta.

The crater population produced by the SFD from [42] mainly ranges in diameter between 10 km and 40 km. The impacts occurring at about 40 km/s results in craters with diameters of 90 km. The bulk of the crater population associated to the SFD from [41] ranges in diameter between 60 km and 200 km, with a tail down to 20 km in diameter in the no migration scenario. Finally, the crater population due to the SFD by [39] ranges between 40 km and 300 km in the scenario where Jupiter does not migrate, while in the scenarios where migration occurs the crater population concentrates between 100 km and 200 km. As can be easily seen, even if the flux of OSS impactors is more limited with respect to the one of the ISS impactors, it is still large enough to saturate the surface of Vesta at least to a 10% level. The cratering and the erosion produced by the OSS impactors therefore appear to globally affect the surface of Vesta.

3.4. Delivery of Volatile Materials to Vesta

Now that we concluded the assessment of the implications of the flux of OSS impactors for the survival of Vesta and of its basaltic crust, we can move to investigating the delivery of water and volatile materials to the asteroid by the JEB. Before proceeding it is important to note that, while water realistically represented the majority of the delivered material, a varying fraction of the mass of the impacting OSS planetesimals was in the form of other ices (mainly carbon monoxide and carbon dioxide, see [63]) and trace compounds (e.g., ammonia and methane, see [63]). However, the available observational constraints are based on the detected emissions of present day comets [63], which are not necessarily representative of the primordial planetesimals considered in this study or even of the bulk composition of the very same comets. Moreover, the abundances of the different volatile materials vary significantly from one comet to the other [63], introducing another large source of uncertainty. Based on the observational evidences, water plausibly represented at least 50% of the material delivered by OSS impactors, so we can constrain the uncertainty on the water delivery to about a factor of two.

The absolute values of the retained masses of volatile materials that we estimated with our model in the four migration scenarios and for the four SFDs of the planetesimals are reported in Table 3. As we can see, the amounts of water delivered to Vesta vary between a few 10^{18} g and a few 10^{20} g and, for a given migration scenario, the retained amount of water varies by a factor of 2–3 between the lowest and the highest value. As a reference, the content of hydrogen and hydrated materials measured by the Gamma Ray and Neutron Detector (GRaND) on-board the Dawn spacecraft in the topmost 1 m of the vestan regolith [2] and suggested to be due to the cumulative flux of impactors on Vesta over the last 4 Ga [11] is about 2.7×10^{15} g. In the case of the SFD by [40] the single value reported refers to the case of no migration of Jupiter, as in the other cases the cumulative impact probability of all OSS impactors is not enough to produce even a single impact. It must be noted, however, that the single impact associated to this SFD would produce major effects on the surface of Vesta and would likely destroy the basaltic crust, if it was already formed.

In order to better illustrate the effects of the JEB in terms of the delivery of water and volatile materials to Vesta, in Figure 10 we show the quantities of Table 3 expressed in units of the present mass of Vesta. If we compare these enrichment factors with the fraction of the Earth's mass represented by water, *i.e.*, 5×10^{-4} [64], we can immediately see that the scenario of no migration of Jupiter could result in a primordial Vesta as enriched in water as the Earth. In all the other migration scenarios, Vesta would be about one order of magnitude less enriched. As we mentioned in Section 2.3 not all this volatile material delivered to Vesta will survive to the subsequent collisional evolution of the asteroid. In particular, as discussed in Paper III the flux of ISS impactors caused by the JEB is intense enough to strip Vesta of the equivalent of at least a few hundreds meters (if not a few kilometres) of surface material. It is therefore plausible that the fraction of volatile elements delivered during the JEB that remained nearer to the surface of the asteroid was removed by the JEB itself through the more abundant ISS impactors. However, if the impacts of the OSS planetesimals can excavate the solid crust of Vesta, part of the volatile material they deliver could be trapped into the molten interior and be incorporated into the crystallizing minerals as in the samples studied by [4]. As discussed by [14,15], in fact, depending on the accretion time and the initial porosity of Vesta the thickness of its solid crust across the JEB could have ranged from a minimum of 7–10 km to a maximum of 20–30 km.

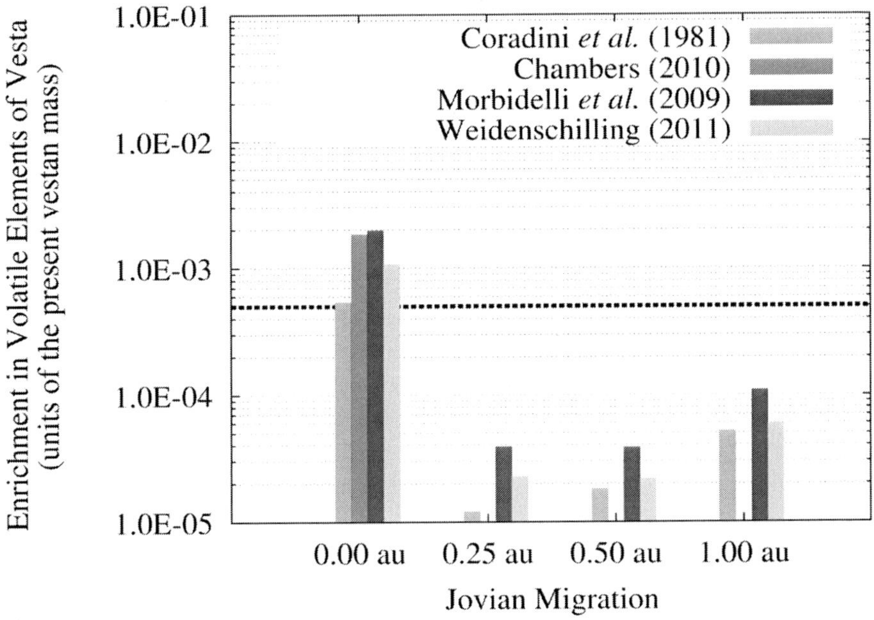

Figure 10. Amounts of volatile materials (expressed in units of the present vestan mass, [32]) delivered by the OSS impactors to Vesta in the four migration scenarios and for the four SFDs we considered. Also indicated, as the horizontal dotted line, is the fraction of the Earth's mass represented by water (5×10^{-4}, [64]).

Table 3. Amounts of volatile materials delivered by OSS impactors to Vesta in the four migration scenarios and for the four SFDs we considered. As a reference value, to produce the same abundance of water of the Earth [64], OSS impactors should deliver 1.3×10^{20} g.

Migration (au)	Coradini *et al.* (1981)	Chambers (2010)	Morbidelli *et al.* (2009)	Weidenschilling (2011)
0.00	1.4×10^{20} g	4.7×10^{20} g	5.1×10^{20} g	2.8×10^{20} g
0.25	3.0×10^{18} g	-	1.1×10^{19} g	5.9×10^{18} g
0.50	4.7×10^{18} g	-	1.0×10^{19} g	5.5×10^{18} g
1.00	1.3×10^{19} g	-	2.8×10^{19} g	1.5×10^{19} g

In order to assess whether the OSS impactors on Vesta would be capable of penetrating the solid crust and reach the molten interior, we estimated the excavation depths of the crater populations we discussed previously. Using the results of Figure 9, we expressed the excavation of the vestan surface in the form of an R-plot of the crater depths across the JEB, as shown in Figure 11. As we can see from Figure 11, the crater populations produced by the OSS planetesimals characterized by the SFDs from [39,41] always saturate the surface of Vesta at least to a 5% level with craters that can excavate at least 7 km (the minimum thickness estimated by [14,15]). The same is not true for the SFD from [42]: in all migration scenarios but one, the crater population excavate between 2 km and 4 km. In the scenario where Jupiter migrates by 1 au, the SFD from [42] can saturate the surface of Vesta to a 5% level with craters that can excavate about 6 km. However, it must be noted that this SFD is characterized by the highest number of impact events among all the SFDs we considered, due to the high abundance of 1–2 km wide planetesimals. Specifically, the number of impact events is high enough to cover the surface of Vesta several times with craters capable to excavate 2–4 km of the vestan crust. Even in the migration scenario characterized by the lowest flux of OSS impactors on Vesta, *i.e.*, the one where Jupiter migrates by 0.5 au, the impacts events are abundant enough to cover the surface of Vesta seven times with craters excavating about 2 km, plus about twice the vestan surface with craters that can reach down to 3–4 km. Moreover, we need to take into account the effects of the contemporary flux of ISS impactors. As we showed in Paper III, even if the bulk of the ISS impactors is characterized by lower impact velocities than OSS impactors, they can also saturate (or cover several times) the surface of Vesta with craters capable of excavating between 2 km and 4 km. When all these effects are taken into account together with the fact that impacts affected the vestan crust not only by excavating it but also by creating fractures and uncompensated negative gravity anomalies [17], it is not implausible that a fraction of the volatile materials delivered by the OSS impactors could reach the molten interior of Vesta or be incorporated into the solidifying magma brought to the surface by impact-triggered effusive phenomena.

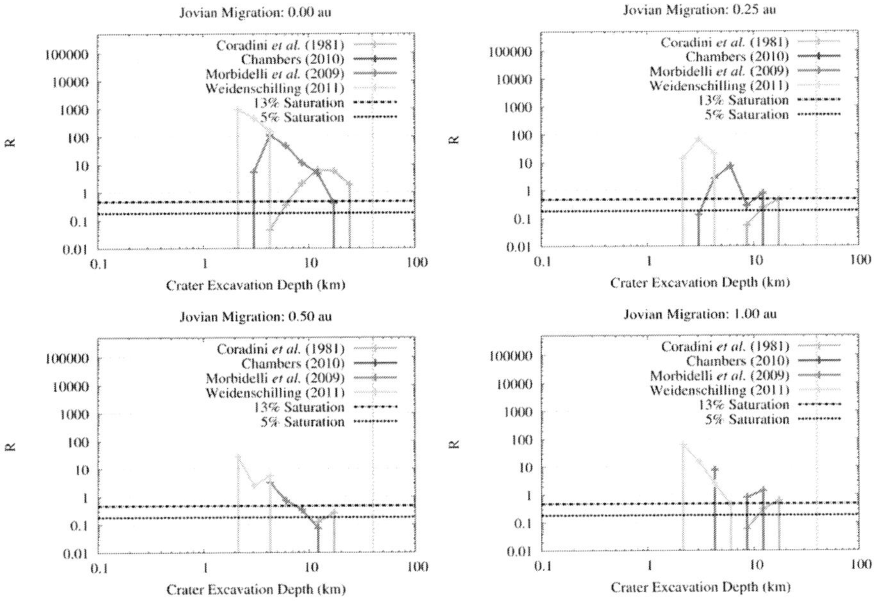

Figure 11. R-plot of the excavation depth of craters on Vesta across the JEB for the different SFDs and migration scenarios considered in this work. The 5% and 13% saturation levels are shown for reference. The vertical light blue dashed line indicates the thickness of the crust of Vesta (*i.e.*, the eucritic and diogenitic layers) as estimated by [12].

4. DISCUSSION AND CONCLUSIONS

In this work we explored the delivery of water and volatile materials to Vesta during the JEB, to assess whether the flux of OSS impactors triggered by the formation of the giant planet could be responsible for the presence of water at the time of the crystallization of some of the eucritic samples studied by [4]. In order for this to be true, one of the following two conditions needs to be fulfilled:

(1) The water is delivered to the vestan surface and is incorporated into eucritic magma brought on the surface by impact-triggered effusive phenomena;

(2) The water is delivered directly into the molten interior of Vesta by impactors capable of excavating the solid crust of the asteroid.

These two conditions are not mutually exclusive and they could have both occurred on Vesta during the JEB. However, while the first condition is plausibly easier to fulfill thanks to the combined effects of the ISS and OSS impactors, the interpretation of the meteoritic data by [4] seems to favor the second condition, at least for the samples investigated to date. An additional constraint provided by the HED meteorites is that, whatever mechanism delivered the water to the asteroid, it should have influenced the eucritic magma only locally and should have preserved the global volatile-depleted nature of Vesta [4].

Before we can compare the results of this study with the picture supplied by the meteoritic data, however, we need to put them into the context of the results obtained in Paper III when studying the erosive effects on Vesta of the ISS impactors caused by the JEB. According to Paper III, the SFD from [40] is (barely) consistent with the survival of the basaltic crust of Vesta and with the observational data supplied by the Dawn mission only if Jupiter did not migrate or migrated by 0.25 au or less. However, if Jupiter did not migrate the OSS impactors would cause a large-scale impact across the JEB, with major implications for the composition of the surface of the asteroid. If instead Jupiter migrated by about 0.25 au, the flux of OSS impactors would be too low to cumulatively produce one impact: the delivery of water to Vesta would therefore be due to stochastic, low-probability events.

Also in the case of the SFD by [39] the results of Paper III would indicate a limited (0.25 au) migration of Jupiter as the scenario most compatible with the observational data. The results of this study mostly confirm the compatibility of the scenario, as the erosion caused by the OSS impactors would be limited. The delivery of water would be due to a few impact events that would be capable of penetrating the solid crust even if its thickness was of the order of the highest values reported by [14,15]. However, these impacts would create 100–200 km wide craters and could in principle bring to the surface significant quantities of diogenitic material, which is currently not observed outside the Rheasilvia basin at the south pole of Vesta [1]. Dedicated impact simulations and geophysical studies, aiming to assess the interior state of Vesta at the time of the JEB and how it affects the outcome of the impacts, are needed before it is possible to assess whether this SFD is compatible with the observational data or not. As an example, the mechanical behavior of a solid crust of limited thickness floating on top of a molten mantle (e.g., the crust can break and/or sink more easily or, conversely, can be eroded less efficiently) can affect the delivery efficiency of water to the interior of Vesta. Alternatively, the existence of an undifferentiated primordial crust like the one suggested by the results of [14,15] can prevent the diogenitic material from being brought to the surface even by such large impact events.

According to Paper III, the most compatible SFDs are those from [41,42]. The former is compatible with the observational data on the composition of the basaltic surface of Vesta for Jupiter not migrating or migrating up to 0.25 au. However, in the scenario where Jupiter did not migrate, the SFD from [41] would be associated to a significant erosion of the vestan surface and it would be compatible with the observational data only if the asteroid possessed an undifferentiated primordial crust as suggested by [14]. If Jupiter did not migrate, moreover, the results of this study indicate that the flux of OSS impactors would be large enough to saturate the surface of Vesta to at least a 50% level with craters 100–200 km wide and capable to excavate more than 10 km in the solid crust. Given the larger flux of OSS impactors associated to the no migration scenario, Vesta could receive enough water and volatile elements to become as enriched in water as the Earth, even when we account for the factor of two uncertainty in the water delivery discussed in Section 3.4. Unless the loss process we discussed above significantly reduced the budget of volatile elements of the asteroids, this scenario would plausibly result in a more frequent signature for the presence of water at crystallization in eucrites. If the presence of water was a local phenomenon as discussed by [4], then also for the SFD from [41] the most plausible scenario would appear the one where Jupiter migrated by about 0.25 au.

Finally, Paper III argued that the SFD most compatible with the observational data is the one from [42]. If Vesta already formed its basaltic crust at the time of the JEB, Paper III indicates as the most plausible scenarios those where Jupiter migrated by 0.25 au or less. If instead Vesta still possessed an undifferentiated crust as discussed by [14,15], Paper III favors a Jovian migration between 0.50 au and 1.00 au. In all the cases where Jupiter migrates the erosion caused by the OSS impactors is limited and the joint action of ISS and OSS impactors should allow for the solid crust to be penetrated locally to deliver water to the molten interior of the asteroid. As discussed for the previous SFD, also in the case of the SFD from [42] the scenario where Jupiter did not migrate could possibly be associated to too high values of the erosion of the surface and of the efficiency in the delivery of water to Vesta (again, even accounting for the factor of two uncertainty discussed in Section 3.4). In all the other scenarios, Vesta would receive enough water to be about one order of magnitude less water-rich than the Earth.

The meteoritic data presently available on the interior of Vesta seem to indicate that the water delivery process in the asteroid belt was efficient (*i.e.*, it needed to deliver to Vesta enough water to allow for a reasonable chance of implanting some of it into its molten interior) but not too efficient (*i.e.*, Vesta should retain its volatile-depleted mineralogical signature). This would translate, according to our results, in the requirement that Jupiter migrated by a significant fraction of au. Once we take into account the global distribution of the material from the impactors on the vestan surface as shown in Figure 1 and Figure 2, our results indicate that the surface of the asteroid could have contained significant quantities of exogenous water and volatile materials mixed with the regolith at the time of the JEB in all considered migration scenarios. Moreover, the results of Paper I indicate that close encounters with the forming core of Jupiter could cause OSS planetesimals to impact Vesta even before the giant planet started accreting its gaseous envelope (albeit with lower frequency than during the JEB), strengthening the link between the appearance of volatile materials on Vesta and the formation of the giant planet. The data supplied by the Dawn mission [2,7,8,9] indicate that hydrated materials have been delivered in significant quantities to Vesta over the last 4 Ga and presently affect about 30% of the surface of the asteroid [2,9,11]. As we said, we don't expect the hydrated and H-rich material presently found of Vesta to date back to the JEB, as the more abundant ISS impactors would plausibly remove the vast majority of the water and volatile elements that did not reach the molten interior of the asteroid. However, it is interesting to note that these results globally suggest that the presence of volatile materials and, likely, of pitted terrains on the surface of Vesta could have been a continuous phenomenon during the life of the asteroid.

We must take into account, however, that the results we present in this work can be significantly affected by the following sources of uncertainty. The first source of uncertainty is linked to the unknown composition of the primordial planetesimals that impacted Vesta, which as we said before should introduce at least a factor of 2 uncertainty in the estimated amounts of water delivered to the asteroid. The second source of uncertainty is due instead to the distribution of the material from the impactors on the surface of the asteroid, which depends on the unknown composition and porosity of the impacting bodies and on their impact angle, and which affects the quantity of water that can penetrate the crust. Another important source of uncertainty in the interpretation of our results is the thickness of the solid crust of Vesta at the time of the JEB. The range of values reported by [14,15] spans between 7 and 30 km:

if the JEB took place when the solid layer was toward the high-end tail of these values, more OSS impactors would be required to have a reasonable chance of a few impactors reaching the molten interior of Vesta. Moreover, the hypothesized presence of an undifferentiated crust would favor more erosive scenarios, and therefore higher fluxes of impactors, than those requested to preserve an already formed basaltic crust. The presence of a thick solid layer would therefore favor the scenario where Jupiter did not migrate, while the existence of an undifferentiated crust would probably require larger displacements of Jupiter (between 0.50 au and 1 au) to allow for its removal. Finally, the dynamical model on which the simulations of Paper I were based did not account for the effects of the perturbations of planetary embryos and of gas drag. These two competing effects act respectively to rise and to damp the orbital eccentricities and inclinations of the planetesimals and therefore affect both the intensity of the fluxes of impactors and the impact velocities (see Papers I, II and III for further discussions).

Addressing these sources of uncertainties, to shed light on the details and the implications of the flux of cometary objects in the asteroid belt caused by the formation of Jupiter, will require further dedicated studies. Notwithstanding these uncertainties, however, the qualitative picture described by the results of this work appears robust, as the formation of Jupiter can deliver water and volatile elements to Vesta at a time when the thickness of the vestan crust was limited and part of the eucritic layer was still in a molten state as required to explain the samples studied by [4]. More generally, the results here presented highlight the fact that the Jovian Early Bombardment could represent a viable mechanism for the delivery of water into the asteroid belt and possibly up to the orbital region of Mars [26]. Such mechanism would be more efficient across the asteroid belt (particularly beyond about 2.5 au, see Papers I and II) than at Mars [26], but the results of theoretical studies indicate that part of the planetary material residing between Mars and the outer bound of the asteroid belt could be later accreted by the forming terrestrial planets (see [65,66] and references therein). This two-phases mechanism for delivering water to the forming terrestrial planets would be in agreement with the global picture supplied by the results of [67], which attributes an important role to comets in the delivery of water to the inner Solar System. The investigation of the efficiency of this mechanism in the Solar System, in turn, could help us improve our understanding of the formation and early evolution of Jupiter and, therefore, of the Solar Nebula. Finally, as pointed out by Paper II the Jovian Early Bombardment is due to physical processes that are general to planetary systems harbouring forming giant planets. The Jovian Early Bombardment therefore represents a specific case of a more general class of events, the

Primordial Heavy Bombardments [16,28], and the results of this work can be applied also to the study of extrasolar planetary systems.

ACKNOWLEDGMENTS

The authors would like to thank Stuart Weidenschilling, Guy Consolmagno, and Maria Teresa Capria for their assistance and suggestions in defining the initial conditions and the observational constraints for this work, and the two anonymous reviewers whose comments helped improve the quality of this work. This research has been supported by the Italian Space Agency (ASI) through the ASI-INAF contract I/010/10/0, by the Russian Fund of Basic Research through Grant 13-05 -00694-a, and by the International Space Science Institute in Bern through the International Teams 2012 project "Vesta, the key to the origins of the Solar System" [68]. The computational resources used in this research have been supplied by INAF-IAPS through the projects "HPP—High Performance Planetology" and "DataWell".

CONFLICTS OF INTEREST

The authors declare no conflict of interest.

REFERENCES

1. De Sanctis, M.C.; Ammannito, E.; Capria, M.T.; Tosi, F.; Capaccioni, F.; Zambon, F.; Carraro, F.; Fonte, S.; Frigeri, A.; Jaumann, R.; et al. Spectroscopic characterization of mineralogy and its diversity across Vesta. Science 2012, 336, 697–700.
2. Prettyman, T.H.; Mittlefehldt, D.W.; Yamashita, N.; Lawrence, D.J.; Beck, A.W.; Feldman, W.C.; McCoy, T.J.; McSween, H.Y.; Toplis, M.J.; Titus, T.N.; et al. Elemental mapping by dawn reveals exogenic H in Vesta's regolith. Science 2012, 338, 242–246.
3. McCord, T.B.; Adams, J.B.; Johnson, T.V. Asteroid Vesta: Spectral reflectivity and compositional implications. Science1970, 168, 1445–1447.
4. Sarafian, A.R.; Roden, M.F.; Patiño-Douce, A.E. The volatile content of Vesta: Clues from apatite in eucrites. Meteorit. Planet. Sci 2013, 345.
5. Bizzarro, M.; Baker, J.A.; Haack, H.; Lundgaard, K.L. Rapid timescales for accretion and melting of differentiated planetesimals inferred from 26Al-26Mg chronometry. Astrophys. J 2005, 632, L41–L44.
6. Schiller, M.; Baker, J.; Creech, J.; Paton, C.; Millet, M.-A.; Irving, A.; Bizzarro, M. Rapid timescales for magma ocean crystallization on the howardite-eucrite-diogenite parent body. Astrophys. J 2011, 740.
7. De Sanctis, M.C.; Combe, J.P.; Ammannito, E.; Palomba, E.; Longobardo, A.; McCord, T.B.; Marchi, S.; Capaccioni, F.; Capria, M.T.; Mittlefehldt, D.W.; et al. Detection of

widespread hydrated materials on Vesta by the VIR imaging spectrometer on board the dawn mission. Astrophys. J 2012, 758.
8. Denevi, B.W.; Blewett, D.T.; Buczkowski, D.L.; Capaccioni, F.; Capria, M.T.; de Sanctis, M.C.; Garry, W.B.; Gaskell, R.W.; le Corre, L.; Li, J.-Y.; et al. Pitted terrain on Vesta and implications for the presence of volatiles. Science 2012, 338, 246–249.
9. McCord, T.B.; Li, J.-Y.; Combe, J.-P.; McSween, H.Y.; Jaumann, R.; Reddy, V.; Tosi, F.; Williams, D.A.; Blewett, D.T.; Turrini, D.; et al. Dark material on Vesta from the infall of carbonaceous volatile-rich material. Nature 2012, 491, 83–86.
10. Reddy, V.; le Corre, L.; O'Brien, D.P.; Nathues, A.; Cloutis, E.A.; Durda, D.D.; Bottke, W.F.; Bhatt, M.U.; Nesvorny, D.; Buczkowski, D.; et al. Delivery of dark material to Vesta via carbonaceous chondritic impacts. Icarus 2012, 221, 544–559.
11. Turrini, D.; Combe, J.-P.; McCord, T.B.; Oklay, N.; Vincent, J.-B.; Prettyman, T.H.; McSween, H.Y.; Consolmagno, G.J.S.J.; de Sanctis, M.C.; le Corre, L.; et al. The contamination of Vesta by impacts and the delivery of the dark material. Icarus 2014. submitted for publication.
12. Ruzicka, A.; Snyder, G.A.; Taylor, L.A. Vesta as the HED parent body: Implications for the size of a core and for large-scale differentiation. Meteorit. Planet. Sci 1997, 32, 825–840.
13. McSween, H.Y.; Ammannito, E.; Reddy, V.; Prettyman, T.H.; Beck, A.W.; de Sanctis, M.C.; Nathues, A.; le Corre, L.; OBrien, D.P.; Yamashita, N.; et al. Composition of the Rheasilvia basin, a window into Vesta's interior. J. Geophys. Res. Planets 2013, 118, 335–346.
14. Formisano, M.; Costanzo, F.; Turrini, D.; Coradini, A.; Capaccioni, F.; de Sanctis, M.C.; Pauselli, C. The heating history of Vesta and the onset of differentiation. Meteorit. Planet. Sci 2013.
15. Tkalcec, B.J.; Golabek, G.J.; Brenker, F.E. Solid-state plastic deformation in the dynamic interior of a differentiated asteroid. Nat. Geosci 2013, 6, 93–97.
16. Coradini, A.; Turrini, D.; Federico, C.; Magni, G. Vesta and Ceres: Crossing the history of the Solar System. Space Sci. Rev 2011, 163, 25–40.
17. Turrini, D.; Magni, G.; Coradini, A. Probing the history of Solar System through the cratering records on Vesta and Ceres. Mon. Not. R. Astron. Soc 2011, 413, 2439–2466.
18. Turrini, D. The primordial collisional history of Vesta: Crater saturation, surface evolution and survival of the basaltic crust. Planet. Space Sci 2013.
19. Bouvier, A.; Wadhwa, M. The age of the Solar System redefined by the oldest Pb-Pb age of a meteoritic inclusion. Nat. Geosci 2010, 3, 637–641.
20. Meyer, M.R. Circumstellar disk evolution: Constraining theories of planet formation. Proc. Int. Astron. Union 2008, 4, 111–122.
21. D'Angelo, G.; Durisen, R.H.; Lissauer, J.J. Giant Planet Formation. In Exoplanets; Seager, S., Ed.; University of Arizona Press: Tucson, AZ, USA, 2011; pp. 319–346.
22. Bottke, W.F.; Durda, D.D.; Nesvorný, D.; Jedicke, R.; Morbidelli, A.; Vokrouhlický, D.; Levison, H.F. The fossilized size distribution of the main asteroid belt. Icarus 2005, 175, 111–140.
23. Bottke, W.F.; Durda, D.D.; Nesvorný, D.; Jedicke, R.; Morbidelli, A.; Vokrouhlický, D.; Levison, H.F. Linking the collisional history of the main asteroid belt to its dynamical excitation and depletion. Icarus 2005, 179, 63–94.
24. Scott, E.R.D. Meteoritical and dynamical constraints on the growth mechanisms and formation times of Asteroids and Jupiter. Icarus 2006, 185, 72–82.
25. Safronov, V.S. Evolution of the Protoplanetary Cloud and Formation of the Earth and Planets; Israel Program for Scientific Translations: Jerusalem, Israel, 1972; p. 212.
26. Weidenschilling, S.J. Mass loss from the region of Mars and the Asteroid belt. Icarus 1975, 26, 361–366.
27. Weidenschilling, S.J.; Davis, D.R.; Marzari, F. Very early collisional evolution in the asteroid belt. Earth Planets Space2001, 53, 1093–1097.
28. Turrini, D.; Coradini, A.; Magni, G. Jovian early bombardment: Planetesimal erosion in the inner Asteroid belt. Astrophys. J 2012, 750, 8–43.

29. Svetsov, V. Cratering erosion of planetary embryos. Icarus 2011, 214, 316–326.
30. Thomas, P.C.; Binzel, R.P.; Gaffey, M.J.; Zellner, B.H.; Storrs, A.D.; Wells, E. Vesta: Spin pole, size, and shape from HST images. Icarus 1997, 128, 88–94.
31. Michalak, G. Determination of asteroid masses—I. (1) Ceres, (2) Pallas and (4) Vesta. Astron. Astrophys 2000, 360, 363–374.
32. Russell, C.T.; Raymond, C.A.; Coradini, A.; McSween, H.Y.; Zuber, M.T.; Nathues, A.; de Sanctis, M.C.; Jaumann, R.; Konopliv, A.S.; Preusker, F.; et al. Dawn at Vesta: Testing the protoplanetary paradigm. Science 2012, 336, 684–686.
33. Lissauer, J.J.; Hubickyj, O.; D'Angelo, G.; Bodenheimer, P. Models of Jupiter's growth incorporating thermal and hydrodynamic constraints. Icarus 2009, 199, 338–350.
34. Coradini, A.; Magni, G.; Turrini, D. From gas to satellitesimals: Disk formation and evolution. Space Sci. Rev 2010, 153, 411–429.
35. Tsiganis, K.; Gomes, R.; Morbidelli, A.; Levison, H.F. Origin of the orbital architecture of the giant planets of the Solar System. Nature 2005, 435, 459–461.
36. Levison, H.F.; Morbidelli, A.; Tsiganis, K.; Nesvorný, D.; Gomes, R. Late orbital instabilities in the outer planets induced by interaction with a self-gravitating planetesimal disk. Astron. J 2011, 142, 152–163.
37. Papaloizou, J.C.B.; Nelson, R.P.; Kley, W.; Masset, F.S.; Artymowicz, P. Disk-Planet Interactions During Planet Formation. In Protostars and Planets V; Reipurth, B., Jewitt, D., Keil, K., Eds.; University of Arizona Press: Tucson, AZ, USA, 2007; pp. 655–668.
38. Davis, D.R.; Chapman, C.R.; Weidenschilling, S.J.; Greenberg, R. Collisional history of asteroids: Evidence from Vesta and the Hirayama families. Icarus 1985, 62, 30–53.
39. Coradini, A.; Magni, G.; Federico, C. Formation of planetesimals in an evolving protoplanetary disk. Astron. Astrophys 1981, 98, 173–185.
40. Chambers, J.E. Planetesimal formation by turbulent concentration. Icarus 2010, 208, 505–517.
41. Morbidelli, A.; Bottke, W.F.; Nesvorný, D.; Levison, H.F. Asteroids were born big. Icarus 2009, 204, 558–573.
42. Weidenschilling, S.J. Initial sizes of planetesimals and accretion of the asteroids. Icarus 2011, 214, 671–684.
43. Weidenschilling, S.J. Accretion of planetary embryos in the inner and outer Solar System. Phys. Scr 2008.
44. Davis, D.R.; Chapman, C.R.; Greenberg, R.; Weidenschilling, S.J.; Harris, A.W. Collisional Evolution of Asteroids—Populations, Rotations, and Velocities. In Asteroids; Gehrels, T., Ed.; University of Arizona Press: Tucson, AZ, USA, 1979; pp. 528–557.
45. Goldreich, P.; Ward, W.R. The formation of planetesimals. Astrophys. J 1973, 183, 1051–1062.
46. Weidenschilling, S.J. Dust to planetesimals—settling and coagulation in the solar nebula. Icarus 1980, 44, 172–189.
47. Cuzzi, J.N.; Hogan, R.C.; Shariff, K. Toward planetesimals: Dense chondrule clumps in the protoplanetary nebula. Astrophys. J 2008, 687, 1432–1447.
48. Cuzzi, J.N.; Hogan, R.C.; Bottke, W.F. Towards initial mass functions for asteroids and Kuiper belt objects. Icarus2010, 208, 518–538.
49. Öpik, E.J. Interplanetary Encounters: Close-Range Gravitational Interactions; Elsevier Scientific Publishing: Amsterdam, The Netherlands, 1976; pp. 7–155.
50. Holsapple, K.A.; Housen, K.R. A crater and its ejecta: An interpretation of deep impact. Icarus 2007, 187, 345–356.
51. Holsapple, K.A. The scaling of impact processes in planetary sciences. Annu. Rev. Earth Planet. Sci 1993, 21, 333–373.
52. Russell, C.T.; Raymond, C.A.; Jaumann, R.; McSween, H.Y.; de Sanctis, M.C.; Nathues, A.; Prettyman, T.H.; Ammannito, E.; Reddy, V.; Preusker, F.; et al. Dawn completes its mission at 4 Vesta. Meteorit. Planet. Sci 2013.

53. Vincent, J.-B.; Schenk, P.; Nathues, A.; Sierks, H.; Hoffmann, M.; Gaskell, R.W.; Marchi, S.; O'Brien, D.P.; Sykes, M.; Russell, C.T.; et al. Crater depth-to-diameter distribution and surface properties of (4) Vesta. Planet. Space Sci 2013.
54. Melosh, H.J. Impact Cratering: A Geologic Process; Oxford Monographs on Geology and Geophysics, No. 11; Oxford University Press: New York, NY, USA, 1989; p. 253.
55. Benz, W.; Asphaug, E. Catastrophic disruptions revisited. Icarus 1999, 142, 5–20.
56. Shuvalov, V.V. Multi-dimensional hydrodynamic code SOVA for interfacial flows: Application to thermal layer effect. Shock Waves 1999, 9, 381–390.
57. Svetsov, V.V. Numerical simulations of very large impacts on the Earth. Planet. Space Sci 2005, 53, 1205–1220.
58. Dienes, J.K.; Walsh, J.M. Theory of Impact: Some General Principles and the Method of Eulerian Codes. In High-Velocity Impact Phenonena; Kinslow, R., Ed.; Academic Press: New York, NY, USA, 1970; pp. 46–104.
59. Thompson, S.L.; Lauson, H.S. Improvements in the Chart D Radiation-Hydrodynamic CODE III: Revised Analytic Equations of State; Report SC-RR-71 0714; Sandia National Laboratory: Albuquerque, NM, USA, 1972; p. 119.
60. Pierazzo, E.; Vickery, A.M.; Melosh, H.J. A reevaluation of impact melt production. Icarus 1997, 127, 408–423.
61. Tillotson, J.H. Metallic Equations of State for Hypervelocity Impact; General Atomic Report GA-3216; Advanced Research Project Agency: San Diego, CA, USA, 1962; p. 140.
62. Crater Analysis Techniques Working Group, Standard Techniques for Presentation and Analysis of Crater Size-Frequency Data; NASA Technical Memorandum Series; National Aeronautics and Space Administration, Scientific and Technical Information Office: Hampton, VA, USA, 1978.
63. Mumma, M.J.; Charnley, S.B. The chemical composition of comets—emerging taxonomies and natal heritage. Annu. Rev. Astron. Astrophys 2011, 49, 471–524.
64. Morbidelli, A.; Chambers, J.E.; Lunine, J.I.; Petit, J.M.; Robert, F.; Valsecchi, G.B.; Cyr, K.E. Source regions and timescales for the delivery of water to the Earth. Meteorit. Planet. Sci 2000, 35, 1309–1320.
65. Righter, K.; O'Brien, D.P. Terrestrial planet formation. Proc. Natl. Acad. Sci. USA 2011, 108, 19165–19170.
66. Morbidelli, A.; Lunine, J.I.; O'Brien, D.P.; Raymond, S.N.; Walsh, K.J. Building terrestrial planets. Annu. Rev. Earth Planet. Sci 2012, 40, 251–275.
67. Hartogh, P.; Lis, D.C.; Bockele-Morvan, D.; de Val-Borro, M.; Biver, N.; Kppers, M.; Emprechtinger, M.; Bergin, E.A.; Crovisier, J.; Rengel, M.; et al. Ocean-like water in the Jupiter-family comet 103P/Hartley 2. Nature 2011, 478, 218–220.
68. Turrini, D.; Consolmagno, G. J.; Golabek, G.; Jutzi, M.; Sirono, S.; Svetsov, V.; Tackley, P.; Tsiganis, K. Vesta, the key to the origins of the Solar System, Available online: http://www.issibern.ch/teams/originsolsys (accessed on 1 November 2013).

Index